OXFORD WORLD'S CLASSICS

SELECTED WRITINGS

GALILEO GALILEI (1564–1642) is one of the most fascinating and controversial figures in the history of science. He was appointed professor of mathematics at the University of Pisa in 1589, and in 1592 he moved to the University of Padua where he taught for the next eighteen years. In 1610 he published a number of sensational telescopic discoveries, including that the lunar landscape is like that of a barren Earth, and his book, *A Sidereal Message*, was sold out in less than a week. The next year he was awarded the prestigious position of personal mathematician and philosopher to the Grand Duke of Tuscany in Florence. He next found that there are dark spots on the face of the Sun and that gave rise to a lively international controversy that is recorded in his *Letters on the Sunspots*.

He argued for a non-literal interpretation of the Bible, and became involved in a dispute over the nature of comets with a Jesuit professor whom he lampooned in a witty essay, *The Assayer*. When the Roman Inquisition banned the Copernican theory in 1616, he refrained from writing about the motion of the Earth until a Florentine friend became Pope Urban VIII in 1623. His *Dialogue on the Two Chief World Systems*, which is not only a scientific masterpiece but an outstanding literary work, appeared in 1632. Summoned to Rome, he was put on trial and condemned to house arrest in 1633. He nonetheless went on to write his *Discourse on Two New Sciences*, the work for which he is remembered as the forerunner of Newton. He died in Florence in 1642.

WILLIAM R. SHEA was Galileo Professor of History of Science at the University of Padua where Galileo taught for eighteen years. He has written extensively on the Scientific Revolution of the seventeenth century, and is currently working on a biography of Galileo.

MARK DAVIE has taught Italian at the Universities of Liverpool and Exeter, and has published studies on various aspects of Italian literature, mainly in the period from Dante to the Renaissance. He is particularly interested in the relations between learned and popular culture, and between Latin and the vernacular, in Italy in the Renaissance.

OXFORD WORLD'S CLASSICS

*For over 100 years Oxford World's Classics have brought
readers closer to the world's great literature. Now with over 700
titles—from the 4,000-year-old myths of Mesopotamia to the
twentieth century's greatest novels—the series makes available
lesser-known as well as celebrated writing.*

*The pocket-sized hardbacks of the early years contained
introductions by Virginia Woolf, T. S. Eliot, Graham Greene,
and other literary figures which enriched the experience of reading.
Today the series is recognized for its fine scholarship and
reliability in texts that span world literature, drama and poetry,
religion, philosophy, and politics. Each edition includes perceptive
commentary and essential background information to meet the
changing needs of readers.*

OXFORD WORLD'S CLASSICS

GALILEO GALILEI

Selected Writings

Translated by

WILLIAM R. SHEA and MARK DAVIE

With an Introduction and Notes by

WILLIAM R. SHEA

OXFORD
UNIVERSITY PRESS

OXFORD
UNIVERSITY PRESS

Great Clarendon Street, Oxford OX2 6DP

Oxford University Press is a department of the University of Oxford.
It furthers the University's objective of excellence in research, scholarship,
and education by publishing worldwide in

Oxford New York

Auckland Cape Town Dar es Salaam Hong Kong Karachi
Kuala Lumpur Madrid Melbourne Mexico City Nairobi
New Delhi Shanghai Taipei Toronto

With offices in

Argentina Austria Brazil Chile Czech Republic France Greece
Guatemala Hungary Italy Japan Poland Portugal Singapore
South Korea Switzerland Thailand Turkey Ukraine Vietnam

Oxford is a registered trade mark of Oxford University Press
in the UK and in certain other countries

Published in the United States
by Oxford University Press Inc., New York

A Sidereal Message translation, editorial material © William R. Shea 2012
All other translations © Mark Davie 2012

British Library Cataloguing in Publication Data

Data available

Library of Congress Cataloging in Publication Data

Data available

Typeset by RefineCatch Limited, Bungay, Suffolk
Printed in Great Britain
on acid-free paper by
Clays Ltd, Elcograf S.p.A.

ISBN 978–0–19–958369–0

10

CONTENTS

CONTENTS

INTRODUCTION

GALILEO made towering and lasting contributions to the Scientific Revolution and the reader will find in this volume all that is essential to grasp his achievements, barring the more technical and mathematical aspects of his work. The selection of texts begins with Galileo's first bestseller, the *Sidereal Message*, which appeared in Venice in March 1610 when Galileo was already forty-six years old. What happened before this date will tell us something about his personality and the science of his age.

The Training of a Renaissance Scientist

Galileo Galilei was born in Pisa on 16 February 1564,[1] the eldest son of Vincenzio Galilei and Giulia Ammannati. His father was a musician and the author of an influential *Dialogue on Ancient and Modern Music* that contains a fierce attack on his former master, Gioseffo Zarlino, and shows a gift for polemics that his son was to display in his own writings. The Galilei were an old Florentine family, and Galileo's great-great-granduncle, also named Galileo Galilei, was a famous physician who had been twice elected officer of the Governing Body of Florence and, in 1445, filled the high office of Minister of Justice. Galileo was proud of his ancestry and he described himself as a 'noble Florentine' on the title-page of his first printed book, a handbook that appeared in 1606 on how to use a geometrical and military compass that he had invented.

Vincenzio Galilei had settled in Pisa when he married in 1562 but the family moved to Florence in 1574. In 1580 his son returned to Pisa to attend the university. At the time philosophy and science were still deeply influenced by the writings of the fourth-century BC Greek philosopher Aristotle, whose works were rediscovered in the Middle Ages. Aristotle maintained that things on Earth were made of four basic elements (earth, fire, air, and water) that were mixed in

---

[1] Italians counted days from sunset before the calendar reform of 1582, and Galileo recorded his date of birth as follows: 15 February at 22.30 hours. The Sun set at 5.30 p.m. on that day according to our modern way of reckoning, which means that he was born on 16 February at 4.00 p.m.

ever-changing proportions. He thought, however, that celestial bodies were made of an entirely different kind of material, which was unchangeable or, to use the old expression, 'incorruptible'. This physics was well adapted to the concept that the Earth is at rest, and it was seen as compatible with Christian theology. Young men studying for the priesthood were expected to grasp the main lines of Aristotelian cosmology, but this did not hinder the development of observational astronomy or mathematics to which the Jesuits, a religious teaching order founded in 1534, made significant contributions.

Galileo attended the University of Pisa for four and a half years, but he left before getting his degree. This practice was not uncommon and it was not held against him when he later applied for a university post. For a short time Galileo attended the Florentine Academy of the Arts of Drawing, where he studied with Ostilio Ricci, who taught not only painting and design but also perspective and geometry. Galileo displayed considerable proficiency and soon composed a short treatise, *The Little Balance*, in which he reconstructed the reasoning that he believed had led the Greek scientist Archimedes to devise a way of detecting whether a goldsmith had substituted baser metal for gold in a crown he had made. In 1587 Galileo went to Rome to meet Christopher Clavius, the celebrated mathematician and professor at the Roman College staffed by the Jesuits. In the same year he applied for the lectureship in mathematics at the University of Bologna, but it was awarded to the astronomer Giovanni Antonio Magini. In 1588 Galileo delivered two lectures to the Florentine Academy on the shape, situation, and size of Dante's Inferno. He ably turned the literary challenge into a mathematical exercise and proceeded to apply geometry to Dantesque space.

Galileo went on to tackle problems related to the determination of the centre of gravity of solids, a hot topic in mathematics and physics. He sent some of his results to Giuseppe Moletti, the professor of mathematics at Padua, who praised them highly. Clavius and an influential nobleman, the Marchese Guidobaldo del Monte, who was also a mathematician, were equally enthusiastic and they recommended Galileo for the lectureship of mathematics at Pisa, a post to which he was appointed in 1589.

First Steps in an Academic Career

Galileo's first biographer, Vincenzio Viviani, was keen on showing that his subject was always way ahead of his time, and it is to him that we owe the famous Tower of Pisa story. Viviani would have us believe that Galileo, then a young professor, ascended the Leaning Tower and, 'in the presence of all the lecturers and philosophers, and of all the student body', dropped balls of different weight. These fell at the same speed, thereby refuting Aristotle who had taught that heavy bodies fall faster.[2] Now Galileo was indeed interested in the problem, and he wrote an extended essay on motion around 1591 in which he mentions towers, but his claim is not that bodies fall at the same speed regardless of their weight but that their speed is proportional to the differences between their specific gravity and the density of the medium through which they descend. In other words, he reached the erroneous conclusion that bodies of different sizes but of the same material fall at the same rate while bodies of the same size but of different materials do not. It was not until many years later that he realized his mistake and made experiments that enabled him to formulate the correct law of free fall.

In 1591 Galileo's father died, leaving him, as the eldest son, with serious financial responsibilities. He could expect little help from his brother Michelangelo, who had become court musician to the Grand Duke of Bavaria but whose income barely enabled him to support his own large family. Galileo had to think of his mother as well as paying the dowry of his sister Virginia, who had recently married, and he would soon have to provide a dowry and find a husband for his second sister, Livia.

The Chair of Mathematics at Padua, vacant since the death of Giuseppe Moletti in 1588, was more lucrative than the one in Pisa and Galileo managed to get it with the renewed assistance of the Marchese Guidobaldo del Monte. Galileo took up his appointment in 1592 and soon became acquainted with Paolo Sarpi, the friar who later became the champion of Venetian civic liberties, and was at that time the leading scientific light in Venice. Together they discussed the theory of the motion of the Earth that had been put forward

---

[2] Viviani wrote a *Historical Account of the Life of Galileo* in the form of a letter to Prince Leopold de' Medici in 1654, twelve years after Galileo's death (Galileo, *Opere*, 19, p. 606).

by the Polish astronomer Nicolaus Copernicus in his book *On the Revolutions of the Heavenly Spheres*, which had appeared in 1543. Notes that Sarpi jotted down between 1595 and 1596 mention the important argument from the tides that Galileo later developed to prove that the Earth moves.[3] The faculty of universities was still small in the sixteenth century, and Galileo was acquainted with all his colleagues. All in all there were only forty-seven of them, and Galileo was the only one to teach mathematics, astronomy, and physics.

Galileo's teaching duties were light: he had to give no more than sixty lectures per year. His classes were attended mainly by medical students who wanted to learn how to make horoscopes for their eventual patients. Without them Galileo would have had an empty classroom, as we know from the formal protest he lodged when a professor of medicine decided to give his course at 3.00 p.m., the time that Galileo had chosen because no one else taught at that hour. Outside the university he gave private lectures on fortifications and military engineering to young noblemen bent on a military career. As was not uncommon for professors, he also rented a large house and let out rooms to rich foreign students.

The first indication of Galileo's commitment to heliocentrism dates from 1597. In August of that year he received a book by the German astronomer Johann Kepler, who argued that Copernicus had been right.[4] After reading the preface Galileo wrote back to voice his long-standing sympathies for the view that the Earth is in motion, but also to express his fear of making his position known to the public at large.[5] When Galileo read Kepler more carefully he became aware of their profound methodological differences, and later reproached him in his *Dialogue on the Two Chief World Systems* because he 'still listened and assented to the notion of the Moon's influence on the water, and occult properties, and similar childish ideas'.[6] This statement refers to Kepler's belief that the tides are caused by the Moon, a phenomenon that Galileo thought he could

---

[3] See William R. Shea, *Galileo's Intellectual Revolution* (New York: Science History Publications, 1977), 173.

[4] Copernicus' *On the Revolutions of the Heavenly Spheres* had appeared in 1543, but heliocentrism was generally considered a mere conjecture rather than a probable fact until Kepler published his *Cosmographic Mystery* in 1597.

[5] Letter to Kepler, 4 August 1597 (*Opere*, 10, pp. 67–8).

[6] *Opere*, 7, p. 486; p. 356 in this volume.

explain on purely mechanical grounds as the result of the combination of the Earth's diurnal and annual revolutions.

Galileo never married, but his common-law wife Marina Gamba bore him two daughters and a son. For all his offspring Galileo carefully cast horoscopes, as he had done for himself at least a couple of times. The two girls were placed in a convent in Arcetri near Florence in 1614; his son Vincenzio was legitimized in 1619.

Around 1602 Galileo began making experiments with falling bodies in conjunction with his study of the pendulum. He first expressed the correct law of freely falling bodies, which says that distance travelled is proportional to the time squared, in a letter to Paolo Sarpi in 1604,[7] but he claimed to have derived it from the erroneous assumption that speed is proportional to distance whereas, as he realized later, it is proportional to the square root of the distance. In the autumn of 1604 the appearance of a particularly bright star, a supernova, revived the debate on the incorruptibility of the heavens that had been so lively a generation earlier, when the Dane Tycho Brahe had argued that another very bright star, which had appeared in 1572, proved that heavenly matter was subject to change. Galileo took his side against Aristotle in a book written in the Paduan dialect under the pseudonym of Cecco di Ronchitti. This was his first attempt to reach an audience outside university circles and poke fun at his opponents.

Galileo supplemented his income by manufacturing and selling a mathematical instrument, a forerunner of the slide-rule, which he called the military and geometrical compass. It was expensive but in great demand, and Galileo hired a skilled artisan to help produce it. Sales of the compass went on increasing, and in 1606 he published an Italian handbook for users. Shortly thereafter Simon Mayr, a German astronomer working in Padua, and his Italian student Baldassare Capra published a similar Latin treatise on the compass and claimed that Galileo had stolen the idea from them. Since Mayr had left Padua, Galileo brought an action against Capra and issued a scathing rebuttal giving his account of the incident. In 1605 he was employed to instruct the young Cosimo de' Medici, heir to the Grand Duchy, in the use of his compass, confirming contacts with Florence and its rulers that were to last for the rest of his life.

---

[7] Letter to Paolo Sarpi, 16 October 1604 (*Opere*, 10, p. 115).

A New World in the Heavens: A Sidereal Message

In July 1609, while visiting friends in Venice, Galileo heard that a Dutchman had invented a device to make distant objects appear nearer, and he immediately attempted to construct such an instrument himself.[8] Others were at work on similar devices, but by the end of August 1609 Galileo had managed to produce a nine-power telescope that was better than those of his rivals. He returned to Venice, where he gave a demonstration of his spying-glass from the city's highest bell-towers. The practical value for sighting ships at a distance greatly impressed the Venetian authorities, who confirmed Galileo's appointment at Padua, which was the official university of the Venetian republic, for life and raised his salary from 520 to 1,000 florins, an unprecedented sum for a professor of mathematics.

Galileo had put together a convex lens at one end of a tube roughly 1 metre long, and a concave lens at the other end where he applied his eye. He had the great good fortune of having access to the best lenses in Europe, made on the island of Murano, just off Venice. Thanks to these unsung craftsmen and to his own exceptional abilities as a draughtsman Galileo was able to discover what others had failed to see or lacked the ability to record. Early in January 1610, he turned an improved version of his telescope to the skies. What he saw and reported in the *Sidereal Message* (*Sidereus Nuncius*), which appeared on 13 March 1610, was to revolutionize astronomy. First, the Moon was revealed as having mountains and valleys like those on Earth.[9] This was exciting news, because if the Moon resembles the Earth then it might be inhabited! Second, innumerable stars popped out of the sky and untold worlds were suddenly and unexpectedly disclosed. Third, the Milky Way, which looks like a whitish cloud when seen with the naked eye, showed itself to be a mass of starlets. Fourth, the faint luminosity observed on the dark side of the Moon when its illuminated part is only a thin crescent four or five days old was correctly interpreted by Galileo as the reflection of sunlight bouncing off the surface of the Earth. So the Moon has 'earthshine', and the reflected light that reaches it is more

---

[8] Actually the telescope had been invented in Italy around 1590. On the technology and the problems involved, see *Galileo's Sidereus Nuncius or A Sidereal Message*, translated from the Latin by William R. Shea. Introduction and Notes by William R. Shea and Tiziana Bascelli (Sagamore Beach, Mass.: Science History Publications, 2009).

[9] See pp. 10 ff. in this volume.

powerful than moonlight on Earth because the Earth is four times as big as the Moon. The fifth discovery was even more spectacular, for four new celestial bodies were found to orbit Jupiter, something that had never been anticipated in the wildest dreams of philosophers or astronomers. Galileo named them after the Medici, the ruling family of Tuscany, the state where he was born and where he soon hoped to be recalled.

The discovery of four satellites orbiting Jupiter was particularly important in the Copernican debate, for if Jupiter could revolve around a central body (be it the Earth or the Sun) with four attendant satellites, then it was no longer absurd to suggest that the Earth went around the Sun with just one satellite, the Moon. Jupiter's satellites did not prove that Copernicus was right, but they removed a major obstacle to his theory's acceptance by astronomers.

While the night sky provided startling news mundane events continued to matter, such as the one that occurred in Florence in January 1609 when the Grand Duke Ferdinand in Florence died and was succeeded by his son, Cosimo II, Galileo's former pupil. Galileo had been wanting to return to Florence for some time and he realized that his newly won fame might assist him in effecting a change of residence. He christened the satellites of Jupiter 'Medicean stars' in honour of Cosimo and began corresponding with Belisario Vinta, the Secretary of State. In May 1610 he formally demanded employment at his new salary of 1,000 florins and with time to complete the ambitious programme he outlined:

two books on the system and constitution of the universe—an immense conception full of philosophy, astronomy, and geometry; three books on local motion, an entirely new science, since no one else, ancient or modern, has discovered any of the very many admirable properties that I have demonstrated to exist in natural and forced motion, so that I may reasonably call this a new science discovered by me from its first principles; three books on mechanics, two pertaining to the principles and foundations and one on its problems—and though others have been written on this same material, what has been written to date is not one quarter of what I am writing, either in bulk or otherwise. I have also various little works on physical subjects such as On Sound and Voice, On Vision and Colours, On the Tides, On the Composition of the Continuum, On the Motions of Animals, and still more.[10]

[10] Letter to Belisario Vinta (*Opere*, 10, pp. 351–2).

The letter ended with the request that, 'to the name of Mathematician his Highness add that of Philosopher, since I have spent more years studying philosophy than I have spent months on mathematics'.[11] The Grand Duke was suitably impressed and Galileo was granted the title of Mathematician and Philosopher. His salary was levied from the University of Pisa, where he would have, however, no teaching duties, something that did nothing to endear him to his Pisan colleagues.

Galileo's departure from the Venetian Republic has often caused surprise. 'Where can you find the freedom and the independence that you enjoyed in Venice?' wrote his friend Sagredo.[12] But Galileo did not see it in this light, and to a Florentine correspondent he confided that the Republic made too many demands on his time, and that he could only hope to obtain the leisure needed to do his work from 'an absolute Prince'.[13]

The New Science Enters the Public Domain:
Letters on the Sunspots

The *Sidereal Message* created a sensation throughout Europe and was reissued in a pirated edition in Frankfurt at the end of 1610. In a characteristically generous and enthusiastic reaction, Kepler immediately hailed Galileo's achievement. Others wavered on theoretical grounds but mainly, as it seems, because their own telescopes were of poor quality. Martin Horky, a friend of Giovanni Antonio Magini, the professor of astronomy at Bologna, declared that the telescope was wonderful for terrestrial observation but useless for studying the stars. Galileo's former student, a young Scot by the name of John Wedderburn, joined the fray and wrote a devastating reply. An influential Florentine nobleman, Francesco Sizzi, criticized Galileo's discoveries on astrological and Hermetic grounds, but the most serious charge was made in a privately circulated manuscript written by the Florentine Ludovico delle Colombe entitled *Against the Motion of the Earth*. It did not attack Galileo directly but ranted against Copernicans and accused them of going

[11] *Opere*, 10, p. 353.

[12] Letter to Galileo of 13 August 1611 (*Opere*, 11, p. 171).

[13] Letter of Galileo to a correspondent in Florence in February 1619 (*Opere*, 10, p. 233).

against Scripture. Galileo's extensive marginal notes to this tract show how incensed he was and how much he feared that he would be dragged into theological waters.

Meanwhile Galileo continued his telescopic observations in the hope of finding other satellites. The Queen of France, Marie de Medicis, who was a Florentine, wrote to say that she would be obliged if her husband, Henry IV, could find a place in heaven. Galileo did his best, and in July 1610, he noticed that Saturn had an elongated shape. He conjectured that two satellites, one on each side of Saturn, might cause this and, as was sometimes done to ensure priority without making an immediate disclosure, he sent out an anagram of thirty-seven words to announce that he had made a discovery. No one worked out the meaning of the anagram, and Giuliano de' Medici, the Tuscan Ambassador in Prague, was charged by the Emperor Rudolph II to ask for the solution. Galileo then officially announced that Saturn had two satellites. Unfortunately the two objects decreased in size and, at the close of 1612, vanished altogether. Were all of Galileo's celestial discoveries to suffer such a fate? Genuinely embarrassed but still capable of mocking his own discomfiture, Galileo wrote to a friend on 1 December 1612: 'Has Saturn devoured its children? Or was their original appearance an illusion produced by the lenses, which deceived me for so long, as well as the many others who observed it with me on many occasions? . . . Lack of time, the unprecedented nature of the event, my limited understanding, and a fear of error all add to my uncertainty.'[14] What Galileo had seen are the rings of Saturn, that are sometimes seen edgewise, when they are hard to detect, and sometimes slanted, when they can be identified with a better telescope than the one Galileo had, something that was done many years later by the Dutch scientist Christian Huygens.

Galileo's telescope was powerful enough, however, to enable him to make another discovery that was both sensational and this time genuine, namely that Venus has phases, not unlike the Moon. This has to be so if Venus orbits around the Sun, but the phases cannot be seen with the naked eye. Galileo felt the need to double-check, and once again he protected his priority claim by sending an anagram to

---

[14] Letter to Mark Welser, the third *Letter on the Sunspots* (*Opere*, 5, pp. 237–8; p. 53 in this volume).

Giuliano de' Medici in Prague on 13 November 1610. Six weeks later he had convinced himself that he had been right, and on 1 January 1611 he gave the solution of the riddle and disclosed that Venus had phases and undoubtedly went around the Sun.[15]

The Jesuits in Rome confirmed Galileo's observations, and when he returned to the Eternal City in the spring of 1611 they gave him the equivalent of an honorary doctorate in a ceremony that included no fewer than three cardinals. The young Prince Federico Cesi, who was in attendance, immediately asked Galileo to join the prestigious Lincean Academy that he had founded. Everyone in Rome wanted to see the new sky, and Galileo was frequently invited to display the marvels of his telescope in the stately gardens of the most important Roman families. His success was so great that Cardinal Francesco Maria del Monte wrote to the Grand Duke Cosimo II in Florence: 'If we were still living under the ancient Roman republic, I am certain that a statue would have been erected to him on the Capitoline.'[16]

In the summer of 1611 professors on leave from the University of Florence and Florentine gentlemen met at the villa of Filippo Salviati at Le Selve near Florence. Galileo was present when a discussion arose on the qualities of hot and cold and, specifically, why ice is lighter than water since the action of cold is to condense, not rarefy. This led to a debate over the cause of floating bodies in general, with Galileo maintaining, along Archimedean lines, that the cause of floating was the relative density, whereas an Aristotelian opponent held that it was the shape. Shortly thereafter Cardinal Maffeo Barberini (the future Pope Urban VIII) and Cardinal Ferdinando Gonzaga happened to be in Florence, and the Grand Duke invited Galileo and the Aristotelian philosopher to repeat their arguments before the distinguished visitors. Cardinal Barberini not only enjoyed the discussion but also sided with Galileo, and the two men became friends. Unfortunately, they were later to clash over the issue of Copernicanism, as we shall see.

In the autumn of 1611 Christoph Scheiner, a Jesuit professor at the University of Ingolstadt in southern Germany, made a number of observations of sunspots and was eager to make them known. He

---

[15] *Opere*, 10, p. 474 and 11, pp. 11–12.
[16] Letter of 31 May 1611 (*Opere*, 11, p. 119).

had a friend and patron in the person of Mark Welser, a wealthy merchant in Augsburg and enthusiastic amateur of science, and he sent him three letters in which he described his findings. Welser had these letters printed and sent copies abroad, notably to Galileo and members of the Lincean Academy. As Scheiner was forbidden to use his name, lest he be mistaken and bring discredit on the Society of Jesus, he had concealed his identity under the pseudonym of *Apelles latens post tabulam*.[17] Galileo, however, identified him as the Jesuit and, in two letters to Welser, took him to task for suggesting that the spots were small satellites orbiting the Sun. Scheiner wrote a reply that he entitled *A More Accurate Discussion of Sunspots and the Stars that Move around Jupiter*. Galileo retorted with a third letter to Welser in December 1612, and the extension of the debate to planets enabled Galileo to bring up the matter of the Copernican system and, for the first time, to endorse it unequivocally in print. His three letters were printed in Rome the following year, under the auspices of the Lincean Academy.[18] Galileo declared that he had observed the sunspots before Scheiner, and in an age when scientists were hypersensitive about priorities, this did not endear him to his rival. It also opened a breach between Galileo and the Jesuits.

Science and the Bible

Meanwhile the problem of Copernicanism did not lie dormant in Florence. In November 1612 Galileo heard that a well-known Dominican friar, Niccolò Lorini, had criticized the heliocentric theory. When asked for an explanation, Lorini replied that he had meant no harm but that when the motion of the Earth had been mentioned in a discussion he had said, 'a few words just to show I was alive. I said, as I still say, that this opinion of Ipernicus—or whatever his name is—would appear to be hostile to Divine Scripture'.[19] Lorini's ignorance of the very spelling of Copernicus'

---

[17] This is a reference to the famous Greek painter Apelles (fourth century BC) who is said to have hidden behind his paintings to hear the comments of passers-by. When a cobbler found fault with sandals he had drawn Apelles made the corrections that very night. Next morning the cobbler was so proud that he began to criticize how Apelles portrayed the leg, whereupon the painter emerged from his hiding-place and said: 'Shoemaker, don't judge above the sandal!' The source is Pliny's *Natural History*, 35.85.

[18] See pp. 33–54 in this volume.

[19] Letter of Niccolò Lorini, 5 November 1612 (*Opere*, 11, p. 427).

name makes it unlikely that he had much interest in astronomy, and it is a pity that, after attacking the Jesuits, Galileo should have been led to antagonize the Dominicans. But things were to get worse. On 12 December 1613 objections were raised, on Scriptural grounds, against the notion that the Earth was a planet at a dinner the Grand Duke gave in Pisa. Galileo was absent, but his friend and former pupil, the priest Benedetto Castelli, defended the Copernican theory when asked about it by Christina of Lorraine, the mother of the Grand Duke. A couple of days later Castelli wrote to Galileo to inform him that he had behaved as gallantly as a knight, but Galileo felt that the battle could not be entrusted to such an inexperienced defender and immediately sent Castelli a letter in which he argued that the motion of the Earth was not at variance with the Bible.[20] Peace seemed restored, but a year later, in 1614, another Dominican, Tommaso Caccini, delivered an inflammatory sermon against the Copernican system from the pulpit of Santa Maria Novella in Florence on 21 December. News of his outburst soon reached Rome, and Galileo's friend, Monsignor Giovanni Ciampoli, raised the matter with Cardinal Barberini, the future Urban VIII, who urged 'greater caution in not going beyond the arguments used by Ptolemy and Copernicus'. Ciampoli explained to Galileo that great care should be exercised because someone might claim that the telescope provided evidence for life on the Moon, and this could lead to someone else asking if the lunar inhabitants 'were descended from Adam or how they came out of Noah's ark'. It was wise, he added, 'to declare frequently that one placed oneself under the authority of those who have jurisdiction over the minds of men in the interpretation of Scripture'.[21]

Not to be outdone by Caccini, Niccolò Lorini forwarded a copy of Galileo's *Letter to Castelli* to the Inquisition in Rome, and in order to cope with this new challenge Galileo expanded this into the *Letter to the Grand Duchess Christina*, which contains his most detailed pronouncement on the relations between science and Scripture.[22] Borrowing the *bon mot* of Cardinal Cesare Baronio, 'the intention of the Holy Spirit is to teach us how one goes to heaven, not how the

---

[20] See pp. 55–61 in this volume.
[21] Letter of Giovanni Ciampoli to Galileo, 28 February 1615 (*Opere*, 12, p. 146).
[22] See pp. 61–94 in this volume.

heaven goes',[23] Galileo stressed that God speaks through the book of nature as well as the Book of Scripture, and that care must be exercised lest metaphorical expressions in the Bible be interpreted as scientific facts.

Further complications arose from the publication early in 1615 of a book by Paolo Antonio Foscarini, a Carmelite monk, who also wanted to defend Copernicanism from the charge that it was at variance with what the Bible taught.[24] Foscarini sent a copy of his book to Cardinal Robert Bellarmine, requesting his opinion. The Cardinal's reply was friendly but firm: 'it seems to me that both you and signor Galileo are acting prudently in confining yourselves to speaking hypothetically and not in absolute terms', and he reminded him that 'to demonstrate that the appearances are saved by the hypothesis that the Sun is at the centre and the Earth is in the heavens, is not the same as demonstrating that the Sun really is at the centre and the Earth in the heavens. I believe that the first can be demonstrated, but I have very great doubts about the second.'[25] Bellarmine added that should such a proof become available then Scripture would have to be reinterpreted with the utmost care.

A copy of the Cardinal's letter was sent to Galileo, who was not unduly troubled because he believed that he had just such a proof. This was his clever (but unfortunately wrong) theory that the tides could not occur if the Earth were not in motion. The theory eventually became the subject of the Fourth Day of his *Dialogue on the Two Chief World Systems*,[26] originally entitled 'A Discourse on the Tides'. Armed with this argument, which he believed to be decisive, Galileo marched off to Rome at the end of 1615. He talked to all and sundry about his proof, but the only result was that the Holy Office of the Inquisition began to take the heliocentric theory seriously, something it had not done before. The eleven theological experts who acted as consultants to the Holy Office were asked on 19 February 1616 to examine the two following propositions:

[23] See p. 70 in this volume.

[24] On Foscarini's book see the note to p. 94 below.

[25] See pp. 94–6 in this volume. 'Saving the appearances' was frequently used in Galileo's day to indicate the common view that astronomical theories were mere geometrical devices put forward for the sake of calculation and not as a representation of physical reality.

[26] See pp. 309–59 in this volume.

(1) the Sun is the centre of the world and hence completely motionless; and (2) the Earth is not the centre of the world and motionless but moves as a whole and also with diurnal motion. Five days later they reached the following unanimous decision: the first proposition was censured as 'foolish and absurd in philosophy, and formally heretical since it explicitly contradicts sentences found in many places in Sacred Scripture according to the proper meaning of the words and according to the common interpretation and understanding of the Holy Fathers and of learned theologians'. The second proposition received the 'same censure in philosophy' and, with respect to theology, was considered 'at least erroneous'.[27] Their report was approved by the Pope and the Cardinals of the Holy Office on 25 February, and next day, acting on the orders of Pope Paul V, Cardinal Bellarmine informed Galileo that he had to abandon the Copernican theory, and stop teaching it orally or in writing. Galileo assented.

As was current practice, the ruling of the Holy Office was passed to the Congregation of the Index, which was in charge of censuring books, and on 5 March Copernicus' *On the Revolution of the Heavenly Spheres* was proscribed 'until revised' and Foscarini's book banned outright. Galileo himself was not mentioned, but he was worried and, before leaving Rome, he called on Cardinal Bellarmine on 26 May to inform him that it was rumoured in Florence that he had been condemned by the Inquisition. The Cardinal gave him a certificate stating that he had not been condemned but merely informed of the decision of the Congregation of the Index. Galileo kept this precious document, which he showed to no one.

The Problem of Longitude at Sea and the Quarrel over Comets in The Assayer

On his return to Florence Galileo applied himself to a much-discussed but non-controversial topic: the determination of longitudes at sea. He hoped that accurate tables of the periods of revolutions of the satellites of Jupiter would enable him to predict how their relative positions would appear from any point on Earth, and hence allow seamen to know their location merely by looking

[27] Minutes of the meeting of 24 February 1616 (*Opere*, 19, p. 321).

through the telescope. The method was sound, and it kept Galileo busy for a long time, but it could not be used because of the practical difficulties of applying it on a ship at sea.

In the autumn of 1618 great excitement was generated over the appearance, in rapid succession, of three comets. Galileo was bedridden at the time and unable to make observations. He was nonetheless asked to comment on accounts given by others, and he chose to attack a lecture delivered by Orazio Grassi, the Jesuit professor of mathematics at the Roman College, who had located the comet beyond the Moon, as we do today. This irked Galileo, who approached the problem from a very different standpoint since he thought that comets were merely optical phenomena caused by refraction in atmospheric vapour or in the clouds. He set out this view—which caused surprise, since it had been that of Aristotle—in a *Discourse on Comets* published under the name of Mario Guiducci, a young lawyer who enjoyed no scientific reputation. It was clear to Grassi from the outset that Galileo was the real author, and he prepared a rejoinder that appeared in print in 1619. Entitled *The Astronomical and Philosophical Balance*, it purported to weigh the arguments of the *Discourse on Comets*. Galileo prepared a rebuttal in which he claimed to ponder Grassi's own argument with a more delicate weighing instrument, called an assayer, which became the title of the book. *The Assayer* appeared in 1623, and was widely acclaimed as a masterpiece of wit.[28] It is here, for instance, that we find the celebrated passage about the book of nature: 'Philosophy is written in this great book which is continually open before our eyes—I mean the universe—but before we can understand it we need to learn the language and recognize the characters in which it is written. It is written in the language of mathematics, and its characters are triangles, circles, and other geometrical figures, without which it is humanly impossible to understand a word of what it says.'[29] Just before it emerged from the press, Cardinal Maffeo Barberini became Pope Urban VIII and Galileo was able to dedicate the book to him.

---

[28] See the extracts in this volume, pp. 115–21.
[29] See p. 115 in this volume.

The Dialogue on the Two Chief World Systems

The election of Urban VIII looked to Galileo like a turning-point in his career, and he went to Rome in 1624 to pay his respects to the new pontiff. During the six weeks of his stay he was admitted to no fewer than six audiences by the Pope, who gave him a painting, two medals, several Agni Dei,[30] and the promise of a pension for his son. However, it does not seem that Galileo was able to ask Urban VIII whether he could now write about the motion of the Earth, but the friendliness of the Pope led him to believe that he could as long as he presented it not as an ascertained fact but a mere scientific hypothesis. His impression was strengthened by a meeting with Cardinal Frederic Eutel Zollern, who offered to broach the Copernican question with the Pope. When he saw Urban VIII the Cardinal pointed out that German Protestants were all in favour of the new system, and that it was necessary to proceed with the utmost caution before attempting to settle the Copernican question. The Pope replied that the Church had never declared the view of Copernicus to be heretical and would not do so, but that he saw no reason to suppose that a proof of the Copernican system would ever be forthcoming. Unfortunately Cardinal Zollern died in 1625, and Galileo lost someone who could have been a key witness at his trial eight years later.

When Galileo returned to Florence he immediately set to work on his *Dialogue on the Two Chief World Systems*, but poor health meant that between 1626 and 1629 he was unable to work with any regularity, and it was only in January 1630 that he managed to finish his long-awaited book. The three interlocutors of the *Dialogue* are the Florentine Filippo Salviati (1583–1614), the Venetian patrician Giovanfrancesco Sagredo (1571–1620), and the Aristotelian Simplicio, an imaginary character. They are presented as having gathered in Sagredo's palace at Venice for four days to discuss the

[30] An Agnus Dei is the name given to discs of wax impressed with the figure of a lamb and blessed at stated seasons by the Pope. The lamb usually bears a cross or flag, and a figure or the name and arms of the Pope are commonly impressed on the reverse. They are made of the remnants of the preceding year's paschal candle, and in the Middle Ages the Popes sent them as presents to sovereigns and distinguished personages. They were considered a protection against blights and tempests. In the penal laws of Queen Elizabeth Agni Dei are mentioned among 'popish trumperies', the importation of which into England was strictly forbidden.

arguments for and against the heliocentric system. Salviati is a militant Copernican, Simplicio an avowed defender of geocentrism, and Sagredo an intelligent amateur already half-converted to the new astronomy.

The First Day belongs to the long history of anti-Aristotelianism, and Galileo borrowed extensively from his predecessors' criticism of Peripatetic philosophy. What must be considered significant about his attack, however, is the skill with which it is conducted. Never before had any critic of Aristotle been so gifted as a writer, so apt at convincing an opponent by the sheer brilliance of his presentation, and so masterful at laughing him off the stage when he refused to be persuaded. Galileo draws from the literary resources of his native Italian to convey insights and stimulate reflection, but his style does not possess the bare factualness of the laboratory report or the unflinching rigour of a mathematical deduction. Words are more than vehicles of pure thought. They are sensible entities, they possess associations with images, memories, and feelings, and Galileo knows how to use these associations to attract, hold, and absorb attention. He does not present his ideas in the nakedness of abstract thought but clothes them in the colours of feeling, intending not only to inform and teach but also to move and entice to action. He wished to bring about nothing less than a reversal of the 1616 decision against Copernicanism, and the dialogue form seemed to him the most conducive to this end. It is true that the written dialogue is deprived of the eloquence of facial expression and the emphasis of gestures, of the support of modulated tone and changing volume, but it retains the effectiveness of pauses, the suggestiveness of questions, and the significance of omissions. Galileo makes the most of these techniques, and it is important to keep this in mind when assessing his arguments, for too often passages of the *Dialogue* have been paraded without sufficient regard for their highly rhetorical content.

The First Day put an end to Aristotelian cosmology by showing that terrestrial and celestial bodies have to be explained in the same way. The next logical step was to ask whether the Earth moves, and the remaining three days of the *Dialogue* are devoted to this problem. The daily rotation of the Earth is discussed in the Second Day and its annual revolution around the Sun in the Third. The Fourth Day attempts to show that it is only when the diurnal and annual motions are acknowledged that the phenomenon of the tides can be adequately explained.

Galileo hoped that his manuscript would be steered through the shoals of Roman censorship by his friends Giovanni Ciampoli and the Dominican Niccolò Riccardi, who had become master of the Apostolic Palace and whose duty it was to authorize the publication of books. In the spring of 1630 Galileo set off once more for Rome, where he personally handed over his manuscript to Riccardi, who passed it on to a fellow Dominican, Raffaello Visconti, who knew some astronomy but whose interests extended to astrology and the occult sciences as well. He was a personal friend of Orazio Morandi, the abbot of Santa Prassede in Rome, who was considered an authority on horoscopes. Shortly before Galileo's arrival in Rome on 3 May 1630 Morandi had published a number of prophecies based on astrological computations, among them one that predicted the early death of the Pope. Galileo was almost certainly unaware of this incident when he received, on 24 May, an invitation to dine with Morandi in the company of Visconti. Roman gossip lost no time in linking his name with the astrologer, and this was to prove a more delicate matter than Galileo could have foreseen. He left Rome on 26 June, and less than three weeks later Morandi was imprisoned by the Inquisition. When Galileo requested information from a mutual friend, he was told that the trial was so secret that there was no way of knowing what was happening.[31] An 'Astrological Discourse on the Life of Urban VIII' bearing Visconti's name was brought forward at the trial, but Visconti was at least partly successful in his plea of innocence, since he was only banished from Rome, while several others received heavy sentences. Morandi himself died in prison on 7 November 1630, before the completion of his trial.

The next year Urban VIII issued a papal edict against astrologers for claiming that they could know the future and set in motion secret forces for the good or harm of the living. Urban ordered his staff to be on the lookout for prognostications that were directed against his own life and that of his relatives down to the third degree. Guilty parties were to be punished with death and confiscation of property.

---

[31] Letter of Vincenzo Langieri to Galileo, 17 August 1630 (*Opere*, 14, pp. 134–5). On the Morandi affair, see Brendan Dooley, *Morandi's Last Prophecy and the End of Renaissance Politics* (Princeton: Princeton University Press, 2002).

That Galileo's name should have been associated with those of Morandi and Visconti was unfortunate, to say the least. Little did he suspect that his intimacy with another Vatican official, Giovanni Ciampoli, was to prove even more damaging.

Urban VIII was a poet in his leisure hours and he enjoyed the company of intellectuals, one of whom was Ciampoli, who handled his international correspondence. Ciampoli was impatient to secure the Cardinal's hat that Urban VIII distributed to men whom Ciampoli considered his inferiors. In his frustration he became reckless and allowed himself to be drawn into the circle of the Spanish Cardinal Gaspare Borgia, spokesman of Philip IV and a thorn in Urban's flesh. After Cardinal Borgia publicly protested against the Pope's position in the struggle between France and the House of Hapsburg in a stormy consistory on 8 March 1632, Urban decided to purge his entourage of pro-Spanish elements. He was particularly incensed upon hearing of Ciampoli's behaviour. He stripped him of his considerable powers and in August 1632 ordered him to leave Rome and take up residence in the small town of Montalto.

Ciampoli's downfall was to have important consequences for Galileo. Although Visconti had informed Riccardi that Galileo's *Dialogue* only needed a few minor corrections, it was actually Ciampoli who told Riccardi to grant the permission to print because the Pope would have no objection. Ciampoli had no warrant for saying this, and the Pope was incensed when he heard about it later. The outbreak of the plague in 1630 had rendered communications between Tuscany and Rome difficult, and Riccardi allowed the *Dialogue* to be printed in Florence. He insisted, however, that the preface and conclusion be sent to him prior to publication, and he expected Galileo to return to Rome to discuss the final draft. When Galileo objected that the plague made travel between Florence and Rome dangerous, Riccardi agreed that it would be enough if a copy of the manuscript were sent to Rome 'to be revised by Monsignor Ciampoli and himself'.[32] Even this requirement was eventually waived, and thereafter Riccardi heard no more of the book until a printed copy reached him in Rome. Above the Florentine

---

[32] As reported by Benedetto Castelli in his letter to Galileo, 21 September 1630 (*Opere*, 14, p. 150).

imprimatur he discovered, to his horror, his own approbation. As Urban VIII expostulated: 'The name of the Master of the Holy Palace has nothing to do with books printed elsewhere.'[33] Summoned to account for his behaviour, Riccardi excused himself by saying that he had received orders to license the book from Ciampoli himself.[34]

The *Dialogue* had gone to press in June 1631, but Galileo had received no financial assistance and had to agree to pay the publisher for the large run of a thousand copies. Printing was completed on 21 February 1632 but copies did not reach Rome until the end of May, thus bursting onto the Roman scene only a few weeks after the consistory in which Cardinal Borgia had attacked Urban VIII. The Roman imprimatur on a Florentine publication created a stir, and Riccardi was instructed to have a ban placed on the sale of Galileo's book, pending further notice.

An Astronomer on Trial

In the summer of 1632 Urban VIII ordered a Preliminary Commission to investigate the licensing of the *Dialogue*. In the file on Galileo in the Holy Office they found the memorandum of 1616 that enjoined him not to hold, teach, or defend the view that the Earth moves. This was enough to conclude that Galileo had contravened a formal order of the Holy Office and he was summoned to Rome in September 1632, but he pleaded ill-health and managed to postpone his trip for another five months. He arrived in Rome on 13 February 1633 and was allowed to stay with Francesco Niccolini, the Tuscan ambassador and an old friend, while three theologians read his *Dialogue* to determine whether he had presented the Copernican doctrine as a proven fact and not a mere hypothesis. These censors were Niccolò Riccardi, the Master of the Sacred Palace, whose office included the responsibility of licensing books to be printed; Agostino Oreggi, who was the papal theologian; and a Jesuit by the name of Melchior Inchofer, who had had one of his own books placed on the Index. They agreed that there could be no doubt that Galileo argued in favour of the motion of the Earth, although he made an apparent disclaimer in the closing paragraph of the

---

[33] This was said to the Tuscan Ambassador, Francesco Niccolini, who mentions it in his letter to the Secretary of State in Florence, 5 September 1632 (*Opere*, 14, p. 384).

[34] Account given by Giovanfrancesco Buonamici in 1633 (*Opere*, 19, p. 410).

Dialogue, where the highest authority, namely Pope Urban VIII, is quoted as having said that 'it would be the height of presumption to try to limit or restrict the divine power and wisdom to any one particular fantasy'.[35] Unfortunately, the one who says this is Simplicio, the Aristotelian pedant who had cut such a poor figure throughout the whole *Dialogue*. The theologians were quick to spot this, and the Pope, when it was called to his attention, was personally affronted.

On 12 April Galileo was taken to the headquarters of the Holy Office, where he was not placed in a cell but provided with a three-room suite. He could stroll in the garden and his meals were brought in from the Tuscan Embassy, which had one of the best chefs in Rome. It is sometimes said that Galileo's trial was conducted before the Cardinals of the Holy Office, but this is not the case. The 'court' consisted of two officials: Vincenzo Maculano, a Dominican scholar and engineer who had recently been appointed Commissioner-General of the Holy Office, and Carlo Sinceri, who had been working there since 1606. Galileo met them four times: on 12 April, 30 April, 10 May, and 21 June 1633. No one else was present. After the first meeting and formal interrogation that took place on 12 April, Commissioner Maculano decided on a private meeting with Galileo. They met on 27 April and worked out an informal compromise: Galileo would admit that he had presented the evidence for the motion of the Earth in too strong a light and, in return, he would be treated with leniency. In practice, this meant that he would recant his 'errors' and the Inquisition would condemn him to imprisonment, but this would be immediately commuted to house arrest. After the second official meeting on 30 April, at which Galileo confirmed that he now realized that he had overstepped the limits imposed upon him, he was allowed to leave the Vatican and return to the Florentine Embassy. He was summoned back to the Holy Office on 10 May to be informed, as ecclesiastical law required, that he had eight days to reconsider. Galileo said that this was not necessary, and he withdrew to the Tuscan Embassy until 21 June, when he met Maculano and Sinceri for the fourth and last time. He vigorously denied that he had willingly infringed the ban on advocating the heliocentric system, but he was informed that he

[35] *Opere*, 7, p. 489; p. 358 in this volume.

would nonetheless be condemned on the very next day for having contravened the orders of the Church. At no time was he tortured or molested. He spent the night at the Holy Office and on the morning of 22 June 1633 was taken to a hall in the convent of Santa Maria Sopra Minerva in Rome, where he was made to kneel while the sentence was read condemning him to imprisonment. Still kneeling, Galileo formally abjured his errors. As had been agreed, he was not formally imprisoned but allowed to leave for Siena, where the archbishop, Ascanio Piccolomini, had invited him to be his guest. In 1634 he obtained permission to return to Florence, where he was confined to his country house on the outskirts of the city.

Galileo sought comfort in work, and within two years he completed the *Two New Sciences*, the book to which his lasting fame as a physicist is attached. Here Galileo worked out the mathematical and physical implications of ideas he had adumbrated in his *Dialogue*. The most important is his discovery that all bodies fall at the same speed regardless of their weight. This was historically important because it led Newton to realize that new laws of motion were required to explain why this should be the case. The story that Galileo dropped balls from the Leaning Tower of Pisa is probably apocryphal, but he did show great ingenuity in devising experiments with rolling balls along an inclined plane. He carefully measured the distance they travelled and the time it took. The outcome was the law that relates distance to the square of the time (expressed as $s = \frac{1}{2}gt^2$, where s stands for distance, g for the acceleration caused by gravity, and t for time). The insight behind Galileo's reasoning is the surprising fact that the vertical and horizontal components of projectile motion are independents. He illustrated this by showing that when balls are projected horizontally from the same height they go further if impelled with a greater force but that, regardless of the force, they strike the ground at exactly the same time. A ball dropped vertically from the same height when the balls were projected will also strike the ground at the same time.

When Galileo cast about for a publisher he came up against a new problem: the Church had issued a general prohibition against printing or reprinting any of his books. But his manuscript was smuggled out of the country and reached the publisher Louis Elsevier in Holland, a Protestant country over which the Roman Church had no power. Galileo feigned surprise and pretended not to know how the manuscript had left Italy. Although it is unlikely that anyone believed his story, the Church let the publication of the *Two New Sciences* in 1638 go unchallenged. Galileo, however, was never successful in obtaining the pardon he longed for. Urban VIII was adamant and Galileo remained under house arrest even after he became blind in 1638.

Galileo lived for four more years with the young Vincenzio Viviani as his assistant and, for the last few months, with Evangelista Torricelli. He died on 8 January 1642, five weeks before his seventy-eighth birthday. The sternness of Urban VIII was unabated; the Grand Duke wished to erect a suitable tomb for Galileo in the Church of Santa Croce but was warned not to do so. Nearly a century elapsed before Galileo's body was placed in the main body of the church, and it was not until 1822 that the *Dialogue on the Two Chief World Systems* was removed from the Index of Proscribed Books.

NOTE ON THE TEXT AND TRANSLATION

THE texts in this volume are all taken from the standard edition of Galileo's works: Galileo Galilei, *Opere*, edited by Antonio Favaro, 20 vols. (Florence, 1890–1909, with subsequent reprints), referred to below and in the Explanatory Notes as *Opere* followed by volume and page number. The *Sidereal Message* was written by Galileo in Latin and is translated here by William R. Shea. The translation was originally published in Galileo, *Sidereus Nuncius or A Sidereal Message* (Sagamore Beach, Mass.: Science History Publications, 2009), and the editors wish to thank Neale W. Watson, the President of Science History Publications, for allowing it to appear in this volume. All the other texts in this selection were written by Galileo in Italian and are translated by Mark Davie.

Details of the texts selected for this volume are as follows:

Sidereal Message: *Opere*, 3, 1, pp. 53–96. The text included here is complete apart from the detailed description of Galileo's observations of the satellites of Jupiter between 16 January and 2 March 1610, pp. 83–94.

Letters on the Sunspots: the first of Galileo's three letters is taken from *Opere*, 5, pp. 94–113, with the omission of a technical passage on pp. 102–5. The extract from his third letter is in the same volume, pp. 231–8.

The section on science and religion includes the complete texts of Galileo's letters to Benedetto Castelli (*Opere*, 5, pp. 281–8) and to the Grand Duchess Christina (*Opere*, 5, pp. 309–48). The letter from Cardinal Bellarmine is in *Opere*, 12, pp. 171–2. The essays we have entitled *Observations on the Copernican Theory*, which were unpublished in Galileo's lifetime and are untitled in his manuscripts, can be found in *Opere*, 5, pp. 351–70.

The extracts from *The Assayer* are taken from *Opere*, 6, pp. 232–3, 279–81, 339–40, and 347–50.

Dialogue on the Two Chief World Systems: the first and fourth days of the dialogue are included in full, from *Opere*, 7, pp. 26–131 and 442–89. The extracts from the second and third days are from *Opere*, 7, pp. 175–224, 248–61 (day 2), and 347–56 (day 3).

The documents on Galileo's trial can be found in *Opere*, 19, pp. 324–7, 336–47, 361–2, and 406–7.

The extracts from the *Two New Sciences* are taken from *Opere*, 8, pp. 49–51, 201–8, and 210–13.

In choosing to write his main scientific works after the *Sidereal Message* in Italian rather than Latin, Galileo was following a tradition going back to Machiavelli and ultimately to Dante, of writing for an educated lay public rather than for a narrowly academic readership. This tradition provided a model of Italian prose combining analytical rigour with the forceful, sometimes colloquial, directness of the vernacular. Galileo was able to exploit these qualities to the full, especially in the *Dialogue on the Two Chief World Systems*, where his polemic against traditional science is set within the fictional framework of an informal conversation among friends. I have tried to maintain the balance between these contrasting elements in my translation, and have felt free to draw on a correspondingly wide range of English usage in attempting, however imperfectly, to do justice to his prose.

M.D.

SELECT BIBLIOGRAPHY

Galileo's Works

The standard edition is *Le Opere di Galileo Galilei*, edited by Antonio Favaro, 20 vols. (Florence, 1899–1909, reissued several times).

English Translations

Bodies that Stay Atop Water or Move in it, translated with introduction and notes in Stillman Drake, *Cause, Experiment and Science* (Chicago: Chicago University Press, 1981).

Controversies on the Comets of 1618, translated by Stillman Drake (Philadelphia: University of Pennsylvania Press, 1960).

Dialogue on the Two Chief World Systems, translated with introduction and notes by Stillman Drake (Los Angeles, 1962, frequently reissued); an abridged translation by Maurizio Finocchiaro will be found in *Galileo on World Systems* (Berkeley: University of California Press, 1997).

Galileo's Logical Treatises, translation by William A. Wallace of Galileo's notes on Aristotle's *Posterior Analytics* (Dordrecht: Springer, 1992).

Galileo's Sidereus Nuncius or A Sidereal Message, translation with introduction and notes by William R. Shea and Tiziana Bascelli (Sagamore Beach, Mass.: Science History Publications, 2009). Also translated as *The Sidereal Messenger*, with introduction and notes by Albert Van Helden (Chicago: Chicago University Press, 1983), and as *The Starry Messenger* in Stillman Drake, *Telescopes, Tides and Tactics* (Chicago: Chicago University Press, 1983).

Letters on Sunspots, translated with introduction and notes in Eileen Reeves and Albert Van Helden, *On Sunspots* (Chicago: Chicago University Press, 2010).

On Motion and On Mechanics, translated by I. E. Drabkin and Stillman Drake (Madison, Wisc.: University of Wisconsin Press, 1960).

The Assayer, partially translated in Stillman Drake, *Discoveries and Opinions of Galileo* (New York: Anchor, 1957).

The Essential Galileo, contains selections from Galileo's works edited and translated by Maurice A. Finocchiaro (Indianapolis: Hackett, 2008).

The Little Balance, translated by Laura Fermi and Gilberto Bernardini, in *Galileo and the Scientific Revolution* (Greenwich, Conn.: Dover, 1965).

Two New Sciences, translated with introduction and notes by Stillman Drake (Madison, Wisc.: University of Wisconsin Press, 1974).

Works on Galileo

Bagioli, Mario, *Galileo Courtier* (Chicago: Chicago University Press, 1993).

—— *Galileo's Instruments of Credit* (Chicago: Chicago University Press, 2006).

Camerota, Michele, *Galileo Galilei e la cultura scientifica nell'età della Controriforma* (Rome: Salerno Editrice, 2004). In Italian.

Cohen, H. Floris, *How Modern Science Came into the World* (Amsterdam: Amsterdam University Press, 2010).

Drake, Stillman, *Galileo at Work: His Scientific Biography* (Chicago: Chicago University Press, 1978).

—— *Essays on Galileo*, 3 vols. (Toronto: University of Toronto Press, 1999).

Fantoli, Annibale, *Galileo, for Copernicanism and for the Church* (Notre Dame, Ind.: University of Notre Dame Press, 1996).

Finocchiaro, Maurice A., *The Galileo Affair: A Documentary History* (Berkeley: University of California Press, 1989).

—— *Retrying Galileo 1633–1992* (Berkeley: University of California Press, 2005).

Heilbron, J. L., *Galileo* (Oxford: Oxford University Press, 2010).

Machamer, Peter (ed.), *The Cambridge Companion to Galileo* (Cambridge: Cambridge University Press, 1998).

McMullin, Ernan (ed.), *The Church and Galileo* (Notre Dame, Ind.: Notre Dame University Press, 2005).

Numbers, Ronald L., *Galileo Goes to Jail and Other Myths about Science and Religion* (Cambridge, Mass.: Harvard University Press, 2009).

Palmieri, Paolo, *Re-enacting Galileo's Experiments* (Lewiston, NY: The Edwin Mellen Press, 2008).

Redondi, Pietro, *Galileo: Heretic* (Princeton: Princeton University Press, 1987).

Renn, Jurgen (ed.), *Galileo in Context* (Cambridge: Cambridge University Press, 2001).

Roland, Wade, *Galileo's Mistake: The Archeology of a Myth* (Toronto: University of Toronto Press, 2001).

Sharratt, Michael, *Galileo Decisive Innovator* (Cambridge: Cambridge University Press, 1996).

Shea, William R., *Galileo's Intellectual Revolution*, 2nd edn. (New York: Neale Watson Academic Publications, 1977).

—— and Artigas, Mariano, *Galileo in Rome: The Rise and Fall of a Troublesome Genius* (Oxford: Oxford University Press, 2003).

—— —— *Galileo Observed: Science and the Politics of Belief* (Sagamore Beach, Mass.: Science History Publications, 2006).

Sobel, Dava, *Galileo's Daughter* (London: Penguin, 1999).

Vergara Caffarelli, Roberto, *Galileo Galilei and Motion* (Berlin: Springer, 2009).

Wallace, William A., *Galileo and His Sources: the Heritage of the Collegio Romano in Galileo's Science* (Princeton: Princeton University Press, 1984).

Willach, Rolf, *The Long Route to the Invention of the Telescope* (Philadephia: American Philosophical Society, 2008).

Wootton, David, *Galileo Watcher of the Skies* (New Haven, Conn.: Yale University Press, 2010).

Further Reading in Oxford World's Classics

Ariosto, Ludovico, *Orlando Furioso*, ed. and trans. Guido Waldman.

Aristotle, *Physics*, trans. Robin Waterfield, ed. David Bostock.

Bacon, Francis, *The Major Works*, ed. Brian Vickers.

Tasso, Torquato, *The Liberation of Jerusalem*, trans. Max Wickert, ed. Mark Davie.

Vasari, Giorgio, *The Lives of the Artists*, ed. and trans. Julia Conaway Bondanella and Peter Bondanella.

A CHRONOLOGY OF GALILEO

1543   Nicolaus Copernicus (1473–1543) publishes *On the Revolutions of the Heavenly Spheres* and Andreas Vesalius (1514–64), *On the Fabric of the Human Body*.

1545   Council of Trent convenes and will be in session off and on for eighteen years until 1563.

1551   The Roman College (now the Pontifical Gregorian University) founded by the Jesuits in Rome.

1559   First worldwide Index of Prohibited Books promulgated by the Roman Catholic Church.

1564   Galileo is born in Pisa, on 15 or 16 February; Michelangelo Buonarroti dies in Florence on 18 February.

1581   Galileo enrols at University of Pisa.

1585   He abandons studies at Pisa without taking a university degree. He considers becoming a painter but decides to study mathematics.

1587   Galileo goes to Rome to discuss his essay on the centre of gravity of solids with Christopher Clavius, a Jesuit professor at the Roman College and the most famous astronomer of his day. During this period he also writes on the balance and on Archimedes.

1589   Appointed professor of mathematics at the University of Pisa; develops a rudimentary thermometer; begins to study falling bodies. His lecture notes indicate that he read the works of Aristotle and those of Christopher Clavius.

1591   Around this time, Galileo drafts a work, *On Motion*, in which he is critical of Aristotelian philosophy. Galileo's father, Vincenzio Galilei, dies.

1592   Galileo becomes professor of mathematics at the University of Padua, where he will spend eighteen years.

1593   He writes a *Treatise on Fortifications and Architecture* for young noblemen to whom he gives private lessons.

1594   Drafts a *Treatise on Mechanics* that he will revise and enlarge over the next few years.

1597   Writes a *Treatise on the Sphere* for his students.

1600   Giordano Bruno is burnt at the stake in Rome. Galileo's first daughter, Virginia, is born in Padua.

1601   Birth of his second daughter, Livia.

1603   Prince Federico Cesi founds the Lincean Academy in Rome.

1604   Galileo gives three public lectures on a new star that appeared in the heavens.

1605   The teenage Prince Cosimo de' Medici takes instruction from Galileo in the summer in Tuscany.

1606   Galileo publishes sixty copies of the *Operations of the Geometric and Military Compass*, an instrument that he manufactured and sold. His son, Vincenzio, is born in Padua.

1607   Baldessar Capra publishes a pirated Latin edition of Galileo's *Operations of the Geometric and Military Compass*. Galileo sues him and publishes an account of the incident.

1609   The Grand Duke Ferdinando I dies in Florence; Cosimo II succeeds him. Galileo devises a new telescope, observes and measures mountains on the Moon.

1610   He discovers four satellites around Jupiter and publishes *A Sidereal Message*. He is appointed chief mathematician and philosopher to the Grand Duke of Tuscany and leaves Padua for Florence.

1611   Galileo visits Rome and is made a member of the Lincean Academy.

1612   He publishes a book on floating bodies.

1613   Prince Cesi publishes Galileo's *Letters on the Sunspots* in Rome. His daughters, Virginia and Livia, enter the Convent of San Matteo in Arcetri outside Florence.

1614   Galileo writes a *Letter to Benedetto Castelli* in which he argues that the Copernican theory is not at variance with Catholic doctrine.

1615   Galileo expands his *Letter to Benedetto Castelli* into the *Letter to the Grand Duchess Christina* that is widely circulated in manuscript but will only be published in 1636 in Holland. He writes his *Observations on the Copernican Theory* and leaves for Rome where he arrives in December and will stay until June 1616.

1616   Galileo writes his *Discourse on the Tides*. On 26 February Galileo is given a formal warning forbidding him from holding, teaching, or defending Copernicanism. On 5 March Copernicus' *On the Revolutions of the Heavenly Spheres* is placed on the Index of Prohibited Books.

1618   Three comets appear in rapid succession, generating interest and debate. Galileo goes on a pilgrimage to Loreto.

1619 Galileo publishes anonymously a *Discourse on the Comets* criticizing the interpretation given by the Jesuit professor Orazio Grassi who replies in a work entitled *The Philosophical and Astronomical Balance.*

1623 Maffeo Barberini becomes Pope Urban VIII. Galileo dedicates *The Assayer* to him.

1624 Galileo travels to Rome and sees the Pope six times in as many weeks.

1629 Bubonic plague enters northern Italy from Germany.

1630 Galileo returns to Rome to obtain a printing licence for his *Dialogue Concerning the Two Chief World Systems.* Prince Cesi dies. Bubonic plague strikes Florence, where the deaths will amount to ten thousand.

1631 Galileo's brother, Michelangelo, dies of plague in Germany.

1632 The *Dialogue Concerning the Two Chief World Systems—Ptolemaic and Copernican* is published in Florence.

1633 Galileo stands trial in Rome. His *Dialogue* is prohibited and he is sentenced to house arrest.

1634 His daughter Virginia (in religion Suor Maria Celeste) Galilei dies in Arcetri.

1637 Galileo loses his eyesight.

1638 Louis Elsevier publishes Galileo's *Two New Sciences* in Leiden, Holland.

1641 Vincenzio Galilei draws his father's design for a pendulum clock.

1642 Galileo dies in Arcetri on 8 January.

1644 Pope Urban VIII dies.

A SIDEREAL MESSAGE*

intended for everyone but mainly
philosophers and astronomers,
and making known

THE GREAT AND MARVELLOUS SIGHTS that

GALILEO GALILEI,

A FLORENTINE PATRICIAN*

and Professor of Mathematics at the University of Padua
observed with the aid of a SPYGLASS, recently discovered
by him,* on the face of the Moon, in innumerable fixed stars,
in the Milky Way, and in nebulous stars,
but above all

IN FOUR PLANETS*

that revolve around the star JUPITER with amazing speed
at different distances and in different periodic times. Unknown
to anyone up to this day, these were very recently discovered
by the Author, who decided to call them

MEDICEAN STARS*

VENICE, printed by Tommaso Baglioni, MDCX.
With permission of the authorities and copyright.

DEDICATION

To the Most Serene Cosimo de' Medici II, Fourth Grand Duke of Tuscany *

NOBLE and truly public-spirited was the intention of those who determined to protect from envy the great achievements of men of outstanding virtue and to rescue their names, which deserve immortality, from neglect and oblivion. Hence, as a memorial to future ages, likenesses sculptured in marble or cast in bronze; hence, statues on foot or on horseback; hence, columns and pyramids whose cost is sky-high, as the poet says;* hence also, the building of cities to bear the names of those whom posterity deemed worthy of being remembered throughout the ages. For recollection all too easily slips away from the human mind unless it is constantly reminded by outside stimuli.

There were others, however, who looked for more stable and enduring memorials. These did not entrust the eternal praise of great men to building-blocks or strips of metal but to the custody of the Muses and the imperishable monuments of literature.* But why mention this? Could it be that human ingenuity, satisfied with what happens here below, dared proceed no further? No, on the contrary, looking beyond and realizing that warfare, the weather, or the passing of time eventually raze all human monuments to the ground, less corruptible signs were sought over which devouring time and envious age could claim no rights. So betaking itself to the heavens, human ingenuity inscribed on the well-known and eternal orbs* of the brightest stars the names of those who for their eminent and godlike deeds were deemed worthy of enjoying all eternity in company of the stars. Wherefore, the fame of Jupiter, Mars, Mercury, Hercules, and the other heroes by whose names the stars are called, will not fade until the splendour of the stars themselves is extinguished. But this noble and admirable custom went out of fashion ages ago when those glorious seats were occupied by the ancient heroes who now hold them, as it were, in their own right. In vain did the affection of Augustus try to introduce Julius Caesar into their company.* The star that appeared in his day and that he wanted to name 'Julian' belonged to those that the Greeks call comets and we 'hair-like', and it vanished in a short time and mocked his too eager hope.

But, Most Serene Prince, we can augur far more genuine and happy things for your Highness, for no sooner had the immortal greatness of your mind begun to shine on Earth than bright stars presented themselves in the heavens like tongues to tell and celebrate your surpassing virtues for all time. Behold, therefore, four stars reserved for your famous name. They do not belong to the common and less distinguished multitude of fixed stars but to the illustrious rank of the planets. Moving at different rates around Jupiter, the noblest of the planets,* as if they were his own children, they trace out their orbits with marvellous speed while, at the same time, with one harmonious accord, they go round the centre of the world, namely the Sun itself,* and complete their great revolutions in twelve years.*

The Creator of the stars seems to have himself* provided me with clear reasons for dedicating these new planets to Your Highness' famous name in preference to all others.* For just as these stars, like worthy children, never leave the side of Jupiter by any appreciable distance, so everyone knows that clemency, kindness of heart, gentleness of manners, splendour of royal blood, majesty in deportment, wide extent of influence, and authority over others have all fixed their abode and seat in your Highness. Who, I say, does not know that all these qualities emanate from the most benign planet Jupiter,* according to the will of God from Whom all good things flow? Jupiter, Jupiter I say, that at the moment of your Highness' birth had already risen above the misty vapour on the horizon and occupied the mid-heaven. Illuminating the eastern angle* from its royal house, it looked down from its exalted throne upon your blessed birth, and poured out the brightness of its majesty in the pure air in order that your tender body and your mind, already adorned by God with the most noble ornaments, might imbibe with its first breath the whole of that strength and power.

But why should I use merely plausible arguments when I can demonstrate my conclusion with practically absolute certainly? It pleased God Almighty that I should be considered worthy by your Most Serene Parents to teach your Highness mathematics* during the last four years at that time of year when it is customary to rest from more exacting studies. It was clearly God's will that I should serve your Highness and be exposed at close quarters to the rays of your incredible kindness and gentleness. What wonder is it, therefore, if you have so warmed my heart that I, who am your subject not only by choice but by birth and lineage, should night and day scarcely think of anything else but how to make known how grateful I am to you, and how desirous to promote your glory. And so, since it was under your auspices, most serene Cosimo, that I discovered those stars that were unknown to all astronomers before me, I have the right to call them by the most august name of your family. Since I was the first to discover them, who can rightly reproach me if I also give them a name and call them MEDICEAN STARS, hoping that perhaps as much honour may accrue to the stars from this title, as other stars have brought the other heroes? For to say nothing of your most serene ancestors whose everlasting glory is attested by the historical record, your virtue alone, Great Hero, can confer on those stars a name that is immortal. For who can doubt that you will not only maintain and preserve the high expectations that you have aroused at the beginning of your reign, but far surpass them? So when you have conquered your equals, you will nonetheless still vie with yourself and become greater day by day.

Receive then, most clement Prince, this honour that was reserved for you by the stars, and may you enjoy for many years those blessings that descended upon you not so much from the stars, as from God, their Creator and Governor.

Padua, on the fourth day before the Ides of March MDCX.
Your Highness' most devoted servant,

Galileo Galilei

PERMISSION TO PRINT

THE undersigned, their Excellencies the Heads of the Council of Ten,* having been informed by the Overseers* of the University of Padua that the two persons appointed for this task, namely the Reverend Father Inquisitor and the circumspect Secretary of the Senate, Giovanni Maraviglia, declared under oath that in the book entitled *Sidereal Message, etc.** of Galileo Galilei there is nothing that is contrary to the Holy Catholic Faith, principles or good customs, and that it is worthy of being printed, authorize its publication in this city.

Dated 1 March 1610.

D. M(arco) Ant(onio) Valaresso

D. Nicolò Bon   } Heads of the Council of Ten

D. Lunardo Marcello*

Bartolomeo Comino
Secretary of the Illustrious
Council of Ten

1610, on 8 March. Registered in the book on page 39.

Giovanni Battista Breatto
Coadjutor of the Office against
Blasphemy

AN ASTRONOMICAL MESSAGE

that contains and explains
recent observations
made with the aid of a new spyglass of the face of the Moon,
the Milky Way, the nebulous stars, and
an innumerable number of fixed ones, as well as in
four planets,
*called cosmic stars**
that have never been seen before.

GREAT indeed are the things that this small book offers to the consideration and study of anyone interested in nature. I call them great because of the excellence of the subject matter and their absolute and unheard-of novelty, but also on account of the instrument whereby they became known to our senses.

It is surely a great thing to add countless stars to the large number that have already been observed with the naked eye, and to render them clearly visible when they had never been seen before, and are more than ten times as numerous as the old with which we are familiar.*

It is a most beautiful and a very pleasing sight to look at the body of the Moon, which is removed from us by almost sixty terrestrial diameters,* and to see it as if it were only two diameters away.* This means that the diameter of the Moon looks almost thirty times larger, its surface nine hundred times bigger, and its whole body close to twenty-seven thousand times more voluminous than when seen with the naked eye.* The observational evidence is so compelling that anyone can grasp for himself that the Moon's surface is not smooth and polished but rough and uneven. Like the face of the Earth, it is covered all over with huge bumps, deep holes, and chasms.

Furthermore, it is no small matter to have put an end to disputes about the Galaxy, namely the Milky Way, and to have made its nature clear to the senses, let alone the understanding. It is also a fine and a pleasant thing to be able to point out, as with one's finger, the nature of those stars that all astronomers have hitherto called nebulous, and to show that it is very different from what was believed until now.

But what is even more admirable, and what we mainly want to let astronomers and philosophers know, is that we have found four wandering stars* that no one before us had heard about or observed, and that these revolve around one of the conspicuous planets. Like Venus and Mercury, which go around the Sun, they have their own periods of revolution so that they sometimes precede, sometimes follow their planet but in such a way that they never stray beyond certain limits.* All this was found and observed a few days ago with a spyglass that I devised after having been enlightened by divine grace.

Other and perhaps greater things will be discovered in the days to come by me or by others with the aid of a similar instrument, so before giving an account of my observations I will briefly say something about its shape and its construction, as well as how I came to think of it.

About ten months ago* the rumour reached us that a Dutchman had made a spyglass by the aid of which visible objects, although at a great distance from the eye of the observer, were seen distinctly as if near. News of this truly wonderful result spread, and if it was believed by some, it was denied by others. But confirmation arrived a few days later in a letter written from Paris by a noble Frenchman, Jacques Badouère,* and this is what made me concentrate all my energy on finding how this was achieved and by what means I could make a similar instrument. Basing myself on the theory of refraction, I achieved my goal in little time. I first got a tube of lead, and fitted the ends with two lenses that were both plane on one side but, on the other side, convex in one case, and concave in the other. I then applied my eye to the concave lens, and saw objects as fairly large and near at hand. They appeared to be three times closer and nine times larger than when seen with the naked eye alone. I then constructed a better instrument that made objects appear sixty times bigger.* Sparing neither time nor expense, I managed to construct an instrument that was so good that objects seen through it appeared a thousand times bigger, and more that thirty times closer than when viewed by the natural power of sight alone. It would be superfluous to list the number and the importance of the advantages of this instrument on land and at sea.

But leaving terrestrial observation, I turned to the study of the heavens and saw, first of all, the Moon as near as if it were hardly two terrestrial diameters away.* Next I repeatedly observed, with the greatest pleasure, the fixed and the wandering stars. When I realized their huge number, I began to ponder how to measure the intervals between them, and I eventually discovered a way of doing so. It is fitting to warn those who may want to undertake this kind of observation that they must, in the first place, acquire an excellent spyglass that shows objects clearly, distinctly, and free from any haziness. It should magnify at least four hundred times so that the objects appear twenty times closer. Without such an instrument, everything we saw in the heavens and that we list below will be sought for in vain.

In order to determine without great trouble the magnifying power of the instrument, trace on paper the outline of two circles or two squares such that one is four hundred times as large as the other, as will be the case when its diameter is twenty times that of the other. Then, having attached these two figures to the same wall, observe them both simultaneously from a distance, looking at the smaller one through the spyglass, and at the larger with the other, naked eye. This can easily be done by keeping both eyes opened at the same time. The two figures will appear to be of the same size if the instrument magnifies the objects in the said ratio. With such an instrument, we can determine distances, and we proceed in the following way. In the interest of clarity, let ABCD be the tube, and E the eye of the observer. When there are no lenses in the tube, the rays are carried to the object FG along the straight lines ECF and EDG,* but when the lenses are inserted they are carried along the refracted lines ECH and EDI. The rays, which before were directed without constraint towards FG, are now squeezed together and only include the part HI. Next we determine the ratio of the distance EH to the

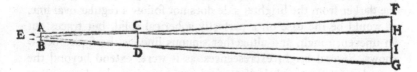

line HI and, with the help of a table of sines, we find that the size of the angle subtended at the eye by object HI is only a few minutes of an arc.* Now if we perforate thin sheets of metal, some with larger and others with smaller holes, and place over the lens CD one size or another, as need may be, we can obtain any number of angles subtending a few minutes of an arc. By this means we shall be able to conveniently measure the intervals between stars a few minutes apart within an error of one or two minutes.

Let it suffice for the present, however, to have touched upon this rapidly and to have given a foretaste of what is to come, for on some further occasion we shall provide a complete theory of this instrument.* But let us now review the observations that we made over the last two months, and here I call upon all those who are eager for true philosophy to witness the beginnings of important considerations.

Let us first consider the face that the Moon turns towards us and, to make things easier, I distinguish two parts, the brighter and the darker one. The brighter seems to surround and illuminate the whole hemisphere of the Moon whereas the darker one, like a cloud, spreads over the face of the Moon and makes it appear covered with spots that are somewhat dark and of considerable size. These have always been observed and are obvious to everyone. We shall call them *great* or *ancient* spots to distinguish them from others of smaller size that are so thickly scattered that they cover the whole surface of the Moon but mainly the brighter part. These were never observed by anyone before us. After examining them repeatedly we were led to a conclusion about which we are certain. The surface of the Moon is not even, smooth, and perfectly spherical, as the majority of philosophers have conjectured that it and the other celestial bodies are, but on the contrary, rough and uneven, and covered with cavities and protuberances just like the face of the Earth, which is rendered diverse by lofty mountains and deep valleys. The appearances that enabled me to reach this conclusion are the following.

On the fourth or the fifth day after new moon, when the Moon presents itself with bright horns, the boundary-line that separates the darker from the brighter side does not follow a regular oval line, as would be the case on a perfectly spherical solid, but traces out an uneven, rough, and altogether sinuous line, as the figure below shows. Several bright excrescences, as it were, extend beyond the boundary of light and darkness and penetrate into the darker part, while on the other hand, patches from the dark side enter the brighter one. Indeed, a great number of small blackish spots, completely separated from the dark part, are scattered over most of the area that is already flooded by sunlight, with the exception of the part occupied by the *great* and *ancient* spots. We note furthermore that the small spots just mentioned always have this feature in common, that their darker side faces the Sun while on the side opposite the Sun their contours are brighter, as if they were crowned with shining peaks. We see exactly the same thing on Earth at sunrise when the sunlight has not yet spread over the valleys although the mountains surrounding them on the side away from the Sun are already shining brightly. And just as shadows in hollows on Earth decrease in size as the Sun rises higher, so these lunar spots shed their darkness as their illuminated parts grow larger.

But not only is the boundary of light and shadow on the Moon seen to be uneven and sinuous, what causes even greater astonishment is that very bright points appear inside the darker portion of the Moon. They are divided and separated from the illuminated part, and removed from it by a considerable distance. After some time, they gradually increase in size and brightness until, after two or three hours, they become joined with the rest of the bright portion, which has now increased in size. In the meantime, more and more bright points light up inside the dark portion, swell in size, and eventually embrace the brighter surface that has extended still further. This is illustrated in the same figure.

On Earth, before sunrise, are not the peaks of the highest mountains illuminated by the Sun's rays while the plains are still in shadow? In a little while, does not the light spread further until the middle and larger parts of these mountains become illuminated and, in the end, when the Sun has risen, the illuminated parts of the plains and the hills are joined? On the Moon, however, the difference between high peaks and depressions appears to be much greater than the one caused by ruggedness on the surface of the Earth, as we shall show below. In the meantime, I cannot pass over in silence something worthy of consideration that I observed when the Moon was hasting towards first-quarter, as can be seen in the same figure above. Near the lower cusp, a great dark gulf extends into the illuminated side. I observed it for a while and saw that it was dark throughout, but after a couple of hours a bright peak began to emerge a little below the centre of the depression. It gradually grew in size and assumed a triangular shape that was still completely detached and separated from the illuminated area. Around it three other small points soon began to shine until, when the Moon was just about to set, this triangular shape, which had become more extended and larger, joined the rest of the illuminated part and penetrated into the dark gulf like a vast promontory, still surrounded by the three bright peaks that we have just mentioned. At the ends of the top and bottom cusps some bright points emerged, completely separated from the rest of the illuminated part, as can be seen in the same figure.

There was also a great number of dark spots in both cusps, but mainly in the lower one. Those nearer the boundary of light and shadow appeared larger and darker while those further removed were fainter and not so dark. But as we mentioned above, the dark portion of each spot was always turned towards the incoming rays of the Sun while the bright rim that surrounded it on the other side always faced the dark region of the Moon. This part of the lunar surface, which is spotted as a peacock's tail is decked with azure eyes, resembles glass vases that are plunged while still hot into cold water and acquire that crackled and wavy surface from which they receive the common name of frosted glass. But the *great* spots on the Moon do not appear to be cracked or crowded with depressions and prominences in this manner, but rather to be even and uniform for only here and there do some bright patches emerge.

So if someone wanted to revive the ancient Pythagorean theory, namely that the Moon is like another Earth, its land surface would be more fittingly represented by the brighter region, and the expanse of water by the darker one. I have never doubted that if the terrestrial globe were observed from afar, bathed in sunlight, the land surface would appear brighter and the expanse of water darker. Furthermore, whether the Moon is waxing or waning, the *great* spots on the Moon appear to be more depressed than the brighter tracts that appear here and there in the vicinity of the *great* spots, and always along the boundary of light and shadow, as we noticed when drawing the figures. The edges of the large spots are not only lower but more even, and free from creases and ruggedness. The bright part stands out particularly near the spots and, before first quarter and approaching last quarter, huge stretches arise above and below a certain spot in the higher and northerly region of the Moon,* as can be seen in the figures reproduced here below.

Before the last quarter this same spot is seen to be surrounded by darker contours that, like the highest ridges of mountains, appear darker on the side that is facing away from the Sun, and brighter on the side that is turned towards the Sun. Now just the opposite happens in the case of the cavities where the side that is facing away from the Sun appears brilliant, while the side that is turned towards

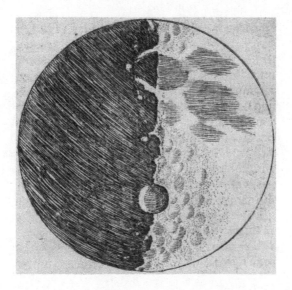

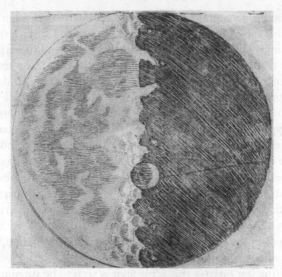

the Sun appears dark and shadowy. When the illuminated portion of the Moon has decreased in size, and the spot we have mentioned is nearly all covered in darkness, the brighter mountain ridges climb above these shadows. This twofold appearance of the spot is illustrated in the following figures:

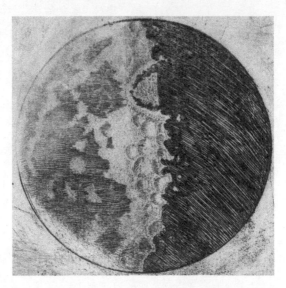

There is another thing that I must not omit, because I found it very striking: near the centre of the Moon there is a cavity that is larger than all the other ones and perfectly round in shape. I observed it near both the first and the last quarter, and I have portrayed it as well as possible in the second of the two figures above. As to light and shade, this cavity offers the same appearance as would a region like Bohemia, if it were enclosed on all sides by very high mountains arranged along the circumference of a perfect circle. For on the Moon this area is walled in by such high peaks that the side adjacent to the dark portion of the Moon is seen to be bathed in sunlight before the boundary between light and shadow reaches halfway across the cavity. Just like other spots, its shaded portion faces the Sun while its lighted part faces the dark region of the Moon. It gives me pleasure, for the third time, to draw attention to this very cogent argument that ruggedness and unevenness are spread over the entire brighter region of the Moon. Of these spots, moreover, the darker ones are always next to the boundary between light and shadow, and those further away appear smaller in size and less dark, so that when the Moon, at its opposition to the Sun, becomes full there remains only a slight and faint difference between the darkness of the cavities and the brightness of the peaks.

The things we have enumerated were observed in the brighter parts of the Moon. In the *great* spots we see no such differences of depressions and heights as those that we are compelled to recognize in the brighter regions on account of the changes in their shape that result from the different ways they are illuminated by rays of the Sun that arrive from a variety of angles. Inside the *great* spots we also find small zones that are less dark, as we have indicated in the illustrations, but they always have the same appearance, and their darkness neither increases nor decreases, although they sometimes appear a little darker or a little brighter according as the rays of the Sun fall upon them more or less obliquely. Furthermore, they are joined to their neighbouring parts by a very gradual connection so that their boundaries run together and blend. But matters are quite different in the case of the spots that occupy the brighter part of the Moon's surface.

The sharp contrast of light and shadow gives them well defined boundaries as if they were steep walls covered with jagged and projecting rocks. Moreover inside these *great spots* certain small zones are observed to be brighter than the surrounding region, and some are very bright indeed. But their appearance, as well as that of the darker parts, is always the same because there is never any change in their shape, brightness, or darkness. Hence, it is proved beyond doubt that their appearance results from a real dissimilarity of parts and not merely from the different ways the rays of the Sun alter their shape or the size of their shadow, as happens for other smaller spots found in the brighter part of the Moon. These change from day to day, grow, decrease, and vanish because they are only produced by the shadows that are cast by the peaks.

Nevertheless, I am told that many have serious reservations on this point and are so concerned that they feel compelled to doubt a conclusion already explained and confirmed by so many observations. If that part of the lunar surface that reflects the sunlight more brightly is full of chasms, that is, countless bumps and depressions, why is it that the western edge of the waxing Moon, the eastern edge of the waning Moon, and the whole periphery of the full Moon are not seen to be uneven, rough, and sinuous? On the contrary, they appear as precisely round as if they had been drawn with a compass, and undamaged by bumps or cavities. The more so as the whole border is made of the brighter lunar material which, as we have said, is full of bumps and holes. None of the *great* spots extend as far as the circumference, but are always seen gathered together far from the rim.

Of this fact, which has given rise to such serious doubt, let me propose a twofold explanation and, thereby, a twofold solution. First, if the bumps and the hollows on the lunar body extended only along the circumference at the very end of the hemisphere that we see, then the Moon could—indeed, should—show itself somewhat as a toothed wheel, terminated by a bumpy and sinuous edge. But if there is not only one range of peaks, situated just along the circumference, but many ranges of mountains with their hollows and canyons disposed in ranks near the Moon's edge, and if these are not only on the hemisphere that is visible, but also on the other side (close however to the boundary between the hemispheres), then, from afar, the eye will not be able to distinguish the peaks from the depressions because the intervals between the mountains lying along one circle, namely along the same range, will be hidden by the interposition of yet other peaks on yet other ranges. This will be especially true if the observer is placed along the same straight line as the summits of the peaks. This happens on Earth when the summits of several mountains are so close together that they appear to lie on the same plane surface for an observer located at some distance and at the same altitude. Likewise, when the sea is rough the crests of the waves appear to lie along the same plane, although between the billows there are chasms and hollows of such depth that they hide not only the hulls but also the sterns, the masts, and the rigging of tall ships. Now because there are intricate arrangements of prominences and cavities on the Moon and along its circumference, and because the eye, regarding them from afar, lies in nearly the same plane as their summits, it should come as no surprise that they present themselves to the visual ray that grazes them as a regular and unbroken line.*

To this explanation another may be added, namely that there is around the body of the Moon, just as around the Earth, a substance denser than the rest of the aether* and enclosed in an orb that can receive and reflect the Sun's rays, but without being so opaque as to prevent seeing through it (especially when it is not illuminated). The orb, when illuminated by the rays of the Sun, makes the body of the Moon appear as a larger sphere than it is, and, were it thicker, it would stop our sight from penetrating to the actual body of the Moon. But it is indeed thicker near the rim of the Moon. I do not mean thicker in an absolute sense but relative to our visual rays that cut it obliquely. Therefore, it can obstruct our vision and, especially when illuminated, conceal the circumference of the Moon that is exposed to the Sun. This can be understood more clearly in the figure below in which the body of the Moon, ABC, is surrounded by a vaporous orb, DEG:

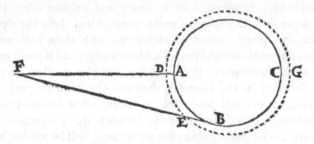

Our vision, coming from F, penetrates to the central region of the Moon, such as at A, through the lesser thickness of vapours, DA, whereas towards the extreme edge a mass of thicker vapours, EB, shuts out its boundary from our sight. Evidence for this is that the illuminated part of the Moon appears to have a larger rim than the rest of the orb that lies in shadow. It can reasonably be conjectured that this is also the cause why the *great* spots on the Moon are not seen to reach the edge of the circumference on any side, although it can be supposed that some should be found there. If they are not visible, it is probably because they are hidden under a deeper and brighter vaporous orb.

That the bright surface of the Moon is dotted all over with bumps and cavities has been made sufficiently clear, I think, by the account of our observations. It now remains to say something about their size, and to demonstrate that terrestrial 'prominences' are much smaller, I mean smaller in absolute terms, not only relative to the size of their respective globes. This is plainly shown as follows.

I have often observed, in various positions of the Moon in reference to the Sun, that some summits in the darker part of the Moon, although fairly removed from the boundary of light, are illuminated. Comparing their distance with the whole diameter of the Moon, I found that it was sometimes greater than one-twentieth of the diameter. Assuming this to be correct, let us represent the lunar globe by the great circle, CAF, with E as its centre:

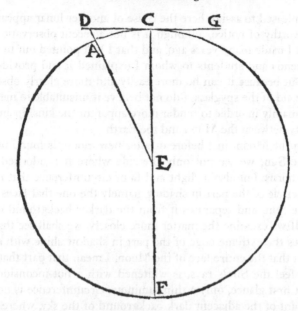

The diameter, CF, is to the diameter of the Earth as two is to seven, and since the diameter of the Earth, according to the most careful observations, contains 7,000 Italian miles,* CF will be 2,000, and CE 1,000, and the one-twentieth part of the whole CF, 100 miles. Now let CF be the diameter of the great circle that divides the brighter part from the darker one (for owing to the great distance of the Sun from the Moon this circle does not differ appreciably from a great one). Let A be separated from point C by one-twentieth of this diameter. Extend the radius EA to meet at point D the tangent GCD (representing the illuminating ray). Then arc CA, or rather the straight line CD, will consist of 100 such units as CE contains 1,000, and the sum of the squares of DC and CE will be 1,010,000.* The square of DE is equal to this. Therefore the whole of ED will be more than 1,004, and AD will be more than 4 such units of which CE contains 1,000. Therefore, the height of AD, which on the Moon indicates a summit reaching up to the ray of the Sun GCD, is removed from C by the distance CD, and is more than 4 Italian miles. Now on the Earth there are no mountains reaching a perpendicular height of even 1 mile high.* It is clear therefore that mountains on the Moon are higher than on Earth.

I am pleased to assign here the cause of another lunar appearance that is worthy of notice, although it is not a recent observation but one that I made many years ago, and that I had pointed out to a few close friends and students to whom I explained it and provided its cause. But because it can be more easily and more clearly observed with the aid of the spyglass, I do not believe it unsuitable to mention it here, mainly in order to render more apparent the kinship and the similarity between the Moon and the Earth.

When the Moon, just before or after new moon, is found not far from the Sun, we see not only the side where it is adorned with shining horns, but also a slight and faint circumference that marks out the circle of the part in shadow, namely the one that faces away from the Sun, and separates it from the darker background of the aether. If we examine the matter more closely, we shall see that not only does the extreme edge of the part in shadow shine with a dim light but that the entire face of the Moon, I mean that part that does not yet feel the Sun's rays, is whitened with a not-inconsiderable glow. At first glance, only a thin luminous circumference is noticed on account of the adjacent dark background of the sky, whereas the rest of the surface appears darker because of the contiguity of the shining horns that obscure our vision. But if someone places himself where a roof, a chimney, or some other object (it must not be too close to the eye) conceals the shining horns while leaving the rest of the lunar globe exposed to view, then he shall see that this tract of the Moon, although deprived of sunlight, glows with a not-inconsiderable light. The more so when the gloom of the night has deepened after the departure of the Sun, since the same light appears brighter against a darker background. It is also the case that this secondary light (so to speak) of the Moon is greater the closer the Moon is to the Sun. It decreases gradually as the Moon moves away from the Sun, so that after the first quarter, and before the second, it is weak and very dim even if observed in a darker sky, whereas at an angular distance of 60 degrees or less it is remarkably bright, even in twilight, so bright indeed, that with the help of the good spyglass the *great* spots can be distinguished in it.

This wonderful brightness has made philosophers marvel, and they have offered various explanations for it. Some have said that it is the inherent, natural brightness of the Moon, some that it is imparted by Venus, others by all the stars, and others still by the Sun whose rays permeate through the solid body of the Moon. Now these opinions are easily refuted and shown to be false. For if this light were the Moon's own or if it were conferred by the stars, it would assuredly be retained during eclipses when it would show itself because it would be left alone in a particularly dark sky. But this goes against experience: the brightness that is seen on the Moon during eclipses is much less, being somewhat reddish and almost copper-coloured, whereas this secondary light is brighter and whiter. Furthermore, the brightness seen during an eclipse is changeable and shifting, for it wanders over the face of the Moon so that the part nearest the circumference of the circular shadow cast by the Earth is brighter, whereas the rest of the Moon is always seen to be dark. Without doubt, this brightness is due to the way the rays of the Sun just graze some denser region that surrounds the Moon. Through this contact, a kind of dawn is spread over the neighbouring regions of the Moon just as twilight spreads in the morning and in the evening on Earth. But I shall deal with this more fully in my book on *The System of the World.**

The suggestion that this kind of light is imparted to the Moon by Venus is so childish that it doesn't deserve an answer,* for who is so ignorant as not to understand that from the time of the new moon to a separation of 60 degrees between Moon and Sun, no part of the Moon that is turned away from the Sun can possibly be seen from Venus? And it is equally untenable that it should come from the Sun penetrating and permeating with its light the solid body of the Moon,* for then it would never lessen since one hemisphere of the Moon is always illuminated by the Sun except at the time of lunar eclipses. But the light does decrease as the Moon hastens towards the first quarter and becomes completely dull when it has passed it. Since, therefore, this secondary light is not inherent and the Moon's own, and is not borrowed from any of the stars or from the Sun, and since there remains in this vast world no other body than the Earth, what, I pray, are we to think? What can we suggest? Can it be said that the Moon, just as would any other opaque and dark body, is flooded with light coming from the Earth? What is marvellous about that? Indeed, the Earth, in fair and grateful exchange, gives back to the Moon an illumination like the one that it receives from her during nearly the whole time in the deepest gloom of the night.

Let us explain the matter more fully. When the Moon is between the Earth and the Sun, as happens at new moon, the rays of the Sun illuminate the side of its hemisphere that is turned away from the Earth. The other side of the hemisphere that faces the Earth is cloaked in darkness, and so the Moon does not illuminate the surface of the Earth at all. But as the Moon moves slowly away from the Sun, the side of its hemisphere that faces us gradually grows brighter, and it turns towards us a slender silvery crescent that slightly illuminates the Earth. The Sun's illumination increases as the Moon approaches her first quarter, and the reflection of its light increases on the Earth. As the illumination on the Moon spreads beyond the first quarter, our nights become brighter. Finally, the whole face of the Moon, on the side that is turned towards us, is bathed in the bright rays of the Sun, which is on the opposite side of the Earth. The surface of the Earth that is covered with moonlight shines forth in all directions. But as the Moon wanes it sends us weaker rays, and the Earth is more faintly illuminated. By the time of new moon, night on Earth is completely dark.

There is thus a monthly period during which the Moon distributes its light, which is alternately brighter and weaker. But the Earth pays the Moon back with an equal benefit, for when the Moon is between the Earth and the Sun it faces the whole surface of the Earth that is exposed to the Sun and is entirely bathed in its vivid rays, which are reflected onto the Moon. And so it happens that the hemisphere of the Moon that faces the Earth, although deprived of sunlight, shines brightly. The same Moon, when it is 90 degrees from the Sun, sees only half of the terrestrial sphere illuminated, namely the western half, for the other, the eastern, is in darkness. Therefore, the Moon is less brightly illuminated by the Earth and, accordingly, its secondary light appears fainter. But if you imagine the Moon when it is on the opposite side of the Sun, it will be facing the hemisphere of the Earth, which is now between itself and the Sun, and will see it as completely dark and steeped in the gloom of night. If this happens when the Moon is in the plane of the ecliptic,* it will receive no light at all, and will be entirely deprived of both solar and terrestrial rays. In its various positions with respect to the Earth and the Sun, the Moon receives more or less light than from the reflection that comes from the Earth according to the greater or smaller portion of the illuminated terrestrial hemisphere that it faces. There is reciprocity between these two globes such that whenever the Earth is more brightly illuminated by the Moon, the Moon is less brightly illuminated by the Earth, and vice versa.

Let these few words suffice here. The matter will be considered more fully* in our *System of the World*, where it will be shown by means of numerous arguments and experiments that the reflection of sunlight from the Earth is indeed very strong. This for the benefit of those who claim that the Earth must be removed from the round of stars,* chiefly for the reason that it has neither motion nor light. We shall demonstrate that it is in motion, that it surpasses the Moon in brightness,* and that it is not the bilge where the rubbish and the refuse of the world have settled down. Furthermore, we shall confirm this with a thousand physical arguments.*

Hitherto, we have spoken of the observations of the lunar body; now we must briefly mention what we have thus far been able to discern about the fixed stars. First of all, it is worth mentioning that when the stars, whether fixed or erratic, are viewed with a spyglass they are by no means magnified in the same ratio as other objects, including the Moon, are enlarged. The increase in the fixed stars is considerably less. For instance, a spyglass that is powerful enough to magnify other objects a hundred times will hardly, I believe, make stars four or five times greater.* The reason for this is that when the stars are looked at with the naked eye they do not present themselves, so to say, in their bare and real size but sparkling all over and fringed with shining rays, especially when night is far advanced. This is why they appear much larger than they would if they were shorn of these adventitious rays, for the angle that they subtend at the eye is determined not by the primary disc of the star but by the brightness that surrounds it.*

This can easily be understood from the fact that when the stars rise at sunset, as twilight is settling in, they appear very small even if they are of the first magnitude, and Venus, if it happens to be visible at midday, appears so small that it hardly seems the size of a star of the last magnitude. Things are different with other objects, including the Moon, that always appears of the same size, whether it is seen in broad daylight or in the depth of night. So the stars are seen unshorn (of their rays) in the dark, but daylight can cut their fringes off, and not only daylight but any slight cloud that passes between them and the eye of the observer. When a dark veil or a piece of coloured glass is placed between the eye and the star, the blaze that surrounds also disappears. The spyglass does the same, for it first removes from the stars their adventitious and accidental splendour before it enlarges their true disc (if indeed they are of that shape), and so they are seen less magnified: a star of the fifth or even the sixth magnitude that is observed through a spyglass is only shown as if it were of the first magnitude.

The difference between the appearance of the planets and that of the fixed stars deserves to be noted. The planets present their discs as perfectly round, just as if they had been traced with a pair of compasses, and they appear as so many little moons, completely illuminated. But the fixed stars never appear bounded by a circular rim, but by blazing light that sparkles all over. Observed through a spyglass, the stars have the same figure as when viewed with the naked eye, but so much larger that a star of the fifth or sixth magnitude seems to equal the Dog Star,* the largest of all fixed stars.

Beyond the stars of the sixth magnitude you will observe through the telescope a host of other stars that escape our natural faculty of sight. They are so numerous as to be almost beyond belief, for you can distinguish more than six new orders of magnitude.* The largest of these, which we may call of the seventh magnitude, or of the first magnitude of invisible stars, appear with the aid of a spyglass larger and brighter than the stars of the second magnitude that are seen with the naked eye. In order that you may see a couple of instances of their inconceivable number, I have described two star-clusters so that from these examples you may decide about the rest. In my first example, I intended to depict the entire constellation of Orion, but I was overwhelmed by the enormous number of stars and the lack of time, and have postponed this until some other occasion. Within the limits of one or two degrees, more than five hundred* are scattered among the old stars. For this reason, I have selected the three stars in the belt of Orion and the six in the sword, which were already known, and I have added in their vicinity another eighty that I recently observed. I have preserved as precisely as possible the intervals between them. In order to distinguish them, I drew the well-known old stars of larger size, and I outlined them with a double line. The others, the invisible ones, I made smaller with one line only. I also preserved the differences of magnitudes as much as possible.

As a second example, I depicted the six stars of the constellation of Taurus, called the Pleiades (I say six since the seventh is hardly ever visible).* These are enclosed within very narrow precincts in the heaven, and near them lie more than forty other invisible ones, not a single of which is much more than half a degree from one of the aforementioned six. Of these I have indicated only thirty-six, and I preserved their intervals, their magnitudes, and the distinction between the old and the new stars, as I did for those in Orion.

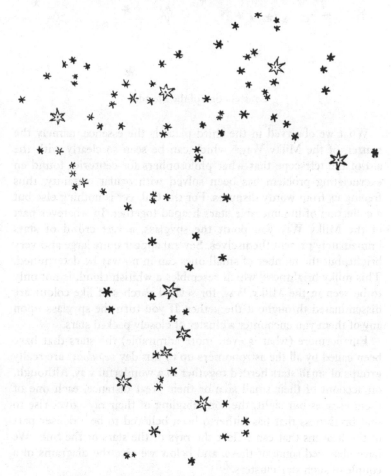

Asterism of the belt and sword of Orion

Constellation of the Pleiades

What we observed in the third place is the essence, namely the matter, of the Milky Way,* which can be seen so clearly with the aid of the telescope that what philosophers for centuries found an excruciating problem has been solved with ocular certainty, thus freeing us from wordy disputes. For the Galaxy* is nothing else but a collection of innumerable stars heaped together. In whatever part of the Milky Way you point the spyglass, a vast crowd of stars immediately present themselves. Several appear quite large and very bright, but the number of small ones can in no way be determined. This milky brightness, which resembles a whitish cloud, is not only to be seen in the Milky Way, for several patches of like colour are disseminated throughout the aether. If you turn the spyglass upon any of them you encounter a cluster of closely packed stars.

Furthermore (what is even more admirable) the stars that have been called by all the astronomers up to this day *nebulous** are really groups of small stars herded together in a wonderful way. Although, on account of their small size or their great distance, each one of them escapes our sight, the commingling of their rays gives rise to that brightness that has hitherto been believed to be a denser part of the heavens that can reflect the rays of the stars or the Sun. We have observed some of these, and below we give the diagrams of a couple of such star-clusters.

In the first, you find the nebula of Orion's Head, as it is called, and in which we counted twenty-one stars. The second contains the nebula known as Praesepe, which is not one star only but a mass of more than forty starlets. Besides the Aselli,* I have indicated thirty-six in the order in which they are arranged:

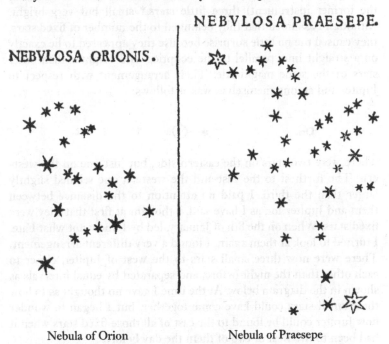

NEBVLOSA ORIONIS.

NEBVLOSA PRAESEPE.

Nebula of Orion   Nebula of Praesepe

I have briefly recounted what I observed in the Moon, the fixed stars, and the Milky Way. There remains what deserves to be considered the most important of all, namely the disclosure of four planets that were never seen from the creation of the world up to our own times, and to declare how they were found and observed, what are their positions, and what observations I made during almost two months concerning their motion and their changes. I call upon all astronomers to devote themselves to the study and the determination of their periods, which so far I have not been able to ascertain because of lack of time. I warn them again, however, in order that they may not undertake such an inquiry in vain, that they will need a very sharp spyglass of the kind I described at the beginning of this account.

On the 7th of January of the present year 1610, at the first hour of night,* when I was observing the celestial bodies with the spyglass, Jupiter came forward. As I had just made an excellent instrument for myself, I saw (what I had not done before because of the weakness of the former instrument) three little stars,* small but very bright. Although I assumed that they belonged to the number of fixed stars, they caused me no little surprise because they appeared to lie exactly on a straight line, parallel to the ecliptic,* and brighter than other stars of the same magnitude. Their arrangement with respect to Jupiter and among themselves was as follows:

Ori.   *   * ◯   *   Occ.

There were two stars on the eastern side, but just one on the western. The furthest to the east and the western one seemed slightly larger than the third. I paid no attention to the distance between them and Jupiter for, as I have said, I thought at first that they were fixed stars. When on the 8th of January, led by I know not what Fate, I turned to look at them again, I found a very different arrangement. There were now three small stars to the west of Jupiter, closer to each other than the night before, and separated by equal intervals as shown in the diagram below. At the time I gave no thought as to how these small stars could have come together, but I began to wonder how Jupiter could be found to the east of all those fixed stars when it had been to the west of two of them the day before.

Ori.   ◯ * * *   Occ.

I therefore asked myself whether, contrary to the computations of astronomers, Jupiter might not be moving eastward at this time and had passed in front of the stars by its own proper motion. I eagerly awaited the next night, but my hopes were dashed: the sky was everywhere covered with clouds.

On the 10th of January the stars appeared in the following position with regard to Jupiter: there were only two and both were to the east:

Ori.   * * ◯   Occ.

The third one, I conjectured, was hiding behind Jupiter. They were, as before, in the same straight line with Jupiter, and located exactly along the line of the Zodiac. When I saw this, and knowing that such changes of position could in no way be ascribed to Jupiter, and moreover, having had to recognize that the stars that I saw had always been the same (there were no other stars within a great distance before or after on the line of the Zodiac), I realized that the changes of position were not due to Jupiter but to the stars that had been observed. My perplexity gave way to amazement, and I therefore resolved to observe them with greater care and attention from then on.

On the 11th of January I saw the following arrangement:

Ori.   *   *   O   Occ.

There were two stars to the east, and the one in the middle was three times as far from Jupiter as it was from the star further east. This last star was almost twice the size of the other whereas, on the previous night, they had appeared about equal in size. I now decided that without the shadow of a doubt there were in the heavens three stars wandering about Jupiter, just as Venus and Mercury go around the Sun. This became clearer than daylight when I later made numerous other observations. Nor were there only three, but four wandering stars revolving around Jupiter. The account that follows will describe the changes of position that I subsequently observed with greater accuracy. I also measured the intervals between them with the spyglass in the manner already explained. Besides this, I have also given the time of the observations, especially when several were made in the same night, for the revolutions of those planets are so swift that it is generally possible to note their hourly changes.

On the 12th of January, at the first hour of night, I saw the stars disposed in this manner:

Ori.   * *O   *   Occ.

The star further to the east was bigger than the westerly one, but both were very conspicuous and bright. The distance of each one from Jupiter was two minutes. A third small star, not visible before, began to appear at the third hour. It almost touched Jupiter on the eastern side and was exceedingly small. They were all on a straight line along the ecliptic.

On the 13th of January, for the first time, I saw four small stars in this arrangement with regard to Jupiter:

Ori.   *   ○ * * *   Occ.

Three were to the west, and one to the east. They made a line that was almost straight, but the one in the middle of those that were to the west deviated a little from the straight line towards the north. The farthest to the west was at a distance of two minutes from Jupiter. The interval between Jupiter and the closest star, and between the stars themselves, was only one minute. All the stars appeared of the same magnitude, and although small were nevertheless very brilliant and far outshone the fixed stars of the same magnitude.

On the 14th of January the sky was overcast.

On the 15th of January,* close to the third hour, the four stars occupied, with respect to Jupiter, the positions depicted below:

Ori.   ○   *   *   *   *   Occ.

All were to the west and situated on a nearly straight line, but the third from Jupiter was raised a little to the north. The nearest to Jupiter was the smallest of all, and the following ones appeared successively larger. The interval between Jupiter and the first star and the intervals between the next two were all equal and each of two minutes, but the one furthest to the west was four minutes away from the one nearest to it. They were very brilliant and did not twinkle at all, and thus they always appeared either before or since. But by the seventh hour there were only three stars in the following configuration with respect to Jupiter:

Ori.   ○   * *   *   Occ.

They lay on a perfectly straight line, but the nearest to Jupiter was very small and at a distance of three minutes. The distance of the second star from this one was one minute, and the third was

4 minutes and 30 seconds from the second. After another hour, the two middle stars were even closer, hardly 30 seconds apart.

[*The observations of the satellites of Jupiter that Galileo went on making until 2 March 1610 are omitted.*]

These relations of Jupiter and its adjacent planets to a fixed star were recorded in order that anyone may grasp that the progress of these planets, both in longitude and latitude, agree exactly with the movements of Jupiter that are calculated from tables.

These are the observations of the Medicean planets recently, and for the first time, discovered by me. Although it has not yet been possible, from these observations, to determine their periods numerically, some things are worthy of mention. And, in the first place, since they sometimes follow and sometimes precede Jupiter by the same interval, and since they remain within very narrow limits to the east or to the west of Jupiter, and since they accompany it whether its motion is retrograde or direct, it is beyond doubt that they make their revolutions around him while at the same time completing, together with him, twelve-year periods around the centre of the world. Moreover, they revolve in circles of unequal size, as is easy to deduce from the fact that two planets are never seen in conjunction at their greatest elongation from Jupiter, although two, three, or indeed all of them have been seen bunched together close to Jupiter.

Furthermore, it has been found that the planets that describe the smallest circles around Jupiter are faster, for those closest to Jupiter are often to be seen to the east when on the day before they had appeared in the west, and vice versa, whereas the planet that traces out the largest orbit appears, on careful examination of its returns, to have a semi-monthly period. Furthermore, we have a particularly strong argument to remove the scruples of those who are willing to examine dispassionately the revolution of the planets about the Sun in the Copernican system, yet are so troubled by the fact that our one and only Moon should go around the Earth while at the same time both carry out an annual revolution around the Sun, that they consider that this theory about the constitution of the universe should be rejected as impossible. But now we have not only one planet revolving about another one, while both trace out an annual circle around the Sun, but our own eyes show us four stars travelling around Jupiter as the Moon travels around the Earth, while, at the same time, they make a grand revolution around the Sun.

Finally, we should not omit the reason why the Medicean stars, in making their very small revolutions around Jupiter, seem sometimes to be more than doubled in size. We can by no means look for the explanation in terrestrial vapours, since Jupiter and the neighbouring fixed stars do change in size when this increase and decrease is taking place. It is utterly untenable that the cause of such a variation should be due to the way they approach and recede from the Earth at the perigee and the apogee of their revolutions, for the narrow circle along which they travel can in no way cause such an effect.* Furthermore, an oval motion (which in this case would be almost rectilinear) seems equally untenable, and would by no means agree with the appearances.* I gladly offer the explanation that occurs to me, and submit it to the appraisal and criticism of all true philosophers. It is well known that the interposition of terrestrial vapours makes the Sun and the Moon appear larger, but the fixed stars and the planets smaller. Thus, the great lights,* when near the horizon, are larger than at other times, whereas the stars appear smaller and often barely visible. They diminish even more if these vapours are flooded with light, and so the stars appear very small by day and in the twilight, whereas the Moon does not,* as I remarked above. Moreover, not only the Earth but the Moon also is surrounded by a vaporous sphere, as is obvious from what we said above, and especially from what will be declared more fully in my *System.** We can reasonably say the same of the other planets, so that it is by no means unthinkable to put around Jupiter a sphere that is denser than the rest of the aether and around which the Medicean planets revolve, as the Moon goes around the sphere of elements.* The planets look smaller at apogee when this sphere is interposed, and when it is removed, at perigee, they look bigger. Want of time prevents me from going further into these matters, but the kind reader may expect more, soon.

LETTERS ON THE SUNSPOTS

FIRST LETTER OF GALILEO GALILEI TO MARK WELSER* CONCERNING THE SUNSPOTS, IN REPLY TO HIS LETTER

Illustrious Sir and worshipful patron,

The letter which you were so kind as to send to me three months ago has long remained unanswered, as various impediments have virtually obliged me to remain silent. In particular a long indisposition, or rather, numerous long indispositions, made all other activities and occupations impossible; above all they prevented me from writing, as indeed to a large extent they still do. However, I am not now so constrained that I cannot at least reply to some of the letters from my friends and patrons, of which I have no small number, all awaiting a reply. I have remained silent also because I hoped to be able to give some satisfaction in replying to your enquiry concerning the sunspots, on which topic you have sent me the brief discourses of the writer calling himself 'Apelles';* but the difficulty of the subject matter and the fact that I have not been able to make many continuous observations have kept me, and still keep me, undecided and in suspense. I have to be more cautious and circumspect than many others in pronouncing upon anything new because, as you know, my recent observations of things far removed from popular and commonly held views have been violently denied and attacked. So I have to conceal and say nothing about any new idea until I have been able to demonstrate it with the most absolute and palpable proof. The enemies of anything new, of whom there are an infinite number, would pounce on any error I make, however trivial, and make it into a capital fault; for it now seems to be the case that it is better to err with the masses than to be alone in speaking correctly. I might add that I would rather be the last to come out with a true statement than forestall everyone else, only to have to withdraw something that I had put forward with more haste and less reflection.

All these considerations, Sir, have made me slow in responding to your questions, and I still am hesitant to offer anything more than negative statements; for I feel more confident of knowing what the sunspots are not than what they really are, and I find it much harder to discover the truth than to refute what is false. But to satisfy your request at least in part, I shall consider those things that seem to me to be worth pointing out in the three letters of 'Apelles', as you ask me to do, as these contain what has been conjectured so far about the essence, location, and motion of these spots.

First of all, I have no doubt that the sunspots are real and not mere appearances or optical illusions, or caused by blemishes in the lenses, as your friend rightly proves in his first *The sunspots* letter. I have observed them for the past eighteen months* and *are real; their* have shown them to several of my friends, and last year at *motion* precisely this time I had many prelates and other gentlemen in Rome observe them. It is true, too, that they do not remain stationary on the surface of the Sun, but appear to move in relation to it with regular motions, as the author also notes in the same letter. It seems to me, however, that this motion is in the opposite direction to what Apelles asserts; that is, they move from west to east, slanting from south to north, and not from east to west and north to south. This can be clearly seen from the observations which he himself describes, which in this respect agree with my own and with those of others that I have seen. The spots observed at sunset are seen to change place from one evening to the next, moving down from the upper parts of the Sun towards the lower, while those seen in the morning move upwards from the lower parts towards the higher, appearing first in the more southerly parts of the solar surface, and disappear or detach themselves from it in the more northerly parts. Thus the spots describe lines on the face of the Sun similar to those that Venus or Mercury trace when they pass in front of the Sun and interpose themselves between the Sun and our sight. So the spots move with respect to the Sun as do Venus, Mercury, and the other planets, moving, that is, from west to east, and obliquely to the horizon from south to north. If Apelles were to assume that these spots did not revolve around the Sun but simply passed beneath it, then their motion could indeed be described as being from east to west. But if we assume that they describe circles around the Sun, and are alternately in front of it and behind it, then we must describe their revolutions as being from west to east, because this is the direction in which they are moving when they are in the more distant part of their circle.

Having established that the spots he has observed are not illusions caused by the telescope or optical defects, the author next tries to reach some general conclusions about their location, and to show that they are neither in the atmosphere of the Earth nor on the surface of the Sun. As regards the first of these, the absence of any noticeable parallax* necessarily shows that the spots are not in the atmosphere, that is, in the space close to the Earth which is commonly assigned to the element of air. That they cannot be on the surface of the Sun, however, does not seem to me to be entirely demonstrated. His first argument, that it is not credible that there should be dark spots on the surface of the Sun because the Sun is entirely bright, is not valid: for we referred to the Sun as pure and entirely bright as long as no impurities or darkness had been seen on it; but if it is shown to have parts which are impure and stained, why should we not call it spotted and not pure? Names and attributes must be accommodated to the reality of things, not the other way round; things existed before names. His second argument would be conclusive if these spots were permanent and unchanging; but I will discuss this further below.

What Apelles says next, namely that the spots that are visible on the Sun are much darker than any that have ever been seen on the Moon, seems to me to be absolutely false. On the contrary, I think that the spots seen on the Sun are not just less dark than the dark spots we see on the Moon, but that they are no less bright than the brightest parts of the Moon even when it is directly illuminated by the Sun. The reason for this is as follows. When Venus appears as an evening star, when it is at its most resplendent, it can only be seen when it is many degrees distant from the Sun, especially if both are well above the horizon. This is because the parts of the aether around the Sun are no less bright than Venus itself. From this we can deduce that if we could place the Moon alongside the Sun,

The spots are no less bright than the light parts of the Moon

even when it is resplendent with light as it is at the full Moon, it would also be invisible, as it would be placed in a field no less bright than itself. Now consider, when we look at the brilliance of the Sun's disc through a telescope, or eyeglass, how much more resplendent it is than the area around it; and then compare the darkness of the sunspots with both the light of the Sun itself and the darkness of the surrounding area. Both comparisons will show that the spots on the Sun are no darker than the area around it. If, then, the sunspots are no darker than the field surrounding the Sun; and if, moreover, the light of the Moon would be invisible against the brightness of this same area; then it necessarily follows that the sunspots are no less bright than the brightest parts of the Moon, even though they appear shadowy and dark against the brilliant brightness of the solar disc itself. And if they yield nothing in brightness to the lightest parts of the Moon, we can imagine how they would compare with the darkest spots on the Moon's surface—especially when we recall that these dark areas are caused by the shadows cast by the lunar mountains, and that they stand out in contrast to the illuminated parts no less than the ink against this sheet of paper. I say this not so much to contradict Apelles as to show that we need not assume the material of the sunspots to be very dense and opaque, as we may reasonably suppose the material of the Moon or the other planets to be. A density and opacity similar to those of a cloud is enough, when interposed between us and the Sun, to produce the required obscurity and darkness.

The material of the sunspots is not very dense

Apelles goes on to suggest that this could be a means of establishing with certainty whether Venus and Mercury revolve around the Sun or between the Earth and the Sun, a point which he develops more fully in his second letter. I am somewhat surprised that he has not heard, or, if he has, that he has not made use of the exquisite and convenient method of determining this which I discovered almost two years ago, and which has been communicated to so many people that it is now generally known, and which can be applied on many occasions. This is the fact that Venus changes its shape in the same way as the Moon. If Apelles will now observe Venus with his telescope, he will see that it is perfectly circular and very small,

The author's observations show that Venus is horned, and of variable size

although not as small as it was when it first appeared in the evening. If he continues to observe it he will see that, at the point of its greatest elongation from the Sun, it appears semicircular, after which it will change into a horned shape, becoming thinner as it approaches the Sun again. Around its conjunction with the Sun it will be as thin as the Moon when it is two or three days old, and its visible circle will be so much increased that its apparent diameter, when it rises as an evening star, is less than a sixth of what it appears when it sets in the evening or rises in the morning. Hence its disc appears almost forty times larger in the latter position than in the former. These things leave no room for doubt about the orbit of Venus, but show with absolute certainty that it revolves around the Sun, which is the centre of the revolutions of all the planets, as the Pythagoreans and Copernicus maintain.* Hence there is no need to wait for bodily conjunctions* to confirm such a self-evident conclusion, or to produce arguments which are open to objections, however feeble, to gain the assent of those whose philosophy is strangely unsettled by this new structure of the universe. For these people, if they are not constrained by other evidence, will say either that Venus shines with its own light, or that it is of a substance that can be penetrated by the Sun's rays, so that it may be illuminated not just on its surface but also in depth. They will shield themselves with this response all the more boldly because there has been no lack of philosophers and mathematicians who have believed this to be the case (*pace* Apelles, who writes otherwise). Copernicus himself had to admit that one of these propositions was possible, indeed necessary, since he was unable to explain how it was that Venus does not appear horned when it is below the Sun; and indeed no other reply was possible until the telescope came to show us that Venus is naturally dark like the Moon, and that it changes its shape as the Moon does.

I have, however, another major doubt about Apelles' investigations. He attempts to see Venus against the disc of the Sun when they are in conjunction, expecting it to appear as a much larger spot than any that has been seen so far, since he takes its diameter to be three minutes of an arc, and therefore

its surface to be just over a hundred-and-thirtieth part of
that of the Sun. This, with respect, is incorrect: the visible
diameter of Venus was less than the sixth part of a minute, and
its surface less than one forty-thousandth of that of the Sun, as
I have established by direct observation and as I will make
plain to all in due course. So you can see how much scope there
would still be for those who continue to insist with Ptolemy
that Venus is lower than the Sun, for they could say that it
would be impossible to see such a small speck against the Sun's
immense and brilliant surface. Finally, I would add that such
an observation would not necessarily convince those who deny
that Venus revolves around the Sun, as they could always fall
back on saying that its orbit is above the Sun's, and indeed
they could claim Aristotle's authority for saying so. Thus, it is
not enough for Apelles to show that Venus does not pass below
the Sun in their morning conjunction, if he cannot also show
that it does pass below the Sun in their evening conjunction.
Such bodily conjunctions in the evening are extremely rare,
and we will not be able to see one. So Apelles' argument is not
sufficient for his purpose.

Venus is very small in comparison with the Sun

I come now to the third letter of Apelles, in which he deals
specifically with the position, motion, and substance of these
spots, and concludes that they are stars close to the surface of
the Sun, and that they revolve around it in the same way as
Mercury and Venus.

To determine their place, he begins with a proof that they
are not on the surface of the Sun, in which case their motion
would be explained by the Sun's turning on its axis, because
their 15-day passage across the visible hemisphere would then
reappear in the same pattern every month; and this does not
happen. This argument would be conclusive provided we could
first be sure that the spots are permanent, that is, that new ones
do not appear while others are erased and vanish; but once it is
admitted that some spots are formed and others disintegrate,
then it can be argued that the Sun turning on its axis carries
them around without necessarily showing us the same spots,
or in the same order, or having the same shape. Now I think
it would be difficult, indeed impossible and contrary to the
evidence of our senses, to prove that they are permanent.

The spots are not permanent

Apelles himself must have seen some of them appear for the first time well within the Sun's circumference, and others fade and vanish before they have finished crossing the Sun, as I too have observed on many occasions. So I neither affirm nor deny that the spots are located on the Sun; I simply state that it has not been sufficiently proved that they are not.

The author goes on to argue that the spots are not in the atmosphere, or in any of the orbs below that of the Sun; and here I find some confusion, not to say inconsistency, for he returns to the old, commonly accepted Ptolemaic system as if it were true, having earlier shown he was aware that it was false. He concluded that Venus does not have an orbit below that of the Sun but revolves around it, as it is sometimes above the Sun and sometimes below it; and he stated the same of Mercury, whose maximum elongation is much smaller than that of Venus and which therefore must be placed closer to the Sun. But then he appears to reject the arrangement which he had earlier accepted as true, as indeed it is, and to adopt the false one; for he places Mercury after the Moon, followed by Venus. I was inclined at first to excuse this small contradiction by saying that he thought it insignificant whether, after the Moon, he named Mercury or Venus first, on the grounds that it did not matter in which order they were named as long as he knew their correct order in reality. But then he undermined my excuse for him by arguing on the basis of parallax that the sunspots are not in the sphere of Mercury, and then adding that this method would not apply to Venus because of its small parallax, which is like that of the Sun; whereas Venus can sometimes have a parallax much greater than those of either Mercury or the Sun.

It seems to me, therefore, that Apelles has a free and not a servile mind; he is well able to grasp true teaching, and now, prompted by the strength of so many new ideas, he is beginning to listen and to assent to true and sound philosophy, especially as regards the arrangement of the universe. But he is not yet able to detach himself completely from the fantasies he absorbed in the past, and to which his intellect sometimes returns and lends assent by force of long-established habit. This is apparent again when, in this same passage, he tries to

prove that the spots are not in any of the orbs of the Moon,
Venus, or Mercury. In doing so, he continues to adhere to
eccentrics, deferents, equants, epicycles, etc.,* as, wholly or in
part, real and distinct entities, which were merely assumed by
mathematical astronomers to facilitate their calculations.
These are not retained by philosophical astronomers whose
aim is not just to preserve the appearances by any means, but
to investigate the true arrangement of the universe—the
greatest and most marvellous problem of all. Such an
arrangement exists, and is unique, true, real, and could not
possibly be otherwise; and its grandeur and nobility are such
that it is worthy to be given pride of place above all other
scientific questions by those who think about such matters.

[. . .]

It remains for us to consider Apelles' conclusion concerning the essence and the substance of these spots, which, briefly
stated, is that they are neither clouds nor comets, but stars
which circle around the Sun. On this point I confess that I am
not yet sufficiently confident to be able to reach or state any
firm conclusion. What I am certain of is that the substance of
the spots could be any one of a thousand things which are unknown and unknowable to us, and that the phenomena which
we discern in them—their shape, their opacity, and their
motion—commonplace as they are, give us little or no general
information about them. So I do not think any philosopher
could be blamed for confessing that he did not know, and had
no means of knowing, the material of which the sunspots are
made. If, however, we were to put forward some suggestion of
what they might be, based on a degree of analogy with
materials that are familiar to us, my opinion would be the very
opposite of what is proposed by Apelles. They do not seem to
me to have any of the essential qualities which define a star;
whereas in contrast, all the qualities which I find in them can
be said to be similar to those which we see in our clouds.
This can be explained as follows.

Sunspots are formed and disintegrate in longer or shorter
periods of time; some condense and others expand greatly
from one day to the next; they change their shape, the majority
of them being completely irregular, and they can be darker in

*The material
of the sunspots
may be
unknown and
unknowable
to us*

*Similarity
between the
sunspots and
our clouds*

one part than another. Being either on or very close to the surface of the Sun, they must necessarily be of enormous bulk. Their uneven opacity means that they can block out more or less of the light from the Sun; and they are sometimes found in great numbers, sometimes few, and at other times none at all. Now where in our experience do we find shapes of enormous bulk which are produced and dissolved in a short time; which last sometimes for a long time, sometimes hardly at all; which expand and contract, easily change their shape, and are more dense and opaque in some parts than others, if not in the clouds? In fact, no other material comes close to meeting all these conditions. And there can be no doubt that if the Earth shone with its own light and did not receive external light from the Sun, it would have a similar appearance to anyone who could observe it from a great distance: as first one region and then another was obscured by clouds, it would appear covered with dark spots which, depending on their greater or lesser density, would block out more or less of the terrestrial splendour, and so they would appear more or less dark; they would appear to be now more numerous, now less, now expanding, now contracting; and if the Earth turned on its axis they would also follow its motion. Since clouds are not very deep in comparison to the breadth to which they commonly expand, those at the centre of the visible hemisphere would appear quite broad, while those nearer the edges would look narrower. In short, I do not think there would be anything in their appearance which is not likewise seen in sunspots. However, since the Earth is dark and receives its light externally from the Sun, an observer looking at it from a great distance would not see any dark spots caused by the distribution of the clouds, because the clouds too would receive and reflect the light of the Sun.

I offer you now these two examples of the changing shape, the irregularities, and the differing densities of sunspots.

Observations of the changes in the density and shape of sunspots, and their irregularity

Spot A was observed at sunset on 5 April last, and was seen to be very faint and not very dark. The following day, also at sunset, it appeared as spot B, having become darker and having changed shape. On 7 April its appearance was similar to figure C. These spots were always well away from the circumference of the Sun.

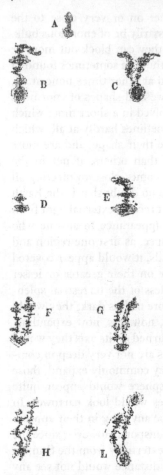

On 26 April, at sunset, there began to appear in the upper part of the Sun's circumference a spot similar to D. On 28 April it appeared similar to E, on the 29th to F, on the 30th to G, on 1 May to H, and on 3 May to L. All the changes in F, G, H, and L occurred well away from the circumference of the Sun, so that their changed appearance cannot be explained by the different aspect they would have when close to the circumference, when the surface of the globe is moving.

It is clear from these and other observations that I have made, and from others that can be made day by day, that no material known to us has as many of the properties of the sunspots as do clouds; and the arguments put forward by Apelles to the contrary seem to me to carry little weight. To his question 'Who would ever place clouds around the Sun?', I would reply, 'Anyone who saw these spots and wanted to say something plausible about their nature, for nothing known to us more resembles them.' When he asks how big they are, I would reply, 'As big as we see them to be in comparison with the Sun; as big as those clouds that sometimes cover a whole region of the Earth'—or if that is not enough, then I would say two, three, four, or ten times as big. Finally, to the third 'impossibility' which he adduces—how they could produce such dark shadows—I would reply that they are less dark than the thickest of our clouds would appear if they were interposed between the Sun and our eyes. This can be clearly seen when one of our darkest clouds covers part of the Sun, and one of the sunspots can be seen in the part that remains visible: it will be clear that there is a significant difference in darkness between the two, even though the edge of the cloud crossing the Sun cannot be very deep. From this it can be deduced that a really thick cloud would produce a much darker shadow than the darkest of the spots. And even if that were not the case, what reason is there to think that some of the solar clouds could not be thicker and denser than terrestrial ones?

This is not to say that I consider sunspots to be clouds of the same material as ours, which are made up of water vapour drawn up from the Earth and attracted by the Sun; I say only that there is nothing of which we have knowledge that more resembles them. Whether they are vapours, or exhalations, or clouds, or fumes produced by the body of the Sun or attracted to it from elsewhere, I have no idea, for they could be any of a thousand other things which are imperceptible to us.

It follows that the term 'stars' is not appropriate for these spots. Stars, whether fixed or wandering, are seen always to maintain the same shape, that of a sphere; we do not see some disintegrating and others forming, but they always remain the same; and their motions are periodic, returning after a fixed interval of time. We do not see the spots returning unchanged; on the contrary, we can sometimes see them disintegrate on the Sun's surface, and I believe one would wait in vain to see the return of the spots which Apelles thinks may revolve in very close orbits around the Sun. In short, they lack the main qualities of those natural bodies to which we give the name of stars. Nor should the spots be called stars because they are opaque bodies, denser than the material of the sky and blocking the rays of the Sun, so that they are illuminated on the side exposed to the Sun's rays and cast a dark shadow on the other; for these are qualities which apply to a stone or a piece of wood, or to a dense cloud, in short to any opaque body. A piece of marble is opaque and therefore blocks the light of the Sun, which it reflects on one side, as the Moon and Venus do, casting a shadow on the other side. On these grounds it could also be called a star; but since it lacks other and more essential

It is not appropriate to call the sunspots 'stars'

properties of a star, as do the sunspots, the name of star seems inappropriate.

I would have preferred it if Apelles had not put the companions of Jupiter (by which I assume he means the four Medicean planets) in the same category, for they show themselves to be constant like any other star. They shine constantly, except when they pass into Jupiter's shadow, when they are eclipsed, like the Moon in the Earth's shadow. They each have their own distinct and orderly period, which have been exactly determined by me; and they do not move in a single orbit, as Apelles seems to believe or to think that others believe, but they have their own distinct orbits with Jupiter as their centre, each of a different size, which I have also discovered. I have likewise detected the reasons why and when one or another of them occasionally tilts towards the north or the south in relation to Jupiter; and I may be able to reply to the objections which Apelles hints that he has on this subject when he specifies what they are. Apelles says he is sure that there are more of these planets than the four which have been observed so far: he may well be right, and such a confident statement by someone who appears to understand the subject very well makes me think that he must have some great new theory which I lack. So I will not be so bold as to say anything definite now, in case I should have to retract it later.

For the same reason, I am reluctant to place anything around Saturn except what I have already observed and discovered: namely, that there are two small stars touching it, one on the east side and the other on the west, which have not so far shown any change, and which I am confident will not show any in the future, barring some extraordinary event quite remote from any motion known to us or even which we could imagine. As regards Apelles' suggestion that Saturn is sometimes oblong in shape and sometimes accompanied by two stars alongside it, I assure you that this is due to an imperfection either in the instrument or in the eye of the observer. The shape of Saturn appears thus, oOo as can be seen when vision and instruments are both perfect. Where vision is impaired it appears thus, ⬭ the separation and shape of the three stars being indistinct. I have observed Saturn at

The Medicean planets are constant; they are subject to eclipse; they have fixed periods which have been discovered by the Author.

They each move in their own distinct orbit

Description of stars alongside Saturn discovered by the Author.

Differences in the appearance of Saturn caused by defective vision

many different times with an excellent instrument and I can assure you that no change whatever has been found in it; and reason, based on our experience of all other stellar motions, renders us certain that none will be found in the future.* For if there were indeed any motion in these stars that was similar to those of the Medicean planets or other stars, they would by now either be separated from or completely conjoined to the body of Saturn, even if their motion had been a thousand times slower than that of any other star that wanders in the heavens.

Apelles puts forward, in conclusion, the suggestion that the spots are planets rather than fixed stars, and that there is a great number of them between the Sun, Mercury, and Venus, of which only those which are interposed between us and the Sun are visible to us. To this I respond, as regards the first part, that I do not believe that they are stars, either wandering or fixed, or that they move around the Sun in circles separate and at some distance from it. If I were to give my own opinion in confidence to a friend and patron, I would say that the sunspots are produced and disintegrate upon the surface of the Sun, with which they are contiguous, and that they are carried along by the motion of the Sun itself, as it turns on its axis in about one lunar month. Some which last for longer than the Sun's rotation period may reappear, but they are so changed in their shape and configuration that it is hard for us to recognize them. This is as far as I am able to conjecture at present, and I hope that you will consider the matter closed with what I have said here.

The sunspots are not stars; what the Author believes them to be

As regards the possibility that there could be another planet between the Sun and Mercury, which goes around the Sun but is invisible to us because its digressions from the Sun are very small, and because it would become visible to us only if it passed directly across the face of the Sun, this seems to me not at all improbable. In fact, I find it equally credible that such planets may exist or that they may not. I do not think there can be a large number of them, however, because if there were it would be reasonable to expect that we would frequently see one of them between us and the Sun, and thus far I have not seen any; in fact I have not seen anything apart from the spots

Only few stars can be between the Sun and Mercury, and between Mercury and Venus

themselves. It is not likely that any such star should have
passed across the Sun's surface in the form of a dark spot, for
if it did, its motion would have to be uniform, and very much
faster than that of the spots. It would have to be much faster
because it would have a smaller orbit than that of Mercury, and
if its motion were analogous to that of all the other planets, its
orbital period would have to be shorter than Mercury's. Now
Mercury passes across the Sun's disc in approximately six
hours, so a planet which moves more quickly could not remain
in conjunction with the Sun for longer than this—unless it
were assumed that its orbit was so small that it almost touched
the Sun's surface, which seems highly fanciful. But if its orbit
was two or three times the diameter of the Sun the effect
would be as I have described, whereas the sunspots remain in
conjunction with the Sun for many days; so it is not credible
that a planet should be among them, or should appear like
them. Moreover, apart from its velocity, the motion of a planet
would have to be almost uniform, even if it were a significant
distance from the Sun, because only a small part of its revolution would be across the Sun, and that small part would be
directly and not obliquely visible to our sight. So any inequality in the angle at which it was visible to us would be so
imperceptible as to be virtually equal, with the result that its
motion would appear to be uniform. But the motion of the
spots is not like this, as they pass quickly across the middle
part of the Sun and slow down when they approach the circumference. So there could only be a small number of planets
between the Sun and Mercury, and fewer still between Mercury and Venus; for they would necessarily have a maximum
digression greater than that of Mercury, and so, like Venus and
Mercury itself, they would have to be brightly visible, especially as they would be relatively close to the Sun and the
Earth. Both their closeness to us and the effective illumination
of the Sun would mean that they shone so brightly that they
would be visible despite their small size.

I recognize that I have taxed your patience, Sir, by writing
at such length and so inconclusively. Please take my long-
windedness as a sign of the pleasure I have in speaking to you
and my desire to oblige you, as far as my powers permit; and so

pardon my excessive loquacity, and accept the warmth of my affection. Let my inconclusiveness be excused by the novelty and difficulty of the subject matter; for so many different thoughts and opinions have gone through my mind, some gaining my assent, others provoking rejection and contradiction, that I have become so timid and uncertain that I hardly dare open my mouth to pronounce on anything at all. This does not mean that I intend to give up and abandon the task; I hope rather that these new ideas may serve a wonderful purpose in enabling me to tune some pipes in this great discordant organ which is our philosophy. For I see many organists striving in vain to bring it into perfect harmony, which they cannot because they ignore three or four of the principal pipes and leave them discordant, so that it is impossible for the others to harmonize with them.

I hope that, as your servant, I may share the friendship you have with Apelles, who seems to me a person of great intelligence and a lover of the truth. Therefore I beg you to greet him warmly in my name, and to tell him that in a few days I shall send him some observations and drawings of the sunspots, made with absolute accuracy as regards both their shape and their changing location from day to day. They are drawn without a hairsbreadth of error, using a highly refined method discovered by a pupil of mine,* and I hope they may be of use to him in his further theorizing about their nature.

I must presume on your patience no longer; so I kiss your hand with all reverence and commend myself to your kind favour, praying that God may grant you all happiness.

From the Villa delle Selve, 4 May 1612.
Your devoted servant,
Galileo Galilei, Lincean Academician.

FROM THE THIRD LETTER OF GALILEO
GALILEI TO MARK WELSER
CONCERNING THE SUNSPOTS

The opinion that the sunspots are clusters of very small stars, examined and refuted

There are not a few Aristotelian philosophers on this side of the Alps who do not pursue philosophy from a desire to know the truth and its causes, for they deny all these new discoveries without distinction and dismiss them as illusions. Now I think it is time for us to dismiss them in their turn, treating them as both invisible and inaudible. They seek to defend the immutability of the heavens which Aristotle himself, if he were alive today, might well abandon. I understand that an opinion is circulating among them which is similar to that of Apelles, except that where he supposes a single star for each spot, they make the spots clusters of very small stars whose different motions make them agglomerate in greater or smaller numbers and then separate again, thus forming larger or smaller spots of irregular and varying shapes.* Since I have already gone well beyond the bounds of brevity in this letter, so that you will have to read it in more than one sitting, I shall take the liberty of offering some comments on this point.

My first reaction is that those who uphold this opinion cannot have had the opportunity to make careful or prolonged observations, for I am sure that they would have encountered several difficulties in working out how to accommodate their view to the phenomena. For while it is generally true that a large number of objects which are too small and distant to be visible individually can, when joined together, form an agglomeration which is visible to our sight, nonetheless we need to go beyond the general rule and take account of the specific properties of stars and what we observe in the spots, and then consider carefully how far they could join together. We should not act like the defender of a castle who rushes with all his small number of soldiers to defend some point which he sees under attack and leaves other parts exposed and undefended. In the same way, when trying to defend the immutability of the heavens, we should not neglect the other perils to which other propositions, just as necessary for the preservation of Aristotelian philosophy, might be exposed. So the first point to be noted if the integrity and solidity of this philosophy is to be maintained is that stars are of two kinds, fixed and wandering. The fixed stars are those which are all in one sphere and move with its motion while remaining unmoved in relation to each other, while wandering stars are those which each have their own individual motion.* Moreover, the motions of both kinds of star must be constant, since it would not be appropriate for the intelligences which move them to have to exert themselves more at some times than at others, which would be inconsistent with their own nobility and unchangeability as well as that of the spheres. With this in mind, first, we cannot say that these 'solar stars' are fixed, because if they did not move in relation to each other it would be impossible to see the continual changes we observe in the spots, and they would always reappear in the same configurations. So they would have to be wandering stars, each with its own motion which is different from all the others but constant in itself. This would explain how some of them collide and separate, but not how they can form spots, as will be clear if we consider some of the things that are observed in the spots. One is that we see some very large spots appear and disintegrate. These must perforce be made up not of two or three stars but of fifty or a hundred, because there are other smaller spots which are less than a fiftieth of the size of the large ones. So if one of the large ones disintegrates so as to disappear completely from our sight, it must divide into at least fifty small stars, each with its own specific motion which is constant and different from all the others, because two with the same motion would never combine or separate on the Sun's surface. If all this is true, surely it is absolutely impossible for spots to form—especially if they last not just for many hours but for many days—just as it would be impossible for fifty boats all moving at different speeds to join and move together for a long space of time? If, on the other hand, these small stars were separate and therefore invisible, they would have to be strung out one behind the other in long lines, according to the length of the parallels at which they are, all travelling in the same direction, as can be seen with the visible sunspots. But then, far from forty, fifty, or a hundred of them being able to come together so frequently and stay together for such long stretches of time, it would be very rare for so many unequal motions to bring such a large number of stars together in one place at all—and if they did they would inevitably disperse again almost immediately. But as we have seen, many spots can remain for many days at a time with very little change in their shape. To maintain, therefore, that the sunspots are clusters of very small stars requires introducing into the heavens and the stars innumerable motions which would be chaotic, unrelated, and completely irregular, something which is not compatible with any plausible philosophy.

What is more, we would have to suppose that these small stars are more numerous than all the other visible stars put together; for if we consider the number and size of all the spots that have been seen over time on the surface of the Sun, and then break them down into particles so small that they cannot be seen, it will be clear that there must be many hundreds of them. And since it is reasonable to suppose that there must be others not only behind the Sun but on either side of it as well, it is hard to escape the conclusion that there must be more than a thousand. But how then could we maintain some proportion between the distances of the wandering stars and their orbital periods, if between the huge orbit of Saturn and the very small one of Mercury we find only ten or twelve stars and not more than six different orbital periods around the Sun, and we now have to find room for hundreds and thousands more within such a small sphere? They would all be contained within the limits of the revolution of Mercury, because they never shine visibly as stars away from the Sun. But why say that they must be within the sphere of Mercury, when we have already seen that the spots must necessarily be either on the surface or at an imperceptible distance from the surface of the Sun? So anyone who would have us believe that the sunspots are clusters of very small stars must first find a way of convincing us that there are many hundreds of dark, dense globes revolving over the Sun's surface, all at different speeds, often colliding and blocking each other's way, so that the course of the fastest is held back for days at a time by the speed of the slowest; and that in this great crowd there frequently form distinct groups, big enough to be visible to us, until the pressure of the throng coming along behind breaks them up and disperses them to force its way through.

Such collisions and crowding together of stars would be absurd

What lengths the advocates of this position have to go to—and how effectively do they demonstrate it, and to what purpose? To keep celestial matter untainted by the condition of the elements, even down to the smallest change. If what is known as 'corruption' was equivalent to annihilation, then the Aristotelians would have good reason to be so resistant to it; but it hardly deserves such hostility if it is nothing more than change. It seems unreasonable to me to call 'corruption' in an egg that which produces a chicken. In any case, if what we call generation and corruption are no more than a tiny mutation in a small part of the elemental world, which cannot be seen even from the Moon, close as it is, why should we deny that it can take place in the heavens? Do they imagine, arguing from the part to the whole, that the Earth itself is going to disintegrate and decay, so that there will come a time when the universe will have the Sun, the Moon, and the other stars, but not the Earth? I do not believe that they have any such fear. So if the Earth's small mutations do not threaten it with total destruction, and indeed are an adornment to it rather than a defect, why should we deny that they occur elsewhere in the universe, or fear that the heavens might disintegrate because of changes no more damaging to the preservation of nature than these? I suspect that our wish to measure everything by our own small standards leads us into strange fantasies, and that our special hatred of death makes the thought of fragility odious to us. But would we really want to escape from mutability by encountering a Medusa's head which would turn us into a block of marble or a diamond, depriving us of our senses and all the other qualities which we could not experience if we did not undergo bodily change?

Change is not alien to the heavens,* or prejudicial to them

I do not wish to go further now in examining the strength of the Aristotelians' arguments, which I hope to do on another occasion. I would simply add that it seems to me not entirely

worthy of a true philosopher to persist obstinately, if I may say so, in defending Aristotelian positions which have clearly been shown to be false, perhaps in the belief that Aristotle himself would do the same if he were living in our own time—as if it were more a sign of clear judgement and great learning to defend a false position than to be persuaded of the truth. I cannot help wondering whether those who act in this way are less interested in weighing the strength of the arguments for and against the Aristotelian position than in maintaining the authority of Aristotle himself, as a much easier way of fending off any dangerous arguments, given that citing and comparing texts is so much easier than investigating truth and formulating new and conclusive proofs. It seems to me, indeed, that we are selling ourselves short and showing disrespect for nature and even for divine Goodness (who to help us understand his great creation has given us two thousand more years of observations and sight twenty times more acute than Aristotle's), if we now attempt to learn from Aristotle what he had no means of knowing rather than relying on the evidence of our own eyes and our own reason. But as I do not

want to stray any further from my main intention, I am content for now to have demonstrated that the sunspots are neither stars nor permanent materials, and that they are not located at a distance from the Sun, but that they are produced and disintegrate upon it in a way not unlike clouds and other vapours around the Earth.

This was all I had to say to you, Sir, on this matter, for the present, as I imagined that this would set the seal on all my new discoveries in the heavens, and that I could now have leisure to return uninterrupted to my other studies. I have already succeeded, after long and painstaking researches, in establishing the periodicity of all four of the Medicean planets,

and drawn up tables for calculating their motions and their individual features; and I shall shortly publish these, together with my considerations on the other celestial discoveries. But my expectations were confounded by an unforeseen marvel which has recently come to concern me regarding Saturn, of which I shall give you an account.

I have already written* of my discovery about three years ago that, to my great astonishment, Saturn is three-bodied: that is, it is an agglomeration of three stars arranged in a straight line parallel to the ecliptic, the central star being much larger than those on either side. I believed that they were fixed in relation to each other, which was not unreasonable given that they had remained without any apparent change for more than two years since I had first observed them, so close that they seemed almost to be touching. It was hardly surprising that I considered their relative positions to be completely fixed, as even a single second of an arc—a motion incomparably smaller than any other motion even in the largest spheres—would have become apparent in that time, either by separating or completely uniting these three stars. Saturn still appeared with its threefold body when I observed it this year, about the time of the summer solstice, after which I did not observe it again for more than two months, as I was quite sure that it did not change. But when I came to observe it again in the last few days I found it unaccompanied, without its usual companion stars, and perfectly round and sharply bounded like Jupiter; and so it still remains. What can we make of such a strange metamorphosis? Have the two smaller stars disintegrated like the sunspots? Have they disappeared and suddenly fled? Has Saturn devoured its children?* Or was their original appearance an illusion produced by the lenses, which deceived me for so long, as well as the many others who observed it with me on many occasions? Is this the moment to revive the fading hopes of those whose profound learning has enabled them to see through all the new discoveries and pronounce them to be fallacies that could not possibly exist? I have no firm conclusion to offer on such a strange and unexpected development: lack of time, the unprecedented nature of the event, my limited understanding, and a fear of error all add to my uncertainty.

But let me be rash for once, and I hope that you will pardon my rashness since I confess it as such. I do not put forward what I am about to say as something based on established principles and secure conclusions, but simply on plausible conjectures,* which I will publish when I need either to show

A new and unexpected marvel concerning Saturn.

Saturn without its accompanying stars

why the opinion to which I now incline was justifiable, or to confirm my position if I should prove to be correct. So my propositions are these: that the two smaller stars of Saturn, which at present are hidden, are likely to reveal themselves a little for about two months around the time of the summer solstice next year, 1613. They will then disappear, and will remain hidden until the winter solstice of the year 1614. They may well be visible again for a few months during that period, but they will disappear again close to the winter solstice. I am more confident that they will reappear at that time and will remain visible thereafter, although at the following summer solstice, in 1615, they will begin to conceal themselves again; but I do not believe that they will disappear altogether, but that after a short time they will show themselves again, and we shall be able to see them not only distinctly, but brighter and bigger than ever. Then, I dare to say with confidence that they will remain visible to us for many years without any interruption at all. I have no doubt that they will return, and so I am cautious in asserting any other specific details, which for now are based only on probable conjecture. But I say with all confidence that, whether their motions turn out as I have described or otherwise, this star will, perhaps no less than the horned appearance of Venus, agree marvellously with the great system of Copernicus, which we can see to be set fair to gaining universal acceptance, so that there is now little fear of its being blown off course.

SCIENCE AND RELIGION

LETTER TO DON BENEDETTO CASTELLI

Very Reverend Father and most worthy Signore,*

I received a visit yesterday from signor Niccolò Arrighetti,* who brought me news of your Reverence. I was delighted to hear what I never doubted, namely the high opinion in which you are held by the whole university,* both the governors and the teachers and students of all nations. This approval, far from increasing the number of your rivals as usually happens among those who work in the same field, has rather reduced their number to very few—and these few will have to desist if their rivalry, which can sometimes be seen as admirable, is not to degenerate and seem rather to be blameworthy and to do more harm to their reputation than to anyone else's. But what set the seal on my pleasure was hearing his account of the arguments which you were able to put forward, thanks to the great kindness of their Serene Highnesses, first at their dinner table and later in Madame's* drawing-room, in the presence of the Grand Duke and the Archduchess* and the distinguished and excellent gentlemen Don Antonio* and Don Paolo Giordano and other excellent philosophers. What greater favour could you wish for than that their Highnesses should be pleased to hold conversation with you, to put their doubts to you, to hear you resolve them and finally to be satisfied with your Reverence's replies?

The points which you made, as signor Arrighetti reported them to me, have prompted me to think afresh about some general principles concerning the citing of Holy Scripture in disputes on matters of natural science, and in particular on the passage in Joshua* which was put forward by the Dowager Grand Duchess, and to which the Archduchess offered some rejoinders, as evidence against the motion of the Earth and the fixed position of the Sun.

As regards the first general question raised by Madame, it seems to me that both she and you were entirely prudent when she asserted and you agreed that Holy Scripture can never lie or be in error, but that its decrees are absolutely and inviolably true. I would simply have added that, although Scripture cannot err, nonetheless some of its interpeters and expositors can, and in various ways. One error in particular, which is especially serious and frequent, is to insist always on the literal meaning of the words, for this can lead not only to many contradictions but also to grave heresies and blasphemies; for it would mean attributing to God feet and hands and eyes, not to mention physical human affections such as anger, repentance, hatred, and sometimes even forgetfulness of past events and ignorance of the future. So, since Scripture contains many statements which, if taken at their face value, appear to be at variance with the truth, but which are couched in these terms so as to be comprehensible to the ignorant, it is up to wise expositors to explain their true meaning to those few who deserve to be set apart from the common herd, and to point out the particular reasons why they have been expressed as they have.

Given, then, that Scripture in many places not only admits but necessarily requires an interpretation which differs from the apparent meaning of the words, it seems to me that it should be brought into scientific disputes only as a last resort. For while Holy Scripture and nature proceed alike from the divine Word—Scripture as dictated by the Holy Spirit, and nature as the faithful executor of God's commands—it is agreed that Scripture, in order to be understood by the multitude, says many things which are apparently and in the literal sense of the words at variance with absolute truth. But nature never trangresses the laws to which it is subject, but is inexorable and unchanging, quite indifferent to whether its hidden reasons and ways of working are accessible to human understanding or not. Hence, any effect in nature which the experience of our senses places before our eyes, or to which we are led by necessary demonstrations,* should on no account be called into question because of a passage of Scripture whose words appear to suggest something different, because not every statement of Scripture is bound by such strict rules as every effect of nature. Indeed, if Scripture has not hesitated to veil some of its most important dogmas and to attribute to God himself qualities contrary to his very essence, solely so as to be accessible to the ignorant and uneducated masses, who would wish to insist that it had set aside the need to be accessible and confined itself strictly to the narrow literal meaning of the words when speaking incidentally about the Earth or the Sun or some other part of creation? This is all the more unlikely when what it says about these parts of creation are things far removed from the primary purpose of Holy Writ, but rather things which, if understood purely in their plain literal sense, would rather have undermined its primary purpose, by making the common people sceptical of its message in matters concerning salvation.

Since this is so, and since moreover it is clear that two truths can never contradict each other, it is the duty of the wise expositor to seek out the true meanings of Scripture which agree with those scientific conclusions which observation or necessary demonstrations have already established as certain. I would go further and argue that since, as we have seen, the Scriptures, albeit inspired by the Holy Spirit, often allow interpretations far removed from their literal meaning, for the reasons given above, and since moreover there is no guarantee that all interpreters are divinely inspired, it would be prudent not to allow anyone to use Scripture to uphold as true any scientific conclusions which observation and necessary demonstration might show to be false. For who would wish to place limits on human understanding, or claim that we already know all that there is to be known? So perhaps it would be wisest not to add unnecessarily to the articles concerning salvation and the foundations of the faith whose certainty is immune to valid arguments ever being raised against it. It would be even more unwise to add to them at the request of those who may or may not be inspired from above, but who are clearly devoid of the intelligence to understand, let alone to challenge, the demonstrations which the exact sciences use to validate their conclusions.

I believe that the purpose of the authority of Holy Scripture is solely to persuade men of those articles and propositions which are necessary to their salvation and which, being beyond the scope of human reasoning, could not be made credible to us by science or by any other means, but only through the mouth of the Holy Spirit himself. I do not consider it necessary to believe that the same God who has endowed us with senses, and with the power of reasoning and intellect, should have chosen to set these aside and to convey to us by some other means those facts which we are capable of finding out by exercising these faculties. This is especially the case with those sciences of which only a tiny part is to be found in scattered references in Scripture, such as astronomy, of which Scripture contains so little that it does not even mention the planets. For if the sacred writers had intended to persuade the people of the order and motions of the heavenly bodies, they would not have said so little about them—almost nothing compared to the infinite, profound, and wonderful truths which this science contains.

So you can see, Father, if I am not mistaken, how flawed is the procedure of those who, in debating questions of natural science which are not directly matters of faith,* give priority to verses of Scripture—often verses which they have misunderstood. But if they really believe that they possess the true meaning of a particular verse of Scripture, and are therefore convinced that they hold in their hand the absolute truth of the matter which they intend to debate, I would ask them to tell me frankly: do they consider that someone to whom it falls to maintain the truth in a scientific debate has a great advantage over their opponent who has to defend what is false? I know they will answer yes, and will say that the one who is defending the truth will be able to draw on numerous experiments and necessary demonstrations to support their position, while their opponent can only fall back on sophistry, false arguments, and fallacies. Why then, if they are so confident that their purely scientific and philosophical weapons are so much stronger than their adversary's, do they immediately have recourse, as soon as battle is joined, to an awesome and irresistible weapon the very sight of which strikes terror into the heart of the most skilful and experienced fighter? If the truth be told, I believe that they are the first to be terror-struck, and that when they realize they are unable to resist the assaults of their adversary they try to find a way of not letting him come near them. But since, as I have said, whoever has the truth on their side has a great, indeed a huge, advantage over their opponent, and since it is impossible for two truths to contradict each other, we have nothing to fear from assaults on any side, as long as we are given an opportunity to speak and to be heard by persons of understanding who are not unduly swayed by their own passions and interests.

To confirm this, I come now to the particular case of Joshua, about which you presented three statements to their Highnesses; and specifically to the third of these, which you rightly attributed to me, but to which I now want to add some further considerations, which I do not believe I have yet explained to you.

So let me first concede to my adversary that the words of the sacred text should be taken in exactly their literal sense, namely that God made the Sun stand still in response to Joshua's prayers,* so that the day was prolonged and Joshua was able to complete his victory. But let me claim the same concession for myself, lest my adversary should tie me down while remaining free himself to change or modify the meanings of words; and I will show that this passage of Scripture clearly demonstrates the impossibility of the Aristotelian and Ptolemaic world system, and on the contrary fits perfectly well with the system of Copernicus.

I ask first, does my adversary know in what ways the Sun moves? If he does, he must perforce reply that it has two motions,* an annual motion from west to east, and a daily one in the opposite direction, from east to west.

My second question then is, do these two different and almost contrary motions both belong to the Sun, and are they both proper to it? To this the answer must be no: only the annual motion is specific and proper to the Sun, while the other belongs to the highest heaven or Primum Mobile,* which draws the Sun, the other planets, and the sphere of the fixed stars along with it, making them complete a revolution around the Earth every twenty-four hours, with a motion which is, as I have said, contrary to their own natural and proper motion.

So to the third question: which of these two motions of the Sun produces day and night, the Sun's own real motion or that of the Primum Mobile? The answer has to be that day and night are the result of the motion of the Primum Mobile, and that the Sun's own motion produces not day and night but the changing seasons, and the year itself.

Hence it is clear that, if the length of the day depends not on the motion of the Sun but on that of the Primum Mobile, in order to prolong the day it is the Primum Mobile which must be made to stop, not the Sun. Indeed, anyone who understands these first elements of astronomy will realize that if God had stopped the motion of the Sun, the effect would have been to shorten the day, not to lengthen it. The motion of the Sun being in the opposite direction to the daily revolution of the heavens, the more the Sun moved towards the east, the more its progress towards the west would be held back; and if the Sun's motion were diminished or stopped altogether, it would reach the point where it sets all the more quickly.* This effect can be clearly seen in the case of the Moon, whose daily revolution is slower than the Sun's by the same amount as its own proper motion is faster than the Sun's.* So it is simply impossible, according to the system of Ptolemy and Aristotle, to prolong the day by stopping the motion of the Sun, as Scripture says happened. It follows therefore that either the motions of the heavens are not as Ptolemy says, or we must change the sense of the words of Scripture and say that, when Scripture says that God stopped the Sun, what it meant was that God stopped the Primum Mobile. But in order to accommodate itself to the understanding of those who are hardly able to comprehend the rising and setting of the Sun, Scripture states the opposite of what it would have said if it had been speaking to men of sense.

To this we might add that it is not credible that God should have stopped only the Sun and allowed the other spheres to keep on turning, because this would unnecessarily have changed the whole order and disposition of the other spheres in relation to the Sun, which would have caused huge disruption to the whole course of nature. But it is credible that He stopped the whole system of the heavenly spheres, and that after that period of rest they all returned harmoniously to their normal operation, without any confusion or disruption of any kind.

However, since we have agreed that we should not change the meaning of the words of Scripture, we must have recourse to another arrangement of the world to see whether it agrees with the plain meaning of the words, as indeed we shall see that it does.

I have discovered and rigorously demonstrated that the globe of the Sun turns on its own axis,* making a complete revolution in the space of roughly one lunar month, in the same direction as all the other revolutions of the heavens. Moreover, it is very probable and reasonable to suppose that the Sun, as the instrument and the highest minister of nature—the heart of the world, so to speak—imparts not only light (as it clearly does) but also motion to the planets which revolve around it.* So if we follow Copernicus in attributing first of all a daily rotation to the Earth, it is clear that, to bring the whole system to a stop solely in order to prolong the extent and time of daylight without disrupting all the other relations between the planets, it was enough that the Sun should stand still, just as the words of Holy Writ say. This, then, is how the length of the day on Earth can be extended by making the Sun stand still, without introducing any confusion among the parts of the world and without altering the words of Scripture.

I have written much more than my indisposition allows, so I close by offering myself as your servant, kissing your hand and praying our Lord that you may have a joyful festive season and every happiness.

In Florence, 21 December 1613, your Reverence's devoted servant, Galileo Galilei.

LETTER TO HER SERENE HIGHNESS, MADAME THE DOWAGER GRAND DUCHESS

A few years ago, as your Highness well knows, I discovered many things in the heavens which had been invisible until this present age. Because of their novelty and because some consequences which follow from them contradict commonly held scientific views, these have provoked not a few professors in the schools against me, as if I had deliberately placed these objects in the sky to cause confusion in the natural sciences. They seem to forget that the increase of known truths, far from diminishing or undermining the sciences, works to stimulate the investigation, development, and strengthening of their various fields. Showing a greater fondness for their own opinions than for the truth, they have sought to deny and disprove these new facts which, if they had considered them carefully, would have been confirmed by the very evidence of their senses. To this end they have put forward various objections and published writings full of vain arguments and, more seriously, scattered with references to Holy Scripture, taken from passages they have not properly understood and which have no bearing on their argument. They might have avoided this error if they had paid attention to a salutary warning by St Augustine, on the need for caution in coming to firm conclusions about obscure matters which cannot be readily understood by the use of reason alone. Speaking about a certain scientific conclusion concerning the celestial bodies, Augustine writes:* 'Aware of the restraint that is proper to a devout and serious person, one should not entertain any rash belief about an obscure question. Otherwise, when the truth is known, we might despise it because of our attachment to our error, even though the truth may not be in any way opposed to the sacred writings of the Old or New Testament.'

With the passing of time the truths which I first pointed out have become apparent to all, and the truth has exposed the difference in attitude between those who simply and dispassionately were unconvinced of the reality of my discoveries, and those whose incredulity was mixed with some emotional reaction. Those who were expert in astronomy and the natural sciences were convinced by my first announcement,* and the doubts of others were gradually allayed unless their scepticism was fed by something other than the unexpected novelty of my discoveries or the fact that they had not had an opportunity to confirm them by their own observations. But there are those whose attachment to their earlier error is compounded by some other imaginary self-interest which makes them hostile, not so much to the discoveries themselves as to their author. Once they can no longer deny the facts they pass over them in silence, and, distracting themselves with fantasies and embittered even more by what has pacified and won over others, they try to condemn me by other means. And indeed, I would not pay any more attention to these than to the other criticisms made against me, which I have never taken seriously as I have always been confident of prevailing in the end, were it not for the fact that these new attacks and calumnies do not stop at questioning the extent of my learning—for which I make no great claims—but try to smear me with accusations of faults which are more abhorrent to me than death itself. Even if those who know me and my accusers know that their accusations are false, I have to defend my reputation in the eyes of everyone else.

These people, persisting in their determination to use all imaginable means to destroy me and my works, know that in my astronomical and philosophical studies I maintain that the Sun remains motionless at the centre of the revolutions of the celestial globes, and that the Earth both turns on its own axis and revolves around the Sun. They know, moreover, that I uphold this position not just by refuting the arguments of Ptolemy and Aristotle, but also by putting forward many arguments to the contrary, in particular, some related to physical effects that are hard to explain in any other way. There are also astronomical arguments depending on many things in my new discoveries in the heavens, which clearly disprove the Ptolemaic system, and perfectly agree with and confirm this alternative position.* Perhaps they are dismayed by the fact that other propositions which I have put forward, which differ from those commonly held, have been shown to be true, and so have given up hope of defending themselves by strictly philosophical means; and so they have tried to hide the fallacies in their arguments under the mantle of false religion and by invoking the authority of Holy Scripture, which they have applied with little understanding to refute arguments which they have neither heard nor understood.

They began by doing their best to spread abroad the idea that these propositions are contrary to Holy Scripture, and therefore to be condemned as heretical. Then, realizing how much more readily human nature will embrace a cause which harms their neighbour, however unjustly, than one which justly encourages them, they had no difficulty in finding others who were prepared to declare from the pulpit, with uncharacteristic confidence, that they were indeed to be condemned as heretical. They showed little pity and less consideration for the injury they were doing not just to this teaching and those who follow it, but to mathematics and mathematicians everywhere.* Now, as their confidence has grown and they vainly hope that this seed, which first took root in their own insincere minds, will grow and spread its branches up to heaven, they spread the rumour that it is shortly to be condemned as heretical by the supreme authority of the Church. And since they know that such a condemnation would not just undermine these two propositions, but would extend to all the other astronomical and scientific observations and conclusions which are logically linked to them, they try to make their task easier by giving the impression, as far as they can at least among the general public, that this view is new and mine alone. They pretend not to know that its author—or rather the one who revived and confirmed it—was Nicolaus Copernicus, a man who was not just a Catholic but a priest and a canon,* and so highly esteemed that he was called to Rome from the furthest reaches of Germany to advise the Lateran Council under Pope Leo X on the revision of the ecclesiastical calendar. At that time the calendar was incorrect simply because they did not know the exact length of the year and the lunar month; so the Bishop of Fossombrone,* who was in charge of the revision, commissioned Copernicus to undertake the prolonged study necessary to establish these celestial movements with greater clarity and certainty. Copernicus set about this task and, thanks to a truly Herculean series of labours combined with his great intellect, made such great progress in this science that he was able to establish the period of the celestial movements with such a high degree of precision, that he came to be recognized as the supreme astronomer, and his findings became the basis not just for the regulation of the calendar but for tables showing the movements of all the planets.* He set his findings down in six books which he published at the request of the Cardinal of Capua and the Bishop of Kulm;* and since he had undertaken the work at the request of the Supreme Pontiff, he dedicated his book, *On the Revolutions of the Heavenly Spheres*, to Pope Leo's successor, Paul III. As soon as the book was printed it was received by Holy Church and was read and studied throughout the world, without anyone expressing the slightest scruples about its content. But now that the soundness of its conclusions is being confirmed by manifest experiments and necessary demonstrations, there are those who, without even having seen the book, want to reward its author for all his labours by having him declared a heretic—and this solely to satisfy the personal grudge they have conceived for no reason against someone whose only connection with Copernicus is to have endorsed his teachings.

So I have concluded that the false accusations which these people so unjustly try to make against me leave me no choice but to justify myself in the eyes of the general public, whose judgement in matters of religion and reputation I have to take very seriously. I shall respond to the arguments which they produce for condemning and banning Copernicus' opinion, and for having it declared not just false but heretical. Under the cloak of pretended religious zeal, they cite the Holy Scriptures to make them serve their hypocritical purposes, claiming to extend, not to say abuse, the authority of the Scriptures in a way which, if I am not mistaken, is contrary to the intention of the biblical writers and of the Fathers of the Church. They would have us, even in purely scientific questions which are not articles of faith, completely abandon the evidence of our senses and of demonstrative arguments because of a verse of Scripture whose real purpose may well be different from the apparent meaning of the words.

I hope that I can demonstrate that I am acting with greater piety and religious zeal than my opponents when I argue, not that

Copernicus' book should not be condemned, but that it should not be condemned in the way they have done, without having understood it, listened to its arguments, or even seen it. For he was an author who never wrote about matters of religion or faith, or cited arguments based in any way on the authority of Scripture, which he might have misunderstood; but he always confined himself to scientific conclusions regarding the movements of the heavens, using astronomical and geometrical proofs resting first of all on the experience of the senses and detailed observations. This is not to say that he paid no attention to what Scripture says, but he was quite clear in his mind that if his conclusions were scientifically proven, they could not contradict Scripture if it was properly understood. Hence he wrote, at the end of his dedication of the book to the Supreme Pontiff:

If there should happen to be any idle prattlers who, even though they are entirely ignorant of mathematics, nonetheless take it on themselves to pass judgement in these matters, and dare to criticize and attack this theory of mine because of some passage of Scripture that they have wrongly twisted to their purpose, it is of no consequence to me and indeed I will condemn their judgement for its rashness. It is well known that Lactantius,* in other respects a famous writer, was a poor mathematician, and shows his childish understanding of the shape of the Earth when he mocks those who said that the Earth has the form of a sphere. So we scholars should not be surprised if we too are sometimes made fun of by such people. Mathematics is written for mathematicians, and if I am not deceived, they will recognize that these labours of mine make a useful contribution to the ecclesiastical state of which Your Holiness now holds the highest office.

Such are the people who are trying to persuade us that an author like Copernicus should be condemned without even being read. To suggest that this would be not just legitimate but laudable, they cite various texts from Scripture, from theologians, and from the Councils of the Church. I revere these and hold them to be of the highest authority, and I would regard it as the height of temerity to contradict them, as long as they are used in conformity with the practice of Holy Church. Equally, I do not believe it is wrong to speak out if there is reason to suspect that someone is citing and using such texts for their own ends in a way which is at odds with the holy will of the Church. So I declare (and I believe that my sincerity will speak for itself) my willingness to submit to removing any errors which, through my ignorance in matters of religion, may be found in this letter. I declare, further, that I have no wish to enter into quarrels with anyone on such matters, even on points which may be disputable. My purpose is only that, if in these reflections which are outside my professional competence there is anything, among whatever errors they may contain, which might prompt others to find something useful to Holy Church in reaching a conclusion on the Copernican system, it may be taken and used in whatever way my superiors may decide. Otherwise let this letter be torn up and burnt, for I have no desire for any gain from it which is not in keeping with Catholic piety. Moreover, although many of the points I shall discuss are things which I have heard with my own ears, I freely grant to whoever said them that they did not say them, if that is what they wish, admitting that I could well have misunderstood them. Hence, let my reply be addressed not to them but to those who do hold the opinion in question.

The reason, then, which they give for condemning the view that the Earth moves and the Sun is stationary, is that there are many places in Holy Writ where we read that the Sun moves and the Earth does not; and since Scripture can never lie or be in error, it necessarily follows that anyone who asserts that the Sun is motionless and the Earth moves must be in error, and such a view must be condemned.

The first thing to be said on this point is that it is entirely pious to state, and prudent to affirm, that Holy Scripture can never lie, provided its true meaning has been grasped. But I do not think it can be denied that the true meaning of Scripture is often hidden and very different from the literal meaning of the words. It follows that when an expositor always insists on the bare literal sense, this error can make Scripture appear to contain not only contradictions and statements which are far removed from the truth, but even grave heresies and blasphemies; for it would mean attributing to God feet and hands and eyes, not to mention corporeal and human affections such as anger, repentance, hatred, and sometimes even forgetfulness of past events and ignorance of the future. So since the biblical writers, inspired by the Holy Spirit, stated these things in this way so as to be comprehensible to the untrained and ignorant, it is necessary for wise expositors to explain their true meaning to those few who deserve to be set apart from the common herd, and to point out the particular reasons why they have been expressed in the terms that they have. This principle is so commonplace among theologians that it would be superfluous to cite any authorities to justify it.

From this it seems reasonable to deduce that whenever Scripture has had occasion to speak about matters of natural science, especially those which are obscure and difficult to understand, it has followed this rule, so as not to cause confusion among the common people and make them more sceptical of its teaching about higher mysteries. As I have said and as is clear to see, Scripture has not hesitated to veil some of its most important statements, attributing to God himself qualities contrary to his very essence, solely in order to be accessible to popular understanding. Who then would be so bold as to insist that it had set this aside and confined itself rigorously to the narrow literal meaning of the words when speaking in passing about the Earth, water, or the Sun or some other part of creation? This is all the more unlikely since what it says about these things has nothing to do with the primary intention of Holy Writ, namely divine worship and the salvation of souls, and matters far removed from the understanding of the masses.

This being the case, it seems to me that the starting-point in disputes concerning problems in natural science should not be the authority of scriptural texts, but the experience of the senses and necessary demonstrations. For while Holy Scripture and nature proceed alike from the divine Word—Scripture as dictated by the Holy Spirit, and nature as the faithful executor of God's commands—it is agreed that Scripture, in order to be understood by the multitude, says many things which are apparently and in the literal sense of the words at variance with absolute truth. Nature, on the other hand, never trangresses the laws to which it is subject, but is inexorable and unchanging, quite indifferent to whether its hidden reasons and ways of working are accessible to human understanding or not. Hence, any effect in nature which the experience of our senses places before our eyes, or to which we are led by necessary demonstrations, should on no account be called into question, much less condemned, because of a passage of Scripture whose words appear to suggest something different. For not every statement of Scripture is bound by such strict rules as every effect of nature, and God is revealed just as excellently in the effects of nature as in the sacred sayings of Scripture. This may be what Tertullian meant when he wrote:*
'I conclude that knowledge of God is first to be found in nature, and

then confirmed in doctrine; in nature through his works, and in doctrine through preaching.'

This is not to imply that we should not have the highest regard for the text of Scripture. On the contrary, once we have reached definite conclusions in science we should make use of them as the best means of gaining a true understanding of Scripture, and of searching out the meanings which Scripture necessarily contains, since it is absolutely true and in harmony with demonstrated truth. I believe therefore that the purpose of the authority of Holy Scripture is chiefly to persuade men of those articles and propositions which, being beyond the scope of human reasoning, could not be made credible to us by science or by any other means, but only through the mouth of the Holy Spirit. What is more, even in matters which are not articles of faith the authority of Scripture should prevail over that of any human writings which are not set out in a demonstrative way, but are simply stated or put forward as probabilities. This should be regarded as right and necessary, to the same extent that divine wisdom surpasses human understanding or conjecture. But I do not consider it necessary to believe that the same God who has endowed us with senses, and with the power of reasoning and intellect, should have chosen to set these aside and to convey to us by some other means those facts which we are capable of finding out by exercising these faculties, so that even in scientific conclusions which the evidence of our senses and necessary demonstrations set before our eyes and minds, we should deny what our senses and reason tell us. Least of all do I think this applies in those sciences of which only a tiny part is to be found in scattered references in Scripture, such as astronomy, of which Scripture contains so little that it does not even mention the planets, apart from the Sun and Moon, and once or twice Venus, under the name of Lucifer. For if the sacred writers had intended to teach the people the order and movements of the heavenly bodies, and that therefore we should learn these things from Scripture, I do not believe that they would have said so little about them—almost nothing compared to the infinite, profound, and wonderful truths which are demonstrated in this science.

Indeed, it is the opinion of the holy and learned Fathers of the Church that the biblical authors not only made no claim to teach us about the structure and movements of the heavens and stars, and their appearance, size, and distance, but that they deliberately refrained from doing so, even though these things were perfectly well known to them. In the words of St Augustine:*

It is commonly asked what we have to believe about the form and shape of heaven according to Sacred Scripture. Many engage in lengthy discussions on these matters, but our writers, with their greater prudence, have omitted them. Such subjects are of no profit for those who seek a blessed life, and, what is worse, they take up very precious time that ought to be given to what is spiritually beneficial. What concern is it of mine whether heaven is like a sphere and the Earth is enclosed by it and suspended in the middle of the universe, or whether heaven is like a disc that covers the Earth on one side? But the credibility of Scripture is at stake, and as I have indicated more than once, there is some danger for a man uninstructed in divine revelation. Discovering something in Scripture or hearing something cited from it that seems to be at variance with the knowledge he has acquired, he may doubt its truth when it offers useful admonitions, narratives, or declarations. Hence, let it be said briefly that concerning the shape of heaven the sacred writers knew the truth, but that the Spirit of God, who spoke through them, did not wish to teach men these facts that would be of no avail for their salvation.

The same lack of interest on the part of the sacred writers in laying down what should be believed about these properties of the celestial bodies is shown again by St Augustine in the next chapter, chapter 10, where on the question of whether the heavens should be deemed to be motionless or in motion, he writes:

Concerning the heaven, some of the brethren have enquired whether it is stationary or moving. If it is moving, they say, how is it a firmament? And if it is stationary, how do the heavenly bodies that are thought to be fixed in it travel from east to west, the more northerly performing smaller circles near the pole? So heaven is like a sphere, if there is another pole invisible to us, or like a disc, if there is no other axis. My reply is that a great deal of subtle and learned enquiry into these questions would be required to know which of these views is correct, but I have no time to go into these questions and discuss them. Neither have they time, those whom I wish to instruct for their own salvation and for the benefit of the Holy Church.

Coming to the particular point with which we are concerned, if the Holy Spirit has chosen not to teach us whether the heavens move or stand still, or whether they have the shape of a sphere, a disc, or a flat surface, or whether the Earth is in the middle of the heavens or to one side, it necessarily follows that He had no intention of giving us a definite answer to other questions of the same kind. The question of the motion or rest of the Earth and the Sun is so linked to those mentioned above that it cannot be determined one way or the other without first answering these. If the Holy Spirit has deliberately refrained from teaching us such things as not being relevant to his intention—that is, to our salvation—how can it be claimed that taking one or other view on this question is obligatory, and that one view is an article of faith and the other is an error? Is it then possible for an opinion to be heretical, and yet have no relevance to the salvation of souls? Or can it be claimed that the Holy Spirit has chosen not to teach us something which concerns our salvation? I cannot do better here than quote what I have heard said by a very eminent churchman,* that the intention of the Holy Spirit is to teach us how one goes to heaven, not how the heaven goes.

Returning to the question of how much weight should be given in questions of natural science to necessary demonstrations and the experience of the senses, and how much these have been regarded as authoritative by learned and holy theologians, the following are two statements among many: 'In discussing the teaching of Moses, we should take care to avoid at all costs saying or declaring categorically ourselves anything that goes against what is clear from manifest experience and the reasoning of philosophy or other disciplines. Since any truth always agrees with every other truth, the truth of Holy Scripture cannot contradict the truth of human sciences established through experience and reason.'* And in St Augustine we read:* 'Anyone who invokes the authority of Scripture in opposition to what is clearly and conclusively established by reason, does not understand what they are doing. What they are opposing to the truth is not the meaning of Scripture, which they have failed to grasp, but their own view, which they have found not in Scripture but in themselves.'

This granted, and since, as has been said, two truths can never contradict each other, it is the duty of the wise expositor to seek out the true meanings of Scripture, which undoubtedly will agree with those scientific conclusions which observation and necessary demonstrations have already established as certain. Now the Scriptures, as we have seen, often allow interpretations which differ from the meaning of the words, for the reasons given above. Moreover, there is no guarantee that all interpreters are divinely inspired, for if they were, they would never disagree over the meaning of a given passage. So it would be highly prudent not to allow anyone to use Scripture to uphold as true any scientific conclusion which observation and demonstrative and necessary reasons might at some time show to be false. For who can place limits on the human mind, or claim that we already know all that there is to be known? Will it be those who on other occasions admit, quite rightly, that 'What we know is only a tiny part of what we do not know'?*

Indeed, since we have it from the mouth of the Holy Spirit that 'He has given up the world to disputations, so that no man may find out what God made from the beginning to the end',* I do not think we should contradict this by closing the path to free speculation concerning the natural world, as if everything had already been discovered and revealed with absolute certainty. Nor do I think it should be considered presumptuous to challenge opinions which were formerly commonplace, or that anyone should be indignant if someone does not share their opinion on a matter of scientific dispute—least of all in the case of problems which the greatest philosophers have debated for thousands of years, such as the view that the Sun is fixed and the Earth moves. This was the view held by Pythagoras and all his followers, Heraclides of Pontus, and also Plato's teacher Philolaus, and by Plato himself, as Aristotle tells us* and as Plutarch confirms in his life of Numa,* where he writes that Plato in his old age used to say that it was absurd to believe otherwise. The same view was held by Aristarchus of Samos, as we learn from Archimedes;* by Seleucus the mathematician;* by the philosopher Nicetas, according to Cicero,* and by many others; and it has finally been developed and confirmed with numerous observations and proofs by Nicolaus Copernicus. Seneca too, that most eminent philosopher, exhorts us in his book *On Comets** to make every effort to establish with certainty whether it is in the sky or on Earth that the daily rotation is located.

So perhaps it would be only wise and prudent not to add unnecessarily to the articles concerning salvation and the foundations of the faith whose certainty is immune to valid arguments ever being raised against it. It would be doubly unwise to add to them at the request of those who may or may not be inspired from above, but who clearly lack the intelligence first to understand, and then to discuss, the demonstrations which the exact sciences use to validate their conclusions. If I were allowed to give my opinion, I would go further and say that it might be more appropriate, and more befitting the dignity of Holy Writ, to stop every lightweight popular writer from trying to lend authority to their writings, often based on empty fancies, by quoting verses from Scripture, which they interpret or rather force into saying things which are as far from the true meaning of Scripture as they are near to making complete fools of themselves when they parade their biblical knowledge in this way. I could give many examples of such abuses of Scripture; let two suffice, both relevant to these astronomical questions. The first are the writings which were published attacking the Medicean planets, which I recently discovered, which cited many verses of Scripture to prove that they could not exist.* Now that these planets are plain for everyone to see, I would like to know what new interpretations of Scripture those who opposed me can give to justify their foolishness. The other example I would give is the writer who has recently published a book* arguing, against astronomers and philosophers, that the Moon shines with its own brightness and does not receive its light from the Sun. He confirms—or rather, he persuades himself that he can confirm—this fanciful idea with various passages of Scripture which he thinks make sense only if his opinion is necessary and true. Yet the natural darkness of the Moon is as plain to see as the brilliance of the Sun.

It is clear then that if the authority of these writers had counted for anything, they would have imposed their faulty understanding of Scripture on others and would have made it obligatory to believe as true propositions which run counter to manifest proof and the evidence of the senses. God forbid that such an abuse should ever gain a foothold, for if it did then all the investigative sciences would very soon have to be forbidden; for since there are always far more men who are incapable of properly understanding Scripture and the other sciences than there are men of understanding, they would indulge in their superficial reading of Scripture and claim the authority to pronounce on every question of natural science, on the basis of some verse which they have misunderstood and taken out of the context in which the sacred writers intended it. And they would overwhelm the small number of those who understand such matters, as they would always have more followers, for people will always prefer to gain a reputation for wisdom without the effort of studying than to wear themselves out labouring tirelessly at rigorous scientific disciplines. So we should give thanks to God that in his kindness he has spared us this fear, by denying such people any authority and ensuring that no weight is given to their shallow writings. Rather, the task of consulting, deciding, and legislating on matters of such importance has been entrusted to the wisdom and goodness of prudent Fathers and to the supreme authority of those who, guided by the Holy Spirit, cannot fail to decree wisely. I think that it was against such men that the Church Fathers wrote with well-justified indignation, in particular St Jerome,* who says:

The garrulous old woman, the senile old man, and the long-winded sophist all presume to have their say about Scripture, mangling it and teaching before they have learnt. Some, prompted by pride, bandy fine-sounding words as they hold forth about Holy Writ among ignorant women; others, I am ashamed to say, learn from women what they teach to men, and as if that were not enough, glibly expound to others things that they do not understand themselves. I will not even speak of those of my colleagues who, perhaps having come to the Holy Scriptures after a career in secular letters, gratify people's ears with carefully constructed phrases, and think that whatever they say is the word of God, without bothering to find out what the prophets and apostles taught. They adapt incongruous testimonies to their own purposes, as if it was an admirable rather than a deplorable way of teaching to distort the meaning of Scripture and twist it to their own contradictory ideas.

There are some theologians whom I hold to be men of great learning and sanctity of life, and for whom I therefore have the highest esteem, whom I would not wish to count among such profane writers. But I must confess that I do have some doubts which I would gladly have resolved when I hear that they claim, on the basis of the authority of Scripture, to require others to accept in scientific debates the view which they consider best harmonizes with Scriptural texts, while at the same time not accepting any obligation on their part to answer the reasons or evidence given to the contrary. They explain and justify this position by saying that theology is the queen of sciences, and therefore should on no account stoop to adapt to the teachings of other less exalted sciences which are subordinate to her, but rather that they should defer to her as the supreme ruler, and change their conclusions to conform to the statutes and decrees of theology. They go further, and say that if those who profess a subordinate science reach a conclusion which they consider to be certain, because it can be demonstrated or proved experimentally, but which is contradicted by a conclusion stated in Scripture, then it is up to them to disprove their own demonstrations and expose the fallacies in their own experiments, without bothering the theologians and biblical scholars. For, they say, it is not befitting the dignity of theology to stoop to investigating the weaknesses of its subject sciences; its role is solely to determine the truth of the conclusion, with absolute authority and in the certainty that it cannot err. The scientific conclusions about which they say we should defer to Scripture, without trying to gloss or interpret it in any way other than the literal meaning of the words, are those where Scripture consistently says the same thing and which the Church Fathers all receive and expound in the same way. There are several points about this ruling which I will raise so as to be advised by those who understand these matters better than me, and to whose judgement I submit at all times.

First, I fear there may be some cause for confusion if the pre-eminence which entitles theology to be called the queen of sciences is not clearly defined. It could be because the material taught by all the other sciences is encompassed and demonstrated in theology, but by more comprehensive methods and with more profound learning—in the same way as, for instance, the rules for measuring fields or keeping accounts are contained pre-eminently in arithmetic and Euclid's geometry rather than in the practical methods of surveyors or accountants. Or it could be because the subject matter of theology surpasses in dignity the subject matter of the other sciences, and because it proceeds by more sublime methods. I do not think that theologians who are conversant with the other sciences would claim that theology deserves to be called queen for the first of these reasons, for it is hard to believe that any of them would say that geometry, astronomy, music, and medicine are more comprehensively and precisely expounded in Scripture than in the works of Archimedes, Ptolemy, Boethius, and Galen.* It follows that the regal pre-eminence of theology must be of the second kind, namely on account of its elevated subject matter, its marvellous teaching of divine revelation, which human comprehension could not absorb in any other way, and its supreme concern with how we gain eternal beatitude. And if theology is concerned with the most elevated contemplation of the divine, occupying its regal throne because of its supreme authority, and does not stoop to the baser and more humble concerns of the subordinate sciences but rather, as has been said above, has no interest in them because they do not concern our beatitude, then those who practise and profess it should not claim the authority to lay down the law in fields where they have neither practised nor studied. If they did, they would be like an absolute prince who, knowing he was free to command obedience as he wished, insisted that medical treatment be carried out and buildings be constructed as he dictated even though he was himself neither a doctor nor an architect, thereby causing grave danger to the lives of his unfortunate patients and the evident ruin of his buildings.

Then, to command that professors of astronomy should be responsible for undermining their own observations and proofs as no more than fallacies and false arguments, is to command something quite impossible for them to do. For it amounts to telling them not to see what they see, and not to understand what they understand and, indeed, to find in their research the very opposite of what evidence shows them. If they were to be asked to do this, they would first have to be shown how to make one mental faculty give orders to another, and the lower faculties to command the higher, so that the imagination and the will were made able and willing to believe the opposite of what the intellect understands (in saying this I am still confining myself to purely scientific questions which are not articles of faith, and not to those which are supernatural and articles of faith). So I do beg these most prudent Fathers to consider very carefully the difference between statements which are a matter of opinion and those which can be demonstrated. If they keep in mind the strength of logical deduction, they will better understand why it is not in the power of those who profess the demonstrative sciences to change their opinion at will, applying themselves first to one view then to another, and that there is a great difference between commanding a mathematician or a philosopher and persuading a merchant or a lawyer to change their mind. It is not as easy to change one's view of conclusions which have been demonstrated in the natural world or in the heavens, as it is to change one's opinion on what is or is not permissible in a contract, a declaration of income, or a bill of exchange. The Church Fathers understood this very well, as can be seen from the great care they took to refute many arguments, or rather fallacies, in philosophy. This may be found expressly in some of them; in particular, we have the following words of St Augustine:*

It is unquestionable that whatever the sages of this world have demonstrated concerning physical matters, we can show not to be contrary to our Scripture. But whatever they teach in their books that is contrary to Holy Scripture is without doubt wrong and, to the best of our ability, we should make this evident. And let us keep faith in our Lord, in whom are hidden all the treasures of wisdom, so that we will not be led astray by the glib talk of false philosophy or frightened by the superstition of counterfeit religion.

From these words, it seems to me, the following principle can be derived: that the writings of secular scholars contain some statements about the natural world which are demonstrably true, and others which are simply asserted. As regards the former, it should be the task of wise theologians to show that they are not contrary to Scripture; and as regards the latter—those which are stated but not conclusively demonstrated—if there is anything in them which is contrary to Scripture, they should be regarded as undoubtedly false, and their falseness should be demonstrated by all possible means. Now if scientific conclusions which are demonstrated to be true should not be made subordinate to Scripture, but rather the text of Scripture should be shown not to be contrary to such conclusions, it follows that before a scientific statement is condemned it must be shown that it has not been conclusively demonstrated. And the responsibility for showing this must lie not with those who uphold its truth but with those who believe it to be false: this is only reasonable and natural, for it is much easier for those who do not believe a statement to identify its weaknesses than for those who believe it to be true and conclusive. Indeed, the upholders of an opinion will find that the more they go over the arguments, examining their logic, replicating their observations, and comparing their experiments, the more they will be confirmed in their belief. Your Highness knows what happened when the former mathematician at the University of Pisa* undertook in his old age to examine the teaching of Copernicus, in the hope of finding grounds for refuting it (for he was secure in his conviction that it was false as long as he had not read it): as soon as he grasped the foundations, the logic, and the demonstrations of the argument he became convinced by it, and from being an opponent of Copernicus' theory he became its staunch supporter. I could also name other mathematicians* who, prompted by my latest discoveries, have acknowledged it necessary to change the accepted system of the world, as it was now completely unsustainable.

If all that was needed to suppress this theory and its teaching was simply to gag a single author, as seems to be the impression of those who, measuring other people's judgement by the standards of their own, cannot believe that it could continue to find supporters, this would be easily done. But the reality is quite different. To achieve such an effect it would be necessary not just to ban Copernicus' book and those of the other authors who have followed his teaching, but to forbid the whole science of astronomy itself. More than that, they would have to forbid men to look at the sky, lest they should see Mars and Venus varying so much in their distance from the Earth that Venus appears forty times, and Mars sixty times, larger at some times than at others. Or they would have to prevent them from seeing Venus appear sometimes round and sometimes crescent-shaped with very fine horns, and many other observations of the senses which are completely incompatible with the Ptolemaic system, but provide solid evidence for the Copernican one. To ban Copernicus' book now, when many new observations and the work of many scholars who have read it are establishing the truth of his position and the soundness of his teaching more firmly every day, and after allowing it to circulate freely for many years when it had few followers and less evidence to support it, would in my view seem to be a contravention of the truth. It would be trying all the harder to conceal and suppress it the more it is plainly and clearly demonstrated. Not to ban the whole book, but just to condemn this particular proposition as false, would, if I am not mistaken, be even more harmful to people's souls, for it would allow them to see the proof of a proposition which they were then told it was sinful to believe. And to forbid the whole science of astronomy would be nothing less than contradicting a hundred passages of Holy Scripture, which teach us that the glory and greatness of God is wonderfully revealed in all his works, and made known divinely in the open book of the heavens.* Nor should anyone think that the lofty concepts which are to be found there end in simply seeing the splendour of the Sun and the stars in their rising and setting, which is as far as the eyes of brutes and the common people can see. The book of the heavens contains such profound mysteries and such sublime concepts that all the burning of midnight oil, all the labours, and all the studies undertaken by hundreds of the most acute minds have still not fully penetrated them, even after investigations which have continued for thousands of years. So let even the ignorant recognize that, just as what their eyes see when they look at the external appearance of the human body is as nothing compared to the marvellous complexity which is apparent to the trained and dedicated anatomist and philosopher, who never ceases to be amazed and delighted as he investigates the uses of the muscles, tendons, nerves, and bones, or when he examines the functions of the heart and the other principal organs, seeking out the seat of the vital faculties, observing the wonderful structure of the sensory organs, and contemplating where the imagination, the memory, and the power of reason dwell—in the same way, the heavens as they appear to the naked eye are as nothing compared to the great wonders which, through long and painstaking observations, the minds of intelligent men can discern there. This concludes what I have to say on this point.

Next let us answer those who assert that those scientific propositions of which Scripture consistently says the same thing, and which the Church Fathers have all received in the same way, should be understood according to the bare meaning of the words, without trying to gloss or interpret them, and should be accepted and believed as absolutely true; and that the motion of the Sun and the fixity of the Earth are such propositions and are therefore to be believed as matters of faith, and the contrary opinion is to be considered an error. On this I would make the following observations. First, there are some scientific propositions about which human speculation and reason cannot arrive at securely demonstrated knowledge, but can only supply a probable opinion and a reasonable conjecture—such as, for example, whether the stars are animate beings. There are other propositions of which we have, or can confidently expect to have certain knowledge, by means of experiment, prolonged observation, and necessary demonstrations; such are the questions whether the Earth or the Sun moves, or whether the Earth is a sphere. As far as the first kind of proposition is concerned, I have no doubt that where human reason cannot reach, and where consequently we cannot have certain knowledge but only an opinion or belief, we ought reverently to submit to the pure meaning of Scripture. But as regards the others, I believe that, as I have said above, we must first be certain of the facts, which will reveal to us the true meaning of the Scriptures, which will undoubtedly prove to be in agreement with the demonstrated facts, even if the surface meaning of the words appears to suggest otherwise; for two truths can never contradict each other. This principle seems to me all the more sound and secure because I find it stated in as many words by St Augustine. Writing specifically about the shape of the heavens and what should be believed about it, since astronomers who say that it is round appear to contradict Scripture which states that the heavens are stretched out like a skin,* he says that there is no reason to be concerned if Scripture contradicts the astronomers. The authority of Scripture is to be believed if what they say is false and founded only on fallible human conjecture; but if what they affirm is proved by incontrovertible arguments, he does not say that the astronomers are to be ordered to undermine their own proofs and declare their conclusions to be false. Rather, he says that when Scripture describes the heavens as being like a skin it must be shown that this is not contrary to what the astronomers have demonstrated to be true. These are his words:

But someone may ask: 'Is not Scripture opposed to those who hold that heaven is spherical, when it says, "who stretches out the heavens like a skin"?' It does oppose Scripture if their statement is false, for the truth is rather in what God reveals than in what groping men surmise. But if they are able to establish their position with proofs that cannot be denied, we must show that what is said about the skin is not opposed to the truth of their conclusions.*

He goes on to warn us that we should be no less careful to harmonize a passage of Scripture with a demonstrated scientific truth, as with another passage of Scripture which appears to state the opposite. Indeed, I think we should admire the prudence with which this saint, even when he is dealing with difficult questions about which we may be sure that no certain knowledge can be arrived at by human proof, is very cautious about laying down what should be believed. This is what he says at the end of the second book *On the Literal Meaning of Genesis*, on the question of whether we should believe the stars to be animate:

Although this problem at present is not easy to solve, yet I believe that in the course of our study of Scripture we may come across relevant passages that will enable us to treat the matter according to the rules for interpreting Holy Scripture and arrive at some conclusion that may be held without perhaps demonstrating it as certain. Meanwhile we should always observe that restraint that is proper to a devout and serious person and not rashly believe something about an obscure point. Otherwise, if the truth later becomes known we might despise it because of our attachment to our error, even if what is said is in no way opposed to the sacred writings of the Old or the New Testament.*

From this and from other passages it seems to me, if I am not mistaken, that the view of the Church Fathers was that on questions of natural science which are not matters of faith, we should first consider whether they have been demonstrated beyond doubt or are known from the evidence of the senses, or whether such certain knowledge is possible. If it is, and since this too is a gift of God, we should apply ourselves to understanding the true meaning of Scripture in those places where it appears to state the opposite. Wise theologians will undoubtedly be able to penetrate its true meaning, together with the reasons why the Holy Spirit should sometimes have chosen to veil it under words signifying something different, either to test us or for some other reason which is hidden from me.

As for the point about Scripture consistently saying the same thing, I do not think that this should undermine this principle, if we consider the primary intention of Scripture. If it was necessary on one occasion for Scripture to pronounce on a proposition with words conveying a different sense from its true meaning, as a concession to the understanding of the masses, then why might it not have done the same, and for the same reason, whenever the same proposition was mentioned? Indeed, to do otherwise would only have added to people's confusion and undermined their readiness to believe. Regarding the state of rest or motion of the Sun and the Earth, experience plainly shows that it was necessary for Scripture to state what its words appear to say; for even in our own time, people far less primitive still maintain the same opinion, for reasons which on careful consideration and reflection will be found to be wholly trivial, and on experiences which are either erroneous or completely irrelevant. And there is no point in even trying to persuade them to change their view, since they are not capable of understanding the arguments against it, depending as these do on observations which are too precise, proofs which are too subtle, and abstractions which require too much power of imagination for them to comprehend. So even if the fixity of the Sun and the motion of the Earth were established and demonstrated with absolute certainty among the wise, it would still be necessary to uphold the opposite to maintain one's credibility among the vast number of the masses. For if you were to quiz a thousand men among the common people about their view of this matter, I doubt whether you would find one who did not declare himself firmly convinced that the Sun moves and the Earth stands still. But no one should take this almost universal popular consent as an argument for the truth of what they assert; for if we were to question these same men about their grounds and reasons for believing as they do, and on the other hand to listen to the experiments and proofs which have led a few others to believe the opposite, we would find that the latter are persuaded by solidly based reasons, while the former are influenced by shallow appearances and vain and ridiculous comparisons.

It is clear, then, that it was necessary to attribute movement to the Sun and rest to the Earth so as not to confuse the limited understanding of the masses, making them stubborn and reluctant to believe in the principal articles which are absolutely matters of faith; and if this was necessary, then it is not surprising that it was done, with great prudence, in Holy Scripture. I would go further, and say that it was not only consideration for the incomprehension of the masses but the prevailing opinion at that time which led the scriptural writers to accommodate themselves, in matters not necessary to salvation, more to received opinion than to the essential truth of the matter. Speaking of this, St Jerome writes,* 'as if there were not many things in the Holy Scriptures that were said according to the opinion of the time when they took place, rather than according to the truth contained'; and elsewhere,* 'It is the practice in the Scriptures for the writer to give the view of things as they were universally believed at that time.' And St Thomas, commenting on the words in Job chapter 27,* 'He stretches out the north over the void, and hangs the earth upon nothing', notes that Scripture refers to the space which enfolds and surrounds the Earth as 'void' or 'nothing', while we know that it is not empty but full of air. Nonetheless, he says, Scripture adapts to the view of the masses, who believe this space to be empty, by calling it 'void' and 'nothing'. In St Thomas's own words,* 'What seems to us in the upper hemisphere of the sky to be nothing but space filled with air, the common people consider to be empty; and Holy Scripture speaks of it according to the belief of the common people, as is its wont.' From this example I think it can clearly be deduced that on the same principle, holy Scripture had all the more reason to refer to the Sun as in motion and the Earth at rest; for if we challenge the common people's understanding, we will find them much more resistant to the idea that the Sun is at rest and the Earth in motion than to the space around us being full of air. So, if the Scriptural authors refrained from trying to convince the common people even of a point about which they could be persuaded relatively easily, it seems only reasonable that they should have followed the same policy in other much more difficult questions.

Copernicus himself recognized how much our imagination is influenced by ingrained habit and by ways of conceiving things which have been familiar to us since childhood; so in order not to make these abstract ideas even more confusing and difficult for us, once he had demonstrated that the movements which appear to us to belong to the Sun and the firmament are actually movements of the Earth, he continued to call them movements of the Sun and the heavens when he came to set them down in tables and show how they worked in practice. So he talks about the rising and setting of the Sun and the stars, of changes in the inclination of the zodiac and variations in the equinoctial points, of the mean motion, anomalies, and prosthaphaeresis of the Sun,* and so on. All these are in fact movements of the Earth; but since we are on the Earth and hence share in its every motion, we cannot discern them in the Earth directly, but have to refer them to the heavenly bodies where they appear to be. Hence we speak of them as if they occurred where we perceive them to be. This shows how natural it is to adapt ourselves to our habitual way of seeing things.

As for saying that when the Church Fathers agree in interpreting a statement in Scripture on a matter of natural science in the same way this should give it such authority that it becomes a matter of faith to believe it, I think that this should apply at most to those conclusions which the Fathers have aired and discussed exhaustively, weighing up the arguments on both sides before all agreeing that one view should be upheld and the other condemned. But the motion of the Earth and the fixity of the Sun are not propositions of this kind, for such an opinion was completely buried and far removed from the questions discussed by scholars at that time, and was not even considered, let alone upheld, by anyone. So it is fair to assume that it never occurred to the Fathers to discuss the matter, since Scripture, their own views, and the common consent of everyone all agreed on the same opinion, without anyone thinking to contradict it. So it is not enough to say that because the Fathers all accept the fixity of the Earth, etc., this is to be believed as an article of faith; it must be proved that they condemned the contrary opinion. For I could always say that as they never had any occasion to reflect on the matter or to discuss it, they simply left it and accepted it as the current opinion, not as something which had been resolved and established. Indeed, I think I have firm grounds for saying this; for either the Fathers reflected on this as a matter of controversy, or they did not. If they did not, then they cannot have reached any judgement about it, even in their own minds, and their indifference to it should not place any obligation on us to accept precepts which they did not even consider imposing. If on the other hand they had turned their minds to it and considered it, they would already have condemned it if they had judged it to be erroneous, and there is no record of their having done so. Indeed, once some theologians began to consider the matter, it is clear that they did not deem it to be erroneous: Didachus of Stunica,* for instance, in his *Commentaries on Job*, chapter 9, verse 6, commenting on the words 'Who shakes the earth out of her place', etc., discusses the Copernican position at length and concludes that the motion of the Earth is not contrary to Scripture.

I have, in any case, some reservations about the truth of the claim that the Church requires us to believe as articles of faith such conclusions in natural science as are supported solely by the common interpretation of the Church Fathers. I wonder whether those who argue in this way may have been tempted to extend the scope of the Conciliar decrees in support of their own opinion; for the only prohibition I can find on this matter is against distorting in a sense contrary to the teaching of the Church and the common consent of the Fathers those passages, and those alone, which concern matters of faith or morals or the building up of Christian doctrine. This is what was stated by the Council of Trent in its fourth Session.* But the motion or fixity of the Earth or the Sun are not matters of faith or morals, and no one is trying to distort the meaning of Scripture in ways contrary to the teaching of the Church or the Fathers. In fact those who have written about this matter have never cited passages of Scripture, leaving it to the authority of wise and learned theologians to interpret these passages according to their true meaning. It is clear that the Conciliar decrees agree with the Church Fathers in this respect; indeed, so far are they from making such scientific questions articles of faith and condemning contrary opinions as erroneous that they consider it pointless to try to arrive at certainty in such matters, preferring rather to concern themselves with the primary intention of the Church. Let your Highness hear what St Augustine says in response to those Christians who ask whether it is true that the heavens move or whether they are at rest:

My reply is that it would require a great deal of subtle and learned enquiry into these questions to arrive at a true view of the matter. I do not have the time to go into these questions and nor have those whom I wish to instruct for their own salvation and for what is necessary and useful in the Church.*

But even if it were resolved to condemn or admit propositions in natural science according to passages of Scripture which have been unanimously interpreted in the same way by all the Church Fathers, I do not see that this would apply to the present case, for the Fathers differ in their interpretation of the same passages. Dionysius the Areopagite* says that it was not the Sun but the Primum Mobile which stood still. St Augustine is of the same opinion, namely that all the celestial bodies came to a stop; so too is the Bishop of Avila.* But there are Jewish writers, cited with approval by Josephus, who maintain that the Sun did not really stand still, but only seemed to because the Israelites took so little time to defeat their enemies. Similarly with the miracle at the time of Hezekiah, Paul of Burgos* says that it was not the Sun that moved but the sundial.* But in any case, I will show below that it is necessary to gloss and interpret the meaning of the text of the book of Joshua regardless of the view we take of the structure of the universe.

But let us finally concede to these gentlemen more than they ask, and submit entirely to the judgement of wise theologians; and since there is no record of this particular debate being conducted by the ancient Fathers, let it be undertaken by the wise men of our own age. After hearing the experiences, the observations, the arguments, and the proofs cited by philosophers and astronomers on both sides— for it is a controversy over problems of natural science and logical dilemmas, in which a decision has to be made one way or the other—they will be able to determine the matter positively as divine inspiration dictates. But as for those who are ready to risk the majesty and dignity of Holy Scripture for the sake of defending their own vain imagination, let them not hope that such a resolution as this is to be reached without establishing the facts with certainty and discussing in detail all the reasons on both sides of the argument; nor need those who seek only that the foundations of this teaching should be carefully considered, prompted purely by a holy zeal for the truth, for Scripture, and for the majesty, dignity, and authority in which all Christians are bound to uphold it, have anything to fear from such a procedure. Surely it is plain to see that this dignity is far more zealously sought and secured by those who submit whole-heartedly to the Church, without asking for one or other opinion to be prohibited but only that they should be allowed to bring matters forward for discussion so that the Church can reach a decision with greater confidence, than by those who, blinded by their own self-interest or prompted by the malicious suggestions of others, preach that the Church should wield its sword straight away simply because it has the power to do so? Do they not realize that it is not always beneficial to do what one has power to do? This was not the view of the Church Fathers; on the contrary, they knew how prejudicial and how contrary to the primary intention of the Catholic Church it would be to use verses of Scripture to establish scientific conclusions which experience and necessary demonstrations might in time show to be contrary to the literal meaning of the text. Hence they not only proceeded with great caution, but they also left the following precepts for the guidance of others:

In matters that are obscure or far from clear, if we should read anything in Holy Scripture that may allow of different interpretations that are consistent with the faith we have received, we should not rush in headlong and so firmly commit ourselves to one of these that, if further progress in the search of truth justly undermines this position, we too fall with it. That would be to battle not for the meaning of Holy Scripture but for our own, by wanting something of ours to be the meaning of Scripture rather than wanting the meaning of Scripture to be ours.*

A little further on, to teach us that no proposition can be contrary to the faith if it has not first been shown to be false, he adds: 'Nothing is contrary to the faith until unerring truth gives the lie to it. And if that should happen, it was never taught by Holy Scripture but stemmed from human ignorance.' It is clear from this that any view which we attributed to a passage of Scripture would be false if it did not agree with demonstrated truth. Therefore we should use demonstrated truth to help us discover the correct meaning of Scripture, and not try to force nature or deny the evidence of experience and necessary demonstrations in order to conform to the literal meaning of the words, which our imperfect understanding might think to be true.

But note further, your Highness, how carefully this great saint proceeds before affirming that a particular interpretation of Scripture is correct, and so firmly established that there need be no fear of encountering any difficulty which might undermine it. Not content that a reading of Scripture should agree with a demonstrated truth, he adds:

But when some truth is demonstrated to be certain by reason, it will still be uncertain whether this sense was intended by the sacred writer when he used the words of Holy Scripture, or something else no less true. And if the general drift of the passage shows that the sacred writer did not intend this sense, the other, which he did intend, will not thereby be false. Indeed, it will be true and more worth knowing.

Yet this author's caution is even more remarkable when, not being convinced after seeing the demonstrations, the literal meaning of Scripture, and the context of the passage as a whole all pointing to the same interpretation, he adds: 'But if the context supplies nothing to disprove this to be the mind of the writer, we still have to enquire whether he may not have meant the other as well.' Not even then resolving to accept one interpretation and reject the other, he seems to think he can never be cautious enough, for he goes on: 'But if we find that the other also may be meant, it will not be clear which of the two meanings he intended. And there is no difficulty if he is thought to have wished both interpretations if both are supported by clear indications in the context.' Finally, he justifies this rule of his by showing the dangers to which Scripture and the Church are exposed by those who, being more interested in maintaining their own error than in upholding the dignity of Scripture, seek to extend the authority of Scripture beyond the terms which Scripture itself prescribes. He adds the following words, which alone should suffice to restrain and moderate the excessive licence which some claim for themselves:

It often happens that a non-Christian knows something about the earth, the heavens, and the other elements of this world, about the motion and orbit of the stars and even their size and relative positions, about the predictable eclipses of the Sun and Moon, the cycles of the years and the seasons, about the kinds of animals, shrubs, stones, and so forth, and this knowledge he holds to as being certain from reason and experience. Now, it is a disgraceful and dangerous thing for an infidel to hear a Christian, presumably giving the meaning of Holy Scripture, talking nonsense on these topics. We should take all means to prevent such an embarrassing situation in which the non-believer will scarce be able to contain his laughter seeing error written in the sky, as the proverb says. The shame is not so much that an ignorant individual is derided, but that people outside the household of the faith think our writers hold such opinions, and criticize and reject them as ignorant, to the great prejudice of those whose salvation we are seeking. When they find a Christian mistaken in a field which they themselves know well and hear him maintaining foolish opinions about our books, how are they going to believe those books in matters concerning the resurrection of the dead, the hope of eternal life, and the kingdom of heaven, when they think their pages are full of false-hoods about things which they themselves have learnt from experience and decisive argument?

This same saint shows how much the truly wise and prudent Fathers are offended by those who try to uphold propositions which they do not understand by citing passages of Scripture, compounding their original error by producing other passages which they understand even less than the first; he writes:

Rash and presumptuous men bring untold trouble and sorrow on their wiser brethren when they are caught in one of their false and unfounded opinions and are taken to task by those who are not bound by the authority of our sacred books. For then, to defend their utterly reckless and obviously untrue statements, they call upon Holy Scripture, and even recite from memory passages which they think support their position, although they understand neither what they mean nor to what they properly apply.

This seems to me to describe exactly those who keep citing passages of Scripture because they are unable or unwilling to understand the proofs and experiments which the author of this doctrine and his followers advance in its support. They do not realize that the more passages they cite and the more they insist that their meaning is perfectly clear and cannot possibly admit any other interpretation than theirs, the more they would undermine the dignity of Scripture (if, that is, their opinions carried any weight) if the truth were then clearly shown to contradict what they say, causing confusion at least among those who are separated from the Church and whom the Church, like a devoted mother, longs to bring back to her bosom. So your Highness can see how flawed is the procedure of those who, in debating questions of natural science, give priority in support of their arguments to passages of Scripture—and often passages which they have misunderstood.

But if they really believe and are quite certain that they possess the true meaning of a particular text of Scripture, they must necessarily be convinced that they hold in their hand the absolute truth of the scientific conclusion which they intend to debate, and so must know that they have a great advantage over their opponent who has to defend what is false. The one who is defending the truth will be able to draw on numerous sensory experiences and necessary demonstrations to support their position, while their opponent has to fall back on deceptive appearances, illogical reasoning, and fallacies. Why then, if they are so confident that their purely scientific and philosophical weapons are so much stronger in every way than their adversary's, do they immediately have recourse, as soon as battle is joined, to an awesome and irresistible weapon the very sight of which strikes terror into the heart of their opponent? If the truth be told, I believe that they are the first to be terror-struck, and that when they realize they are unable to resist the assaults of their adversary they try to find a way of not letting him come near them. To that end they forbid him to use the gift of reason which the divine goodness has granted him, and abuse the right and proper authority of Scripture which by common consent of theologians, if it is understood and used properly, can never contradict the evidence of plain experience and necessary demonstrations. But I do not think that their resorting to Scripture to cover up their inability to understand, let alone to answer, the arguments against them will do them any good, since this opinion has never hitherto been condemned by the Church. So if they wish to deal honestly they should either confess by their silence that they are unqualified to discuss such matters, or they should first consider that it is not in their power or that of anyone except the Supreme Pontiff or the Councils of the Church to declare a proposition to be erroneous, although they do have the right to debate whether it is true or false. Then, since it is impossible for any proposition to be both true and heretical, they should concern themselves with what they are entitled to discuss, namely demonstrating that it is false. Once they have established its falsehood, either there will be no more need to prohibit it because no one will subscribe to it, or it can safely be prohibited without any risk of causing scandal.

So let these people first apply themselves to refuting the arguments of Copernicus and others, and leave condemning his view as erroneous and heretical to those who have the authority to do so; but let them not hope to find in the wise and cautious Fathers of the Church or in the absolute wisdom of the One who cannot err those hasty judgements into which they are sometimes drawn by their own desires or vested interests. No one doubts that the Supreme Pontiff always has absolute power to admit or to condemn these and similar propositions which are not directly articles of faith; but it is beyond the power of any created being to make them true or false, in defiance of what they are de facto by their own nature. So they would be better advised first to establish with certainty the necessary and immutable truth of the matter, over which no one has any control, than to condemn either side in the absence of any such certainty. This would only deprive them of their own authority and freedom to choose, by imposing necessity on matters which at present are undetermined and a subject of free choice but still reserved to the authority of the Supreme Pontiff. In short, if it is not possible for a conclusion to be declared heretical while there is still uncertainty over whether it may be true, it is a waste of time for anyone to clamour for the condemnation of the motion of the Earth and the fixity of the Sun before they have demonstrated that such a position is impossible and false.

It remains finally for us to consider how far the passage in Joshua can be taken in the straightforward literal meaning of the words, and how it could come about that the day was much prolonged when the Sun obeyed Joshua's command to stand still.

If the movements of the heavens are taken according to the Ptolemaic system, such a thing cannot happen. For the Sun moves through the ecliptic in the order of the signs of the zodiac, that is, from west to east, and hence in the opposite direction to the motion of the Primum Mobile,* which is from east to west, this being the motion which produces day and night. It is clear therefore that if the Sun were to cease its own proper motion, the effect would be to make the day shorter, not longer. The way to make the day longer would be to speed up the Sun's motion; for to make the Sun remain at the same point above the horizon for some time without declining towards the west, its motion would have to be speeded up until it equalled that of the Primum Mobile, which would be about 360 times its normal speed. So if Joshua had meant his words to be taken in their strictly literal sense, he would have commanded the Sun to speed up its motion so that the Primum Mobile ceased to carry it along towards its setting. But since his words were heard by people whose knowledge of the motions of the heavens was very likely confined to just the universally known movement from east to west, and since he had no intention of teaching them about the structure of the spheres but only that they should comprehend the great miracle of prolonging the day, he adapted his words to their understanding and spoke in the way which would make sense to them.

Perhaps it was this consideration which prompted Dionysius the Areopagite to say* that the miracle consisted in stopping the Primum Mobile, with the consequence that all the other celestial spheres also stood still; St Augustine himself is of the same opinion,* and the Bishop of Avila also confirms it at length.* Indeed, it is clear that Joshua's intention was to make the whole system of celestial spheres stand still, because his command also included the Moon even though the Moon had nothing to do with prolonging the day. By commanding the Moon he implicitly included all the other planets, which are not named here any more than they are elsewhere in Scripture, since it has never been the intention of Scripture to teach us the science of astronomy.

It seems clear to me, therefore, if I am not mistaken, that if we were to accept the Ptolemaic system, we would have to interpret the words of Scripture in a sense different from their literal meaning; but bearing in mind the salutary warnings of St Augustine, I do not say that this is necessarily the correct interpretation, as someone else might come up with a better and more appropriate one. I would, however, like to conclude by asking whether this passage can be understood in a sense closer to what we read in Joshua if we assume the Copernican system, together with a further observation which I have recently made concerning the body of the Sun. I put forward this suggestion always with the reservation that I am not so wedded to my own ideas as to claim they are superior to other people's, or to deny that better interpretations may be forthcoming which would conform more closely to the intention of Holy Scripture.

Let us assume, then, in the first place, that the miracle in the book of Joshua meant bringing the whole system of celestial revolutions to a standstill, as suggested by the authors quoted above; this is because if just one sphere were to stand still it would upset the whole system, introducing unnecessary disruption throughout the whole of nature. Secondly, I take into account that the body of the Sun, while remaining fixed in the same place, nonetheless turns on its own axis, completing one revolution in about a month, as I believe I have demonstrated in my *Letters on the Sunspots*.* We can observe this movement and see that in the upper part of the Sun's globe it is inclined towards the south, and therefore in the lower part it inclines towards the north—in just the same manner as all the revolutions of the planets. Thirdly, if we consider the nobility of the Sun and the fact that it is the source of light, not only for the Moon and the Earth but, as I show conclusively, also for all the other planets, all of which similarly have no light of their own, I think it is not unreasonable to suggest that the Sun, as the chief minister of nature and, in a sense, the heart and soul of the universe, dispenses not only light to the bodies which surround it but also motion, by virtue of its turning on its own axis. This means that, just as if an animal's heart stopped beating all the other parts of the body would also stop moving, so if the rotation of the Sun were to cease, the rotations of the planets would stop as well. Of the many weighty writers I could cite to confirm the wonderful strength and power of the Sun, let it suffice for me to quote one passage from the blessed Dionysius the Areopagite in his book *On the Divine Names*.* Writing of the Sun, he says: 'His light also gathers and converts to itself all the things that are seen, moved, lighted, or heated, in a word everything that is held together by its splendour. Therefore the Sun is called Helios, for it gathers and brings together everything that is scattered.' A little later he goes on to say,

This Sun which we see is one, and although the essences and qualities of those things that we perceive with our senses are many and varied, yet the Sun sheds its light equally on all things, and renews, feeds, protects, completes, divides, unites, fosters, makes fruitful, increases, changes, fixes, builds up, moves, and gives life to them all. Every single thing in this universe, inasmuch as it can, partakes of one and the same Sun, and the causes of many things which partake of it are equally anticipated in it; and for all the more reason, etc.

So, the Sun being the source of both light and motion, if God wished that at Joshua's command the whole system of the world should rest and remain for several hours in the same state, it sufficed to make the Sun stand still. When the Sun stopped, all the other revolutions stopped as well; the Earth, Moon, and Sun remained in the same relationship to each other, as did all the other planets; and as long as this continued the day did not decline towards night, but was miraculously prolonged. In this way, by stopping the Sun it was possible to prolong the day on Earth, without altering or disrupting the other aspects and mutual positions of the stars, which agrees perfectly with the literal sense of the sacred text.

Another point which, if I am not mistaken, is of no small significance, is that the Copernican system makes another detail in the literal account of this miracle perfectly clear: namely, that the Sun stood still 'in the midst of the heavens'. This passage has caused learned theologians some difficulty, because it seems likely that when Joshua prayed for the day to be prolonged it was already near to sunset, not at midday—for if it had been at midday, and given that this happened about the time of the summer solstice when the days are at their longest, it seems unlikely that he would have needed to pray for the day to be prolonged so that he could pursue the battle to victory; the seven hours or more of daylight which remained would have been more than enough. So some very learned theologians have concluded that it must have been near to sunset; and indeed this is implied by Joshua's words, 'Sun, stand still', since if it had been at midday, either he would not have needed to ask for a miracle or it would have been enough to pray for the Sun to slow down. This is the view of Cajetan,* to which Magalhães* also subscribes, and confirms it by pointing out that Joshua had already done so many other things that day before he commanded the Sun that he could not possibly have completed them all in half a day. So they are reduced to interpreting the words 'in the midst of heaven' in a somewhat forced way, saying that they simply mean the Sun stopped when it was in our hemisphere, that is, when it was above the horizon. But I think we can avoid this and any other forced reading if we follow the Copernican system and place the Sun 'in the midst', that is, in the centre of the heavenly orbs and the revolutions of the planets, as indeed we must. Then, regardless of the time of day, whether at midday or at any other time towards evening, the day was prolonged and the revolutions of the heavens stood still when the Sun stopped in the midst of heaven, that is, in the centre, where it belongs. Apart from anything else, this is a more natural reading of the literal sense of the text, for if the writer had wanted to say that the Sun stood still at noon it would have been more correct to say that it 'stood still at midday, or in the circle of the meridian', not 'in the midst of heaven'. For the only true 'midst' of a spherical body like the sky is its centre.

As for other passages of Scripture which appear to contradict the Copernican position, I have no doubt that, if this position were once known to be true and proven, those same theologians who now, believing it to be false, find such passages incapable of being interpreted in a way compatible with it, would find interpretations for them which would accord with it very well, especially if their understanding of Holy Scripture were combined with some knowledge of astronomy. Just as now, believing this position to be false, they read the Scriptures and find only passages which conflict with it, so if they once entertained a different view of the matter they might well find just as many others which agreed with it. Then they might judge it fitting for the Holy Church to proclaim that God placed the Sun in the centre of the heaven and, by turning it on its axis like a wheel, gave the Moon and the other wandering stars their appointed course, when she sings the hymn:*

> O God, whose hand hath spread the sky,
> and all its shining hosts on high,
> and painting it with fiery light,
> made it so beauteous and so bright:

Thou, when the fourth day was begun,
didst frame the circle of the sun,
and set the moon for ordered change,
and planets for their wider range.

They could also say that the word 'firmament' is literally correct for the starry sphere and for everything which is beyond the revolutions of the planets, for in the Copernican system this is totally firm and immobile. And since the Earth moves in a circle, when they read the verse, 'Before he had made the Earth and the rivers, and the hinges of the earth',* they might think of its poles, for it seems pointless to attribute hinges to the terrestrial globe if it does not turn on its axis.

LETTER FROM ROBERTO BELLARMINE TO PAOLO ANTONIO FOSCARINI, 12 APRIL 1615

To the Very Reverend Father Paolo Antonio Foscarini,* Provincial of the Carmelites of the Province of Calabria.

Very Reverend Father,

I was pleased to read the letter in Italian and the essay in Latin which you sent me. They are full of insight and learning, and I thank you for them both. As you ask for my opinion I will reply very briefly, as at present you have little time for reading and I for writing.

First, it seems to me that both you and signor Galileo are acting prudently in confining yourselves to speaking hypothetically and not in absolute terms, as I have always understood Copernicus to have done. It is perfectly proper, and poses no danger, to say that all the appearances are saved* more effectively by the hypothesis that the Earth moves and the Sun is fixed than by postulating eccentrics and epicycles; and this is as far as a mathematician can go. But to say that the Sun actually is at the centre of the universe, that it turns on its axis but does not move from east to west, and that the Earth is in the third heaven and moves at a great speed around the Sun—this incurs a great danger, not only of provoking all scholastic philosophers and theologians, but also of undermining the faith by suggesting that Holy Scripture is in error. For while you have shown many ways in which Scripture can be expounded, you have not applied these to particular cases, and there is no doubt that you would have had great difficulty if you had tried to expound all the passages which you yourself cited.

Second, as you know, the Council forbids expounding the Scriptures in ways contrary to the common consent of the Church Fathers; and if you read not just the Fathers but also modern commentators on Genesis, the Psalms, Ecclesiastes, and Joshua, you will find that they all agree in interpreting the text literally when it says that the Sun is in the heavens and revolves at great speed around the Earth, and that the Earth is immobile at the centre of the universe, at a very great distance from the heavens. So consider now, in all prudence, whether the Church can allow the Scriptures to be interpreted in a way contrary to the Fathers and to all the Greek and Latin expositors. And it is no response to say that this is not an article of faith, because even if the subject matter is not an article of faith, the authority of the speaker is. It would be just as heretical to deny that Abraham had two sons and that Jacob had twelve, as it would be to deny that Christ was born of a virgin, for both are declared by the Holy Spirit speaking through the mouths of the prophets and apostles.

Third, if it were demonstrated to be true that the Sun is at the centre of the universe and the Earth is in the third heaven, and that the Earth goes round the Sun and not the Sun round the Earth, then we would have to consider very carefully how to interpret the Scriptures which appear to state the contrary; and rather than declare a demonstrated truth to be false, we would have to say that we do not understand them. But I will not believe that such a demonstration exists until it is shown to me. To demonstrate that the appearances are saved by the hypothesis that the Sun is at the centre and the Earth is in the heavens, is not the same as demonstrating that the Sun really is at the centre and the Earth in the heavens. I believe that the first can be demonstrated, but I have very great doubts about the second; and where there is doubt we should not abandon the Holy Scriptures as they have been expounded by the Fathers. I will add that it was Solomon who wrote 'The Sun also rises, and the Sun goes down, and hastens to the place where it rises', etc.,* and that Solomon, quite apart from being inspired by God, was a man wiser than all others and most learned in all human sciences and the knowledge of the created world. All his wisdom came from God, so it does not seem likely that he would affirm something which is contrary to demonstrated or demonstrable truth. It may be objected that Solomon was speaking according to appearances, and that the Sun appears to us to be revolving in the heavens when in fact it is the Earth that is revolving, in the same way as it appears to someone on a ship moving away from the shore that it is the shore that is moving away from the ship.* To this I would reply that someone sailing away from the shore, even if the shore appears to be moving away from them, knows that this is an error and corrects it, as they can clearly see that the ship is moving and not the shore; but as far as the Sun and the Earth are concerned, no wise man has ever needed to correct the error, because he clearly perceives that the Earth is not moving and that his eyes do not deceive him when he judges the Sun to be moving, in the same way as he does not err when he judges the Moon and the stars to be moving. So let this suffice for now.

With this, Reverend Father, I send cordial regards to you, and pray that God may grant you every contentment.

From my residence, 12 April 1615.
As a brother,
Cardinal Bellarmine

OBSERVATIONS ON THE COPERNICAN THEORY

In order to remove, as far as God grants me, any obstacle which might deflect those charged with adjudicating in the present controversy from reaching a correct judgement, I shall try to eliminate two notions which some people appear to me to be trying to impress on them, both of which, if I am not mistaken, are at variance with the truth.

The first is that there is no reason to fear that their judgement might cause a scandal; for they say that the fixity of the Earth and the motion of the Sun are so well established in philosophy that they can be regarded as absolutely secure and certain, and that to assert the contrary is such an immense paradox and so patently foolish that there is no reason to fear that it might be demonstrated, or even countenanced by any person of sound judgement, either now or at any time in the future. The other notion which they are trying to establish is that, while Copernicus or other astronomers have taken the contrary view, they have done so purely hypothetically, as better preserving the appearances of the motions of the heavens and the calculations and computations of astrologers,* and that even those who advance this hypothesis do not believe it to be factually true in nature; hence they conclude that it can safely be condemned. I believe that this argument is fallacious and at variance with the truth, as I shall make clear in the following observations, which are of a general nature and such as can be understood without great difficulty or study even by those who are not expert in the natural sciences and astronomy. Should the occasion arise to discuss these points with those who are well versed in these subjects, or at least have the time to consider them with the application that the difficulty of the material requires, I would recommend simply that they read Copernicus' book itself, and the strength of its demonstrations will clearly reveal the truth or falsehood of the two notions we are discussing.

That this is not a view to be dismissed as ridiculous is clear, then, from the many distinguished men, ancient as well as modern, who have held it in the past and who hold it today. Anyone who considered it ridiculous would first have to regard as foolish Pythagoras and all his followers, Plato's teacher Philolaus and Plato himself, as Aristotle testifies in his book *On the Heavens*, Heraclides of Pontus and Ecphantus, Aristarchus of Samos, Nicetas, and Seleucus the mathematician.* Seneca himself, indeed, far from deriding it, mocks those who dismiss it as ridiculous; as he writes in his book *On Comets*:*

It will also help to clarify this if we know whether the universe revolves around the stationary Earth, or the universe is stationary and the Earth revolves. There have been those who have said that it is we who, without realizing it, are carried along by the natural world, and that it is not the motion of the heavens that causes them to rise and set, but that we ourselves rise and set. It is a matter worthy of consideration to know what our true state is: whether we stand still or move very fast, and whether God rotates all things round us or has us rotate.

Among the moderns, Nicolaus Copernicus was the first to revive the idea and confirm it amply in the whole of his book, and others have followed, among them William Gilbert, a distinguished doctor and philosopher, who discusses it at length and confirms it in the last book of his work *On the Magnet*;* Johannes Kepler,* an eminent living philosopher and mathematician in the service of the past and present emperor,* is of the same opinion; David Origanus, at the beginning of his *Ephemerides*,* devotes a long discussion to proving the motion of the Earth; and there is no lack of other authors who have published their reasons for supporting it. What is more, I could name numerous others living in Rome, Florence, Venice, Padua, Naples, Pisa, Parma, and other places, who uphold this position even though they have not published their views in print. So it is not a ridiculous position, but one which is held by very distinguished men; and if they are few in comparison with those who adhere to the commonly held view, this is evidence that it is hard to understand, not that it is unfounded.

Moreover, the weighty and convincing arguments which support this view can be deduced from the fact that all those who now follow it started off by believing the opposite—indeed for a long time laughed at it and considered it ridiculous, as Copernicus himself and I, and all its other living supporters can testify. Is it credible that an opinion which was generally considered to be unfounded, even foolish, and which was not upheld by more than one philosopher in a thousand, indeed was rejected by the prince of present-day philosophy,* could be established by any but the most solid proofs, the clearest experiments, and the most accurate observations? Surely no one will be persuaded to change their mind about an opinion which they have taken in with their mother's milk and their earliest lessons, which almost the whole world considers obvious and which is supported by the authority of the most reputable authors, if the arguments against it are not wholly compelling. If we consider carefully we shall see that the support of one person who upholds the Copernican opinion is worth a hundred of those who oppose it, because those who are persuaded of the truth of the Copernican system all began by believing the opposite. So I reason as follows: those who have to be persuaded are either capable of understanding the reasoning of Copernicus and his followers, or not; this reasoning is either true and demonstrable, or false. If those who have to be persuaded are incapable of following the arguments, they will not be convinced by them whether they are true or false, and those who are capable of following the arguments will equally be unpersuaded if these arguments are false. Therefore neither those who understand nor those who do not will be persuaded by false arguments. So, since no one is going to be persuaded to change their mind by arguments which are false, it necessarily follows that if anyone is convinced of the opposite of what they believed to start with, the reasons which convince them must be true and convincing. And since there are indeed many who are persuaded by the arguments of Copernicus and others, these arguments must be effective and should not be dismissed as ridiculous, but deserve to be considered with the greatest care and attention.

As for judging the plausibility of an opinion by the sheer number of those who follow it, this can easily be shown to be unsound. There is no one who supports the Copernican view who did not originally believe the opposite; whereas, in contrast, there is not a single person who having adopted this view changed their mind again because of any arguments they may have heard. So it seems very likely, even to someone who has not heard the arguments on either side, that the proofs for the motion of the Earth are much stronger than those for the opposite view. If the probability of the two positions were put to a vote, I would gladly concede the point if the opposite view had one vote in a hundred more than mine—indeed, I would go further and would be willing for every single vote of my opponent to be worth ten of mine—provided the vote was conducted among those who had listened to all the arguments fully, had gone into them in detail, and had carefully examined all the reasoning and the evidence on both sides; for it is reasonable that only such people should be eligible to vote. So this is not an opinion which can be dismissed as ridiculous; rather, it is those who seek to place great weight on the universal view of the multitude who have not carefully studied these authors who are on shaky ground. How much less, therefore, should we pay attention to the noisy protests and idle chatter of those who have not even considered the first and most basic principles of these matters, and may not even be capable of understanding them in any case?

There are some who persist in saying that Copernicus only put forward the motion of the Earth and the fixity of the Sun as an astronomical hypothesis, as a more satisfactory way of saving the appearances of the heavens and calculating the motions of the planets, but did not believe this to be actually true in nature. They show—and I mean no disrespect in saying this—that they have been too ready to believe those whose views are perhaps based simply on their own opinion rather on a proper reading of Copernicus' book or an understanding of the nature of the discussion, and so what they say is somewhat wide of the mark.

Turning first, then (and still confining ourselves to general conjectures), to the preface in which he dedicates his book to Pope Paul III, it is clear that, as if to comply with what was expected of him as an astronomer, Copernicus initially carried out his computations following the established philosophy and conforming to the Ptolemaic system, so that nothing was lacking. But then, setting aside the role of a pure astronomer and assuming that of an observer of nature, he undertook to examine whether this supposition which had already been introduced by astronomers, and which fully satisfied the requirements of calculating and accounting for the motions of each individual planet, could really exist in the natural world. He found that it was quite impossible for all the parts to be arranged in such a way, even though each part by itself was well proportioned, but that when they were all put together they formed a monstrous chimera. So, as I have said, he set about considering what could be the real natural structure of the universe, no longer from the point of view of the pure astronomer—whose calculations he had already fully satisfied—but in order to understand this profound problem of nature; for he was confident that, if it was possible to satisfy the mere appearances with hypotheses which were not true, this could be done much better by means of the true physical structure of the world. And so, drawing on a rich collection of genuine and direct observations about the courses of the stars (without which no such understanding is possible), he applied himself tirelessly to discovering this structure. First, prompted by the authority of so many great men of the past, he began to consider whether the Earth might be mobile and the Sun at rest—a possibility which would not have occurred to him if it had not been for the prompting of these authoritative figures, or if it had, he would have considered it (as he confessed he did when it first appeared to him) as a wild idea and a great paradox. But then, through painstaking observations, considerations which all agreed with each other, and conclusive proofs, he found it to be so in keeping with the harmony of the universe that he became completely convinced of its truth. He arrived at this position, therefore, not to satisfy the needs of the pure astronomer, but to satisfy the requirements of nature.

What is more, Copernicus goes on to say in the same place that he realized that publishing this opinion abroad would give him the reputation of a madman among the infinite numbers of followers of the current philosophy, not to mention the general masses; but he published it nonetheless, compelled by the insistence of the Cardinal of Capua and the Bishop of Kulm.* How much more mad would he have been if he had published this opinion believing that it was false, but presenting it as if he believed it to be true, in the certainty that he would be regarded as foolish by the whole world? Would he not rather have said that he was only putting it forward as an astronomer while denying it as a philosopher, thereby avoiding being universally called a fool, and winning praise for his sound judgement?

Besides, Copernicus states in detail the grounds and reasons for the ancients' belief that the Earth was at rest, and then examines the force of each one in turn, showing them to be invalid.* What author with any sense would set out to refute the proofs in favour of a proposition which he deemed to be true? And what would have been the point of disproving and condemning a conclusion which he really wanted the reader to think he believed to be true? Such inconsistencies cannot be attributed to such a man.

We must recognize that the question of whether the Earth or the Sun moves or is at rest is a choice between contradictory positions, one of which must be true; we cannot fall back on saying that perhaps neither of them is correct. If, then, the fact is that the Earth is at rest and the Sun moves, and that the contrary position is absurd, what reasonable person could say that the false position accords better with the visible and observed appearances, and with the motions and positions of the stars, than the true one? Everyone knows that all the truths of nature harmonize with each other, and that false positions clash discordantly with the facts. Can it be, then, that the motion of the Earth and the fixity of the Sun agree in every respect with the position of all the other bodies in the universe, and with the very many observations that we and our predecessors have painstakingly made, and yet this position is false? And can the fixity of the Earth and the motion of the Sun be considered true, and yet not be in agreement with the other truths? If it were possible to say that neither position is true, then it might be the case that one preserves the appearances better than the other; but to say of two positions, one of which is necessarily false and the other true, that the false one better accords with the effects in nature, is beyond my comprehension. Rather, I ask: if Copernicus admits that he has fully satisfied the astronomers working with the hypothesis which is commonly held to be true, how can anyone say that he wanted or was able to satisfy them again with another hypothesis which is foolish and false?

I move on now to examine in detail the nature of the question, and to show how much careful attention is necessary in discussing it.

There are two kinds of supposition which have so far been made by astronomers. Some are primary, and concern the absolute truths of nature; others are secondary, and have been imagined to explain the appearances of the stellar motions, as these seem in some way not to agree with the primary, true suppositions. Ptolemy, for example, before attempting to account for the appearances, makes certain suppositions not as a professional astronomer, but as a thorough-going philosopher—borrowing them, indeed, from the philosophers themselves. He supposes that the celestial motions are all circular and regular, that is, uniform; that the heavens are spherical in shape; that the Earth is at the centre of the celestial sphere, and that it too is spherical, and at rest; and so on. Turning then to the irregularities which we perceive in the motions and the distances of the planets, which appear to clash with the established primary physical suppositions, he goes on to make another kind of supposition, with the aim of explaining, without altering the primary suppositions, why there is such evident and observable irregularity in the motions of the stars as they approach or recede from the Earth. In order to do this he introduces other circular motions which do not have the Earth as their centre, but describe eccentric circles and epicycles. It is this second kind of supposition which the astronomer can be said to posit to meet the needs of his computations, without having to maintain that they are actually true in nature.

Let us now consider which kind of hypothesis Copernicus considered the motion of the Earth and the fixity of the Sun to be. There is no doubt that he placed it among the primary and necessary suppositions about nature; for as I have said, he had already given the astronomers satisfaction by the other route, and he only applied himself to this new one to resolve the greatest problem of nature. In fact, so far was he from adopting this supposition in order to satisfy astronomical calculations that he himself, when he comes to these calculations, sets aside the new position and returns to the old, as being more accessible and easier to learn, and still wholly suitable for these computations. For although both suppositions, namely whether the Earth revolves or the heavens revolve, are inherently equally suitable for particular calculations, nonetheless many geometricians and astronomers had in many books already demonstrated the properties of right and oblique ascension of parts of the zodiac in relation to the equinoctial, the declinations of the ecliptic,* the variety of the angles between it and the oblique horizons and the meridian, and a hundred and one other specific details which are needed for the completeness of astronomical science. So Copernicus himself, when he considers these details of the primary motions, discusses them in the old manner, as made of circles drawn in the heaven around the unmoving Earth, even though in reality the fixity and motionlessness were in the highest heaven, called the Primum Mobile, and the motion was in the Earth. This is why he concludes, at the end of the preface to the second book: 'Let no one be surprised if I still speak of the rising and setting of the Sun and the stars, and similar things. It should be recognised that I am using accustomed figures of speech which everyone can understand; nonetheless I am always conscious that "To us who are being carried along on the Earth, the Sun and Moon pass over us, and the stars return to their former places and move away".'*

There is no doubt, then, that Copernicus posits the mobility of the Earth and the fixity of the Sun solely in order to establish this hypothesis of the first kind; and that in contrast, when he comes to dealing with astronomical computations he reverts to positing the old hypothesis, in which the circles of the basic motions are imagined as occurring in the highest heaven and moving around the stationary Earth, since ingrained habit makes this easier for everyone to grasp. But why should I labour this point? Such is the strength of the truth and the weakness of falsehood that those who argue in this way reveal their own imperfect grasp of these matters, allowing themselves to be persuaded that Ptolemy and the other authoritative astronomers did not believe the second kind of hypothesis to be true in nature, but considered them mere chimerical fantasies introduced for the sake of their astronomical computations. Their only basis for this deluded opinion is a passage where Ptolemy, having been unable to observe more than one simple anomaly in the Sun, writes that this could be explained by positing either a simple eccentric or an epicycle on a concentric, and adds that he will adopt the first hypothesis as being simpler than the second.* On the slender basis of these words some argue that Ptolemy considered both suppositions to be unnecessary, indeed that they were totally fictitious, since he said that either would serve although only one could be a correct explanation of what we observe in the Sun. What a shallow argument! Surely anyone can see that, if the primary suppositions are taken to be true, namely that the motions of the planets are circular and regular, and if, as the evidence of our senses undeniably shows, the planets all move through the zodiac at different speeds, indeed the majority of them sometimes stand still or move in a retrograde direction, and that they appear sometimes very large and close to the Earth and at other times very small and far away—surely, I say, anyone trained in the discipline who understands these basic observations realizes that eccentrics and epicycles must actually exist in nature. A misunderstanding, which is excusable in those who are not trained in these sciences, reveals in those who profess them that they do not even understand the meaning of the terms *eccentric* and *epicycle*: it is as if someone could identify the letters G, O, and D but then denied that when they are combined they form the word GOD, and claimed that they spelt the word SHADOW. But if discursive reasoning is not enough to persuade them that eccentrics and epicycles must necessarily have a real existence in nature, then the evidence of their own senses ought to convince them. The four Medicean planets,* which are very far from circling the Earth, can be seen to describe four small circles around Jupiter, that is, four epicycles; the appearance of Venus, which is sometimes full of light and sometimes has very thin horns, necessarily shows that it revolves around the Sun and not around the Earth, that is, that its course is an epicycle; the same can be deduced of Mercury. And the fact that the three superior planets* are very close to the Earth when they are in opposition to the Sun, and very remote from the Earth at the time of their conjunction, so that Mars appears more than fifty times bigger when it is at its nearest point than when it is furthest away (with the result that it has sometimes been feared that it had got lost and disappeared, when in fact it was simply invisible because it was so distant), can only lead to the conclusion that their revolution is made in eccentric circles, or in epicycles, or in a combination of the two, if we take account of the second anomaly.* So to deny that there are eccentrics and epicycles in the motions of the planets is like denying light to the Sun; it is a self-contradiction.

Let me apply this specifically to the point at issue here. Some argue that modern astronomers introduce the motion of the Earth and the fixity of the Sun purely hypothetically, in order to save the appearances and to facilitate their calculations, in the same way as they accept eccentrics and epicycles for the same reason even though they regard these as illusions which cannot occur in nature. My reply to them is that I will gladly accept their argument provided they accept the same condition, namely that the mobility of the Earth and the fixity of the Sun are true or false in nature in the same way as epicycles and eccentrics are. So let them do all in their power to disprove the real actual existence of these circles, and if they succeed in demonstrating that they do not exist in nature I shall immediately admit defeat, and acknowledge that the mobility of the Earth is a great absurdity. But if they find that they have to accept epicycles and eccentrics, then let them also admit to the mobility of the Earth, and acknowledge that they have been convicted by their own contradictions.

I could make many other points on this question, but since I doubt whether anyone who is not convinced by what I have said already is likely to change their mind because of any number of other arguments, let these suffice. I will simply add what might be the reason which has enabled some to maintain, with at least an appearance of plausibility, that Copernicus himself did not really believe in the truth of his own hypothesis.

On the back of the title-page of Copernicus' book is a preface to the reader, which is not by the author, as it speaks of him in the third person, and is unsigned.* This clearly states that the reader should not believe for a moment that Copernicus considered his position to be true, but that it was simply a fiction which he introduced for the purpose of calculating the motions of the heavens; and it ends by saying that it would be folly to believe that the theory is actually true. This conclusion is stated so firmly that anyone who reads no further, and who assumes that it was at least added with the author's consent, can be excused for their error. But I let everyone judge for themselves how much weight should be given to the view of someone who would condemn a book without reading anything more than a brief preface by the printer or bookseller. It can only be a preface added by the bookseller to make the book sell, for the general public would have considered it sheer fantasy if it had not had some such qualification added to it, and the customer usually glances at such prefaces before buying the book. And that the preface was not only not written by the author, but was added without his consent, is clear from the sheer errors of terminology which it contains, which the author would never have accepted.

The author of this preface writes that only someone wholly ignorant of geometry and optics could think it probable that Venus has such a large epicycle that it can sometimes precede and sometimes follow the Sun by 40 degrees or more, as this would mean that when it is at its highest point its diameter would appear to be only a quarter of what it appears to be at its lowest point, and its volume at its lowest point would appear to be sixteen times greater than at its highest. He says that this is contrary to the observations made throughout the centuries, but in saying this he shows, first, that he does not know that Venus precedes and follows the Sun by just under 48 degrees, not 40 as he says. What is more, he says that its diameter would appear four times, and its volume sixteen times greater at its lowest point than at its highest. This shows, first of all, his ignorance of geometry, since a globe whose diameter was four times larger than another would have a volume sixty-four times greater, not sixteen times as he says. Hence, if he considered such an epicycle to be absurd and wanted to show that it was impossible in nature then, if he had understood geometry, he could have made the absurdity all the greater, since the position which he rejects and which is postulated by astronomers has Venus diverging from the Sun by nearly 48 degrees, and its distance at its furthest from the Earth must be more than six times greater than when it is at its nearest. So its diameter would appear to be more than six times greater, and its volume more than 216 times greater, not just sixteen times as he says.* These are such gross errors that it is unbelievable that they should have been committed by Copernicus, or by anyone who was not completely inexperienced. But in any case, why cite the vastness of an epicycle as so great an absurdity that Copernicus could not have believed his conclusions were true, and that neither should others believe them to be true? He should have remembered that in chapter 10 of book I Copernicus, to refute other astronomers who allege that it is absurd to give Venus such an epicycle, which exceeds the whole lunar orbit by more than 200 times and yet contains nothing inside it, eliminates this absurdity by showing plainly that the orbit of Venus contains both the orbit of Mercury and the body of the Sun itself, which is at its centre. What a shallow argument this is, then, to try to show that a theory is erroneous and false on the basis of a difficulty which the theory in question not only does not introduce into nature, but which it entirely removes—just as it removes all the other enormous epicycles which other astronomers were obliged to posit under the other system! This is as much as I shall say about the author of the preface to Copernicus' book, for it is fair to infer that if he had tried to put forward any other arguments based on the science of astronomy, he would simply have multiplied his errors.

Finally, to remove any shadow of doubt, if the fact that such great differences in the apparent size of Venus are not perceptible to our sight should call into question its revolution around the Sun, in conformity with the Copernican system, then let us make diligent observation with an appropriate instrument, namely a perfect telescope, which will show that these are indeed confirmed by observation and experience. For Venus will be seen when it is nearest to the Earth to be horned, and to be at least six times greater in diameter than when it is at its furthest distance, namely above the Sun, at which point it appears round and very small. And if the fact that these variations are not visible to the naked eye, for the reasons that I have explained elsewhere, should seem to give grounds for denying the Copernican position, so now that we can see that it corresponds in this and every other detail exactly to what is observed, let all doubts be set aside and let it be recognized as true and correct.* And if anyone wishes to have confirmation of the opinion of Copernicus himself on every other part of this marvellous system, let them read the author's work in its entirety, not a vain piece of writing added by the printer; and they will be left in no doubt that Copernicus was firmly convinced that the Sun is at rest and that the Earth moves.

* * *

The mobility of the Earth and the fixity of the Sun, if they are shown to be true in nature by philosophers, astronomers, and mathematicians drawing on the experience of the senses, detailed observations, and necessary demonstrations, can never be contrary to the faith or to the Holy Scriptures; rather, if in this case there are passages of Scripture which appear to state the contrary, we should say that this is because of the weakness of our intellect, and that we have been unable to penetrate the true meaning of Scripture in this matter. This is a clear and well-established principle, since two truths cannot contradict each other. So if anyone wishes legally to condemn such a view they must first show it to be false in nature, by rebutting the arguments which support it.

The question then arises, from which end one should start in order to establish that it is false: by citing the authority of Scripture, or by refuting the experiments and proofs of the astronomers and philosophers. My reply is that one should start from the place that is most secure and least likely to give rise to scandal, that is, from the scientific and mathematical arguments. For if the arguments proving the mobility of the Earth are found to be fallacious and the contrary arguments sound, then we shall have established that this view is false and the contrary proposition, which we now see as being in keeping with the meaning of Scripture, is true; in which case the false view can freely and safely be condemned. If on the other hand the arguments supporting the mobility of the Earth are found to be true and compelling, the authority of Scripture is not thereby undermined, but we shall be more cautious, recognizing that in our ignorance we were not able to penetrate its true meaning. This we shall then go on to find, with the help of the new scientific truth we have discovered. Hence we are always on firm ground by starting with the arguments. But if we condemned such a proposition without having examined the arguments, basing ourselves solely on what we took to be the clear meaning of Scripture, how great would be the scandal if the experience of the senses and good reasons then showed that we were wrong? Who, in that case, would have brought confusion on the Church—those who put forward a considered opinion based on proof, or those who dismissed it out of hand? So it is clear which is the safer route to take.

Moreover, once it is granted that a scientific proposition whose truth is demonstrated by physical and mathematical proofs can never contradict Scripture, but that in such a case it is the weakness of our intellect which has failed to penetrate Scripture's true meaning, anyone who then cited the authority of the same passages of Scripture in order to refute such a proposition would fall into the error called 'begging the question'. For if scientific proofs have cast doubt on the true meaning of Scripture, we cannot then take the meaning of Scripture as a secure basis for refuting the same proposition. Rather, we have to challenge the proofs and expose their fallacies by means of other arguments, experiments, and more accurate observations. Once we have established the true position as it is in nature, then and only then can we be certain of the true meaning of Scripture and use it with confidence. Thus, the secure way of proceeding is to start from the proofs, confirming those which are true and refuting those which are false.

If the Earth really does move, we cannot change nature so that it does not move; but we can easily resolve the inconsistency with Scripture, simply by admitting that we cannot penetrate its true meaning. So the secure way of avoiding error is to begin with natural and astronomical enquiries, not with references to Scripture.

It will be objected that the Church Fathers all agree in interpreting the passages of Scripture relating to this point according to the straightforward plain meaning of the words, and that therefore we should not read them in a different way or challenge the accepted interpretation, as this would be to accuse the Fathers of heedlessness or negligence. I agree that such respect for the Fathers is only right and proper, but I would add that there is a perfectly sound reason for their attitude: they never expounded the Scriptures in any but the literal sense in these matters, because at that time the idea that the Earth moves was completely forgotten, and nobody even spoke about it, let alone wrote it or upheld it; so the Fathers cannot be accused of negligence if they did not reflect on something which was totally hidden from them. That they did not reflect on it is clear from the fact that there is not a word about such an opinion in any of their writings; indeed, to say that they did consider it would make it all the more dangerous to condemn it now, as this would mean that the Fathers had considered it and, far from condemning it, had cast no doubt on it whatever.

The Church Fathers, then, have a clear and ready defence. But it would be much harder, if not impossible, to excuse from a similar charge of negligence the supreme Pontiffs, Councils, and compilers of the Index, if for eighty years they had allowed an opinion and a book to circulate with the approval of the Church—a book written at the command of a Pope, then printed by order of a Cardinal and a Bishop, and dedicated to another Pope, and one moreover which is quite specific in its teaching, so that it cannot be said to have remained hidden—if its teaching was erroneous and to be condemned. So if we are, quite rightly, to avoid the disrespect of accusing our ancestors of negligence, we should also be careful lest in our anxiety to avoid one absurdity we fall into a worse one.

If, however, anyone should think it disrespectful to abandon the common interpretation of the Fathers, even of propositions concerning the natural world which they did not discuss or the contrary of which they never considered, my question is this: what are we to do if necessary proof should show that the actual fact in nature is the contrary of what they upheld? Which of the two principles should we compromise—the principle which states that no proposition can be both true and false, or the one which requires us to hold as articles of faith natural propositions endorsed by the common interpretation of the Fathers? Unless I am mistaken, it seems much safer to compromise on this second principle than to try to impose as an article of faith a natural proposition which conclusive proofs have shown to be false in fact and in nature. It seems to me that the common interpretation of the Fathers should be absolutely authoritative in propositions which they discussed and for which there were no proofs to the contrary, and no possibility of such proofs ever being found. It seems quite clear that the Council* requires agreement with the common interpretation of the Fathers only in matters of faith and morals; but I will not pursue this.

* * *

1. Copernicus posits eccentrics and epicycles; it was not these, which undoubtedly exist in the heavens, but other disproportionate assumptions which prompted him to reject the Ptolemaic system.

2. Philosophers, if they are true philosophers, that is, lovers of the truth, should not be resentful but grateful if someone shows them the truth and they realize that they have been mistaken; and if their opinion withstands a challenge, they should see that as a source of pride and not a cause for indignation. Theologians should not be

resentful if this opinion is shown to be false, because then they will be free to condemn it; if it is shown to be true, they should rejoice that someone has shown them the way to discover the true meaning of Scripture and spared them the disgrace of condemning a proposition which is true.

As for showing the Scriptures to be false, that is not and never has been the intention of Catholic astronomers such as myself. On the contrary, our view is that the Scriptures agree perfectly with demonstrated natural truths. Some theologians who are not astronomers should beware of making the Scriptures appear false by trying to interpret them as contrary to propositions which could be true and proven in nature.

3. We may well encounter some difficulties in expounding the Scriptures, etc.; but this is because of our ignorance, not because there really is, or might be, any insuperable difficulty in showing that they agree with demonstrated truths.

4. The Council speaks of 'matters of faith and morals'. In reply to the claim that a given proposition is a matter of faith 'with respect to the one who speaks it' even if not 'by reason of the object' and that therefore it comes within the terms of the Council's ruling, it should be pointed out that everything in Scripture is a matter of faith 'with respect to the one who speaks it' and therefore should be included in the ruling, which is clearly not the case; if it had been, it would have said 'in every word of Scripture the exposition of the Fathers is to be followed', and not just 'in matters of faith and morals'. Hence it is clear that the intended meaning was 'in matters of faith by reason of the object'. This is why it is much more a matter of faith to believe that Abraham had sons, or that Tubal had a dog, because Scripture says they did,* than to believe that the Earth moves, although this too can be read in Scripture, and it is a heresy to deny the former but not the latter. For there have always been men who had two, four, or six sons, or for that matter none; and there have always been some men who had dogs, and others not; so it is equally credible that one man had sons or dogs and another did not, and there is no reason why the Holy Spirit should affirm anything other than the truth on such matters, since negative and positive statements are both equally credible to everyone. But this is not the case with the mobility of the Earth and the fixity of the Sun, which are ideas far beyond the

comprehension of the masses; and it pleased the Holy Spirit to accommodate the pronouncements of Scripture to the understanding of the masses in these matters which do not pertain to their salvation, even though the true facts are otherwise.

5. As regards saying that the Sun is in the heavens and the Earth is not, as Scripture appears to affirm, this seems to me to be simply a matter of our perception and of speaking in terms which we can understand. The reality is that everything which is surrounded by the heavens is 'in' the heavens, just as everything that is surrounded by a city's walls is in the city; indeed it could be said that nowhere is more 'in' the heavens or 'in' the city than the middle or, as we say, in the heart of the city or the heavens. The difference as far as our understanding is concerned is that we regard the elemental region which surrounds the Earth as being quite distinct from the heavens themselves. This would be true wherever we placed the elemental region, and it will always be the case that we will think of the heaven as being above and the Earth beneath, because all the inhabitants of the Earth have the heaven above their heads—what we refer to as 'up'—and the centre of the Earth beneath their feet—what we call 'down'. So from our point of view the centre of the Earth and the surface of the heaven are the furthest points there are, the extremes of the diametrically opposed points which we call 'up' and 'down'.

6. It is absolutely prudent not to believe that there is proof of the mobility of the Earth until it is demonstrated, and we do not ask that anyone should believe such a thing without proof. Indeed we ask only that, for the benefit of the Church, anything which the upholders of this position can produce should be examined with the greatest rigour, and that nothing should be conceded to them unless the arguments they rely on far outweigh those on the other side; if their arguments are only 90 per cent correct, they should be rejected. But if the arguments produced by the philosophers and astronomers who oppose them are shown to be mostly false and all immaterial, then the other side should not be dismissed as a mere paradox which can never be demonstrably proved. There is no risk in offering such generous terms, for those who uphold the false position can never have any valid argument or evidence on their side, whereas on the side of the truth everything will necessarily come together and agree.

7. It is true that showing that the mobility of the Earth and the fixity of the Sun save the appearances is not the same as proving that these hypotheses are actually true in nature; but it is equally true—indeed more so—that the other commonly accepted system cannot satisfactorily explain these appearances. There is no doubt that the accepted system is false, and it is clear that the new one, which fits the appearances very well, may be true. And no greater truth can or should be expected from a theory than that it should meet all the relevant appearances.

8. It is not suggested that in cases of doubt we should abandon the teaching of the Fathers, but only that where there is doubt we should try to arrive at certainty, and that therefore we should not dismiss out of hand the lines of reasoning which great philosophers and astronomers follow and have followed. Then, having exercised due diligence, we should reach our decision.

9. We believe that Solomon, Moses, and all the other Scriptural authors understood the structure of the universe perfectly well, just as they knew that God does not have hands or feet, and that He does not experience anger, forgetfulness, or repentance; and we have no intention of questioning this. But we follow what the Church Fathers and in particular St Augustine have said on this matter, that the Holy Spirit chose to dictate to them in this way for the reasons we have given, etc.

10. The illusion that the shore appears to move and the ship to stand still is familiar to us because we have often stood on the shore watching the motion of boats, and been on a boat watching the shore; so if we were able to stand now on the Earth and now on the Sun or another star, we might be able to tell from sense impressions which of them was moving. But even then, if we only looked at these two bodies we would always think that the one on which we were standing was motionless, in the same way as someone who only looks at the boat and the water will always think that the water is flowing past and the boat is stationary. Besides, the great disparity between a small boat, isolated from all its surroundings, and a great expanse of shore which all our experience tells us is immobile in relation to both the water and the boat, is quite unlike the comparison between two bodies both of which are solid and equally disposed to motion or

rest. It would be more proper to make a comparison of two ships with each other: in this case we would always think that the one we were standing on was stationary, as long as we were unable to compare the two ships with anything else.

There is, then, a great need to make allowance for illusory appearances when considering whether the Earth or the Sun is in motion, for it is clear that anyone standing on the Moon or any other planet would always have the impression that they were standing still and the other stars were in motion. Those who uphold the common opinion will have to give very clear answers to these and many other more evident arguments before they can claim to be taken seriously, let alone to win approval—just as we have responded in minute detail to all the arguments that have been raised against us. In any case, neither Copernicus nor his followers have ever used this analogy of the shore and the ship to prove that the Earth is in motion and the Sun is at rest; they have simply used it as an example, not to prove the truth of their position, but to show that there would be no contradiction between our sense impressions telling us that the Earth is at rest and the Sun in motion, and the reality being the opposite. Indeed, if this was Copernicus' proof, or if his other arguments were not more conclusive than this, then I am sure that no one would endorse what he says.

And yet I have shown that this is indeed the case, and have demonstrated as much with a perfect telescope to anyone who cares to
look.*

*   *   *

I have noticed from long observation a general rule of human
behaviour, that the less someone knows and understands about some
intellectual matter, the more determined they are to hold forth about
it; and that conversely, the more things a person has knowledge of
the more hesitant they are about making pronouncements on anything new. A man once lived in a very isolated place who was
endowed by nature with a most penetrating mind and a high degree
of curiosity. For a pastime he raised various birds, taking great
pleasure in their song and marvelling at the skill with which they
produced different songs at will, all of them delightful, simply with
the air which they breathed. It happened that one night he heard a
delicate sound close to his house and, imagining that it must be
a bird, he went outside to capture it. When he came to the road he
found a shepherd-boy who was producing these sounds by blowing
into a piece of hollowed-out wood, opening and closing holes in its
surface by moving his fingers over them—sounds similar to those of
a bird, but produced in a completely different way. He was amazed
and, prompted by his natural curiosity, he gave the shepherd a calf in
return for this whistle. Reflecting on this, he realized that if the
shepherd had not happened to pass by he would never have discovered that nature has two ways of producing sweet-sounding
songs; so he decided to leave home to see what other adventures he
might meet. The next day, as he was passing by a small hut, he heard
a similar sound coming from it; so he went inside, wanting to see
whether it was a whistle or a blackbird. There he found a boy holding a bow in his right hand, which he moved back and forth like a
saw over some pieces of gut stretched over a hollowed piece of wood,
supporting the instrument with his left hand and moving his fingers
over it, so as to produce a variety of pleasing sounds without any
blowing of any kind. His astonishment can be imagined by anyone
who shares his wit and curiosity; and having discovered two such
unexpected ways of producing sound and song, he came to believe
that nature might well have yet other means of producing them.
Imagine his wonder, then, when entering a temple he looked behind
the door to see who had made a sound, and found that it had come

FROM THE ASSAYER

Sarsi* seems to be convinced that in philosophy one must always rely on the opinion of some famous author, as if our minds were infertile and sterile until they are married to someone else's reasoning. Perhaps he thinks that philosophy is a work of fiction like the *Iliad* or the *Orlando furioso*,* books about which the least important thing is whether what the author says is true. But, signor Sarsi, that is not how it is at all. Philosophy is written in this great book* which is continually open before our eyes—I mean the universe—but before we can understand it we need to learn the language and recognize the characters in which it is written. It is written in the language of mathematics, and its characters are triangles, circles, and other geometrical figures, without which it is humanly impossible to understand a word of what it says. Without these, it is just wandering aimlessly in a baffling maze. But even granted that, as Sarsi thinks, our intellect must become a slave of someone else's (leaving aside the fact that he thereby makes himself and everyone else into copiers, praising in himself what he condemned in signor Mario*), I do not see why he should have chosen Tycho in preference to Ptolemy or Nicolaus Copernicus, both of whom have given us a complete system of the universe, skilfully constructed and worked out in every detail—something which I do not think Tycho has done, unless Sarsi is satisfied by his having rejected the other two systems and promised to produce another, which he never did. In fact, I do not think Tycho can even be credited with having shown the other two systems to be false: as far as the Ptolemaic system is concerned, neither Tycho nor any other astronomer, nor Copernicus himself, was able to prove that it was mistaken, even though the main argument, based on the movements of Mars and Venus, always pointed against it. But since the disc of Venus appeared to show very little difference in size between the points of its conjunction and maximum digression from the Sun, and Mars appeared to be only three or four times greater at its perigee than at its apogee, he would never have been convinced that Venus could appear forty times larger, and Mars sixty times larger, in one state than in the other, as they must be if they orbit the Sun as the Copernican system states.

from the hinges and brackets as he opened it. Another time his curiosity prompted him to go into an inn, expecting to find someone lightly touching the strings of a violin with a bow, and he saw someone running their finger along the rim of a glass and producing a delightful sound. As he then observed that wasps, mosquitos, and flies produced a continuous sound with the very rapid beating of their wings, unlike the intermittent sounds which birds produced with their breath, his wonder increased as he became less and less sure that he knew how sounds were formed. And nothing he had seen so far could have enabled him to believe or to understand how crickets, since they do not fly, could give out such rich and sonorous sounds, not by breathing but by scraping their wings. But just when he thought that there could not possibly be any more ways of producing sound—after he had observed, besides all the ways I have described, every kind of organ, trumpet, pipes, stringed instrument, and even that thin piece of metal which is held between the teeth and uses the mouth cavity as a sounding-box and the breath to carry the sound—just when he thought he had seen everything, he found himself in greater ignorance and amazement than ever when he found a cicada and held it in his hand. He tried closing its mouth and holding its wings still, but he could not diminish the loudness of its chirping. He could not see it moving its scales or any other part of its body, so he finally lifted the outer shell of its thorax and, seeing that it contained hard but flexible cartilages and imagining that the sound came from their movement, he was reduced to breaking them to make it stop. But it was all in vain, until finally, pushing his needle further in, he pierced it and deprived it of both voice and life, so that even then he could not be sure that this had been the source of its song. So he came to have so little confidence in his knowledge that, if he was asked how sounds were produced, he would say that although he knew a few ways, he was quite certain that there were a hundred and one others which were unknown and beyond our comprehension.

I could give many other examples to illustrate how nature produces its effects in a wealth of ways which we could never work out if they were not revealed by the experience of our senses, and sometimes even this is not enough to make up for our lack of understanding. So I hope no one will hold it against me if I am unable to say precisely how a comet is formed, especially as I have never claimed to be able to explain it, knowing that they could be formed in ways which are unimaginable to us. So the difficulty of understanding how a cicada produces its song while it is sitting in our own hand more than excuses our ignorance about how distant comets are generated.

*   *   *

I cannot help being amazed once more that Sarsi should persist in trying to prove by citing witnesses what I can see perfectly well from my own experience. We examine witnesses when we need to know about things of which we are uncertain and which are past and not permanent. So a judge needs to consult witnesses to find out whether Pietro wounded Giovanni last night, but not whether Giovanni was wounded, since that is still apparent and he can see it for himself. But I would go further and say that even in matters which can only be resolved by reasoning, I would give scarcely more weight to the testimony of many than of few, since we know that those who reason correctly on a difficult subject are far fewer than those who reason badly. If reasoning on a difficult problem were like carrying weights, then I would agree that the reasoning of many counts for more than that of one, just as several horses can carry more sacks of grain than just one can; but reasoning is like running rather than carrying, and a single racehorse will outrun a hundred carthorses. So when Sarsi quotes such a multitude of authors it seems to me that, far from strengthening his case, he makes my own and signor Mario's all the more distinguished, by showing that we have out-reasoned many people of great reputation. If Sarsi wants me to believe Suidas* when he says that the Babylonians cooked eggs by whirling them around rapidly in a sling, I shall do so, but I must say that the cause of this effect was very far from the one that he gives. To discover the true cause I will reason as follows: 'If we are not successful in producing an effect which has been successfully produced on another occasion, then it must be that we lacked something that produced that success; and if only one thing was lacking in our attempt, then this must be the true cause. Now we lack neither eggs, nor slings, nor strong men to whirl them around, and yet they do not cook; in fact, if they were warm to start with they will cool down. Since the only thing we lack is being Babylonians, it must be the fact that they were Babylonians that caused the eggs to cook, and not friction of the air'; and this is what I wanted to find. Has Sarsi never observed how, when he rides post-haste, the continual change of air produces a cooling effect on his face—and if he has, then how can he believe something reported by others as having happened in Babylon two thousand years ago, rather than present events that he himself experiences? I pray you, get him to experiment to see if he can cool wine in summer by shaking it rapidly, when nothing else would make it cool.

*   *   *

It remains for me to fulfil the promise I made above, to give you my views on the proposition 'Motion is the cause of heat' and to show how it might be true. But first I must consider what we mean by 'hot', for I suspect that people in general have a concept of this that is very far from the truth; for they believe it to be a real, accidental quality which affects and resides in the material by which we feel ourselves warmed.

When I think of a physical material or substance, I immediately have to conceive of it as bounded, and as having this or that shape, as being large or small in relation to other things, and in some specific place and at any given time, as moving or at rest, as touching or not another physical body, and as being one in number, or few or many. I cannot separate it from these conditions by any stretch of the imagination. But whether it is white or red, bitter or sweet, noisy or silent, and of a pleasing or unpleasant odour, my mind does not feel compelled to bring this in order to apprehend it; in fact, without our senses as a guide reason or imagination unaided would probably never arrive at qualities such as these. So it seems to me that taste, odour, colour, and so on are nothing more than pure names, as far as the objects in which we think they reside are concerned. Rather, they exist only in the mind that perceives them, so that if living creatures were removed, all these qualities would be wiped away and no longer exist. But since we have given them specific names, distinct from those of the other and real primary qualities, we treat them as if they too were real with a distinct existence of their own.

Some examples may help to explain what I mean. Suppose I run my hand over a marble statue, and then over a living man. As far as my hand is concerned the action is the same in both cases, and consists in the primary qualities of motion and touch, and this is what we call them. But the live body which is the recipient of these actions perceives them in different ways depending on the different

parts of the body which are touched. When it is touched on the
soles of the feet, for example, or under the knees or the armpits, it
will feel, besides the ordinary sensation of touch, another sensation
for which we have a specific name, tickling. This sensation is solely
in us, not in the hand, and it would be quite wrong to suggest that
the hand has, in addition to the capacity for movement and touch, a
separate faculty for tickling, as if tickling were a property inherent in
the hand. Or again, a piece of paper or a feather which is lightly
brushed against any part of our body is always doing the same thing,
that is, moving and touching; but as far as we are concerned, if
it touches us between the eyes, on the nose or under the nostrils, it
provokes an almost intolerable titillation, whereas on other parts of
the body we hardly feel it at all. Now this titillation belongs entirely
to us and not to the feather, and without the live and percipient body
it is nothing more than a name. And I believe that no more solid
existence belongs to many of the qualities which we attribute to
physical bodies, such as taste, odour, colour, and others.

A body that is solid and, as we say, quite material, when it is
moved into contact with any part of my body produces in me the
sensation we call 'touch'; and although this exists over my entire
body it seems to reside especially in the palms of our hands, and most
of all in our fingertips, with which we can perceive tiny differences
between rough and smooth, soft and hard, which other parts of the
body distinguish much less clearly. Some of these sensations are
more pleasant than others, depending on the different shapes of the
tangible bodies we come into contact with, whether they are smooth
or harsh, sharp or blunt, hard or yielding. This sense of touch, being
more material than the others and arising from the solidity of matter,
seems to belong to the element of earth. And since some material
bodies are continually dissolving into tiny particles, some of which
are heavier than air and descend while others, being lighter than air,
rise up, this may be the origin of two other senses which act on two
parts of our body which are much more sensitive than the skin,
which cannot perceive contact with materials which are so fine,
thin, and yielding. The tiny particles which descend are received on
the upper side of the tongue, where they mingle with the tongue's
moisture to produce sensations of taste which may be pleasant or
unpleasant depending on how they touch the tongue and whether
they are many or few, and moving more or less rapidly. Those which

rise may enter the nostrils and strike some small protuberances which are the organs of smell, where similarly their contact is perceived as pleasing or distasteful depending on their shape, their speed, and their number. It is clear how providentially our tongue and nostrils are located, the tongue to receive the descending particles and the nostrils those which are rising; and perhaps there is an analogy between the sense of taste and fluids, which are heavier than air, and between the sense of smell and fire, which rises. This leaves the element of air to transmit sounds, which come to us from all directions, above, below, and from either side; for we ourselves inhabit the air, which moves freely in its own region, and is equally distributed all around us. So our ears, too, are in the best possible place to receive sounds from all sides; sounds are formed and perceived by us when a rapid tremor in the air (without any special quality of sonority or transonority*) compresses it into tiny waves that move a certain cartilage in our eardrum. There are many other external ways in which this compression of the air can be caused, and I think the majority of them can be attributed to the trembling of some body which pushes the air, causing waves which spread out rapidly, producing high-pitched sounds if they are more frequent and low-pitched sounds if they are less so. I believe, however, that to excite our senses of taste, smell, or sound nothing is required in external bodies except their size, shape, number, and slow or rapid movements. I think that these qualities of shape, number, and movement would still exist even if we did not have ears, tongues, or noses, whereas odours, tastes, or sounds would not—these being nothing more than names if there are no living creatures to perceive them, just as tickling or titillation are no more than names without such things as armpits or noses. As these four senses are related to the four elements, so I believe that sight, which is our pre-eminent sense, is related to light, but in proportion as the finite is to the infinite, the temporal to the instantaneous, the quantitative to the indivisible, light to darkness. I claim very little knowledge of the sense of sight and of how it works, and that little would take a very long time to explain, or even to outline on paper; and so I pass over it in silence.

DIALOGUE ON THE TWO CHIEF
WORLD SYSTEMS

Most Serene Grand Duke,*

Great though the difference is between men and other animals,
it may not be entirely misguided to say that differences among
men themselves are hardly less. What comparison is there
between one and a thousand? And yet there is a common saying that one man is worth a thousand, where a thousand are
not worth as much as a single one. The difference derives from
their different intellectual capacities, which I equate with
whether or not they are philosophers; for philosophy being
their food, it separates those who can receive nourishment
from it from the common crowd, in varying degrees depending on the degree of nourishment they take from it. Those who
aspire to the highest degree will distinguish themselves the
most; and the highest aspiration is to turn to the great book of
nature, which is the proper subject of philosophy. And
although everything in that book is perfectly proportioned,
being the handiwork of the omnipotent Maker, nonetheless the
part which is most resplendent and worthy is that in which the
greatness of his artistry is most clearly visible to our sight. The
structure of the universe, among all those things in nature
which we can apprehend, is that which I believe should take
pride of place; for as it exceeds everything else in size, since it
contains the whole universe, it should also exceed everything
else in nobility, since it regulates and maintains all things. So if
anyone has ever stood out intellectually above other men,
Ptolemy and Copernicus are those who have scaled the heights
in studying, scrutinizing and reflecting on the structure of the
world. Since these Dialogues of mine deal principally with
their works, it seemed to me that they should be dedicated to
none other than your Highness; for as their content rests on
these two, whom I consider the greatest thinkers to have left
their works to us on these matters, it was only appropriate that
I should commend them to the favour of the one who has given

me the greatest support, that they might receive glory from his patronage. And if Ptolemy and Copernicus have so illuminated my understanding that this work could be considered theirs as much as mine, it can equally well be said to belong to your Highness, for your generosity has given me leisure and tranquillity in which to write, and your effective help, which has never tired of honouring me, has enabled it finally to see the light of day. So may your Highness accept it with your customary kindness; and if you find in it anything which can bring greater understanding and benefit to lovers of the truth, recognize it as your own, for your rule is so beneficial that no one in your happy dominions is molested by the universal troubles which prevail in the world. Praying that you may prosper so that your pious and magnanimous custom may ever increase, I offer you my humble reverence.

Your Serene Highness's humble and devoted servant and subject,
Galileo Galilei.

TO THE DISCERNING READER*

A salutary edict* was published some years ago in Rome which, as a defence against the dangerous scandals of our present age, imposed a timely silence on the opinion of Pythagoras concerning the mobility of the Earth.* There were some who were so bold as to assert that this decree was the product not of judicious consideration but of ill-informed prejudice, and there were protests that adjudicators who were totally inexperienced in astronomical observations had no business clipping the wings of speculative thinkers by means of an over-hasty ban. I could not remain silent in the face of such presumptuous complaints, and as I was fully informed about this entirely prudent decision I resolved to appear on the stage of world opinion as a witness to the unvarnished truth. I was in Rome at the time, where I was received and indeed applauded by the most eminent prelates of the Curia, and I had prior information about the publication of this decree. So it is my intention in this present work to show to all foreign nations that as much is known about these matters in Italy, and especially in Rome, as has been imagined by anyone working north of the Alps; and by bringing together all the speculations concerning the system of Copernicus, to make clear that they were all already known to the Roman censorship, and that this climate can produce ingenious insights to delight the mind as well as dogmas for the salvation of the soul.

To this end I have taken the side of Copernicus in the argument, purely as a mathematical hypothesis, and I have adopted every artifice to show that it is superior to the hypothesis that the Earth is fixed—not as an absolute position, but as it is defended by some who call themselves Peripatetic philosophers, although they are Peripatetics only in name.* For they do not walk about but are content just to worship shadows, basing their philosophy not on their own insight but on their memory of a few principles which they have only half understood.

Three main topics will be discussed. I shall aim to show, first, that none of the experiments that can be carried out on

Earth are sufficient to prove that the Earth moves, but that they are equally compatible with its moving or being fixed; and here I hope to make known many observations which were unknown in antiquity. Second, I shall examine celestial phenomena so as to reinforce the Copernican hypothesis as if it were absolutely proved to be superior, and I shall add new speculations, but only to facilitate the work of astronomers, not as necessities in nature. In the third place, I shall put forward an ingenious theory. I had occasion to observe many years ago that some light might be shed on the unsolved problem of the cause of the tides* if we were to accept the movement of the Earth. This observation of mine came to be generally known, and some people were so charitable as to take it in and adopt it as a brainchild of their own. So now, to forestall any foreigner who might take over our arguments and then criticize our lack of insight in such a fundamental question, I have decided to set out the suppositions which could explain this on the assumption that the Earth moved. I hope that these considerations will show the world that, if other nations have sailed further, we have speculated no less than them, and that if we submit to asserting that the Earth is fixed and the contrary hypothesis is purely a mathematical invention, this is not because we are not aware of what others have thought on the matter, but, if nothing else, for reasons which piety, religion, and a recognition of divine omnipotence and the limitations of human understanding dictate.

It seemed to me, next, that it would be appropriate to set these ideas out in the form of a dialogue, which is not bound always to follow the strict rules of mathematics and so gives scope for occasional digressions, which are sometimes no less interesting than the main argument itself.

Many years ago, in the marvellous city of Venice, I had several occasions to converse with signor Giovan Francesco Sagredo,* a man of distinguished birth and great intelligence. He was visited from Florence by signor Filippo Salviati,* whose ancient lineage and magnificent wealth were the least of his marks of distinction; a man of sublime intellect, there was nothing he enjoyed more avidly than subtle philosophical speculations. I often discussed these matters with them both, together with a Peripatetic philosopher* for whom the greatest obstacle to perceiving the truth was the fame he had acquired as an interpreter of Aristotle.

Now that cruel death has deprived Venice and Florence of these two great luminaries, both in the prime of their life, I have resolved to perpetuate their fame in these pages, as far as my modest talents allow, by making them the participants in this discussion. The good Aristotelian, too, will have his place here; given his great attachment to the commentaries of Simplicius,* I have thought it tactful to give him the name of his revered author rather than calling him by his own name. So may these two great spirits, who still have an honoured place in my heart, accept this public monument to my undying affection, and let the memory of their eloquence help me to expound these speculations to posterity.

These gentlemen had already had some fragmentary discussions in the course of casual meetings, which had only whetted their appetite to learn more; so they wisely decided to meet on several consecutive days when they could set aside all their other concerns and devote themselves more systematically to admiring and speculating on the wonders of God's creation in heaven and on Earth. So, when they were met in signor Sagredo's palace, after the customary brief formalities signor Salviati began as follows.

FIRST DAY

Speakers: SALVIATI, SAGREDO, AND SIMPLICIO

SALVIATI. We concluded and agreed yesterday that we should meet today to discuss, as systematically and in as much detail as we could, the reasons in natural science which have so far been brought forward by the advocates of the Aristotelian and Ptolemaic system on the one hand, and the Copernican system* on the other, and the effectiveness of each. And since Copernicus, by placing the Earth among the mobile bodies in the heavens, makes it a sphere similar to a planet, we should begin our investigation by examining the force of the arguments of the Peripatetics to show that such an assumption is impossible, on the grounds that nature necessarily has two different kinds of substance, one heavenly, which is impassible and immortal, the other made up of the elements, which is changeable and transient. Aristotle discusses this in his book on the Heavens, drawing first on arguments derived from certain general assumptions, and then backing these up by reference to experience and to detailed demonstrations. I will follow the same procedure, putting forward and then frankly declaring my own opinion, and in doing so I will lay myself open to your criticism and especially to that of signor Simplicio, who is such a vigorous defender and upholder of Aristotelian teaching.

Copernicus considers the Earth to be a sphere similar to a planet.

According to Aristotle, nature requires that there be unchangeable heavenly substances and changeable substances made up of the elements.

The first stage in the Peripatetics' argument is where Aristotle proves the integrity and perfection of the world by showing that it is neither a simple line, nor a pure surface, but a body endowed with length, width, and depth; and since there are only these three dimensions, and the world has all three, it is complete and therefore perfect. That these three dimensions constitute completeness and, so to speak, totality—that, starting with the simple dimension of length in that unit of magnitude which is called a line, adding width to produce a surface area, and finally adding height or depth to produce a body, there is then no possibility of passing from these three dimensions to any other—is something which I wish Aristotle

Aristotle considers the world to be perfect because it has three dimensions.

had shown by necessary demonstration, especially since this could be done quite clearly and easily.

SIMPLICIO. Surely there are splendid proofs in the second, third, and fourth parts of the text, after the definition of what is meant by 'continuity'?* Is there not, first of all, the proof that there is no other dimension beyond these three because the number three is everything, and is to be found everywhere? Is this not confirmed by the authority and teaching of the Pythagoreans, who state that everything has a threefold definition—a beginning, a middle, and an end—and that three is the number of totality? And what of the other argument, that the number three is used in sacrifices to the gods, as if by a natural law? And that, taking our cue from nature, we use the word 'all' to refer to three things but not to less—for we refer to two things as 'both', not 'all' as we do for three? All these arguments can be found in the second part of Aristotle's text. In the third part, pursuing the matter in more depth, we read that 'everything', 'all', and 'perfection' are formally the same, and that therefore the body is the only perfect figure, since it alone is defined by three, which is everything, and is divisible in three ways, that is, in all directions. The others, in contrast, are divisible in either one or two ways, because their division and continuity depend on the number which applies to them—so one is continuous in one direction, another in two, but only the body is continuous in all three. Moreover, in the fourth part, after various other arguments, does he not give another proof of the same point, namely that since passing from one dimension to another implies some lack or deficiency (so, for instance, we pass from a line to a surface area because a line lacks width), and since perfection, being everywhere, cannot lack anything, it is not possible to pass from the body to another unit of magnitude? Now, from all these texts, do you not think it has been sufficiently established that there can be no passing from the three dimensions of length, width, and height to any other dimension, and that therefore the body, which has all three, is perfect?

SALVIATI. To tell the truth, the only thing I feel obliged to concede in all these arguments is that something which has a beginning, a middle, and an end can and should be called

Aristotle's demonstrations that there are three dimensions, and no more.

The number three is celebrated by the Pythagoreans.

perfect; but I see no reason why it should follow that there is therefore anything perfect about the number three, or that the number three can confer perfection on anything which possesses it. Nor do I believe that, for example, there is any more perfection in having three legs rather than four or two; or that because there are four elements they are in any way imperfect, or that the perfection would be greater if there were three. So it would have been better if he had left these fanciful ideas to the rhetoricians and established his case with necessary demonstrations,* which is the proper way to proceed in the demonstrative sciences.

SIMPLICIO. I think you are using these arguments facetiously; but it was the Pythagoreans who attributed such qualities to numbers, and yet you, as a mathematician and, as I believe, a Pythagorean in many of your opinions, now seem to despise their mysteries.

SALVIATI. I'm well aware that the Pythagoreans had the highest regard for the science of numbers, and that Plato himself admired the human intellect and regarded it as sharing in divinity simply because it could comprehend the nature of numbers. My own view is not very different, but I don't accept for a moment that the mysteries which caused Pythagoras and his followers to have such veneration for the science of numbers are the same as the foolish beliefs which are spoken and written about among the common people. Indeed, I know that the Pythagoreans were so concerned that these wonders should not be exposed to the scorn and contempt of the crowd that they condemned it as sacrilege to publish the hidden properties of numbers and the immeasurable and mysterious quantities which they studied, and they declared that anyone who revealed these mysteries would be punished in the next world. I think it was for this reason that one of them said, to satisfy the crowd and free himself from their demands, that their numerical mysteries were the same as the trifles which were then widely repeated among the common people. He showed the same astuteness as the wise young man who, to rid himself of the importunate questions of his mother or his inquisitive wife who wanted him to divulge the secret deliberations of the Senate, made up the story which led to her

Plato's opinion that the human intellect shares in the divine nature because of its comprehension of numbers.

The number mysteries of the Pythagoreans are fables.

and many other women being made a laughing-stock, to the great amusement of the Senate itself.*

SIMPLICIO. I have no wish to be counted among those who would pry into the mysteries of the Pythagoreans; but to come back to the point at issue, I reply that the reasons which Aristotle gives to prove that there are and can be only three dimensions seem to me to be conclusive; and I think that, if any further proof had been necessary, Aristotle would not have failed to provide it.

SAGREDO. You might add, if he had known it, or if it had come to his mind. But signor Salviati, I would very much like to hear some evident proof of this, if you have any which is clear enough for me to understand.

SALVIATI. I do indeed, for you and for signor Simplicio as well; and it is one which you can not only understand, but *Geometrical* which you know already,* although you may not be aware of it. *demonstration* To make it easier to understand, we shall take pen and paper, *that there* which I see are ready here for just this purpose, and do a little *are three* *dimensions.*

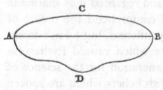

drawing. First let us mark these two points, A and B; and then join them up with these curved lines, ACB and ADB, and this straight line, AB. Now tell me which line in your view determines the distance between A and B, and why.

SAGREDO. I would say it is the straight line, not the curved ones, both because the straight line is the shortest, and also because there is only one, and its position is determined; whereas there is an infinite number of other lines, all longer and of different lengths, and it seems to me that the distance should be determined by the line of which there is only one and whose position is fixed.

SALVIATI. So we have a straight line to determine the

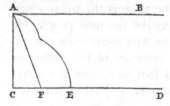

distance between two points. Now let us add another straight line, CD, parallel to the line AB, so that there is a surface area between them, of which I would like you to tell me the width. So, starting from A, tell me where and how you would draw a line to meet the line CD so as to establish the width of the area between these two lines: would you use the curved line AE, or the straight line AF, or . . . ?

SIMPLICIO. I would use the straight line AF, not the curved one, as we have already eliminated curves for this purpose.

SAGREDO. I wouldn't use either of them, because the straight line AF runs obliquely. I would draw a line at a right angle above CD, because this seems to me to be the shortest, and it would be unique among the infinite number of longer lines, all of unequal length, which could be drawn from A to any number of points on the opposite line CD.

SALVIATI. Your choice and the reasoning you give for it are perfect. So we have established that the first dimension is determined with a straight line; the second, namely width, with another line, also straight but, as well as being straight, at right angles to the line which determined length. Thus we have defined the two dimensions of a surface area, namely length and width. What if, now, you had to determine a height—for example, the height of this platform above the floor beneath our feet: given that from any point on the platform you could draw an infinite number of lines, both curved and straight, all of different lengths, to an infinite number of points on the floor underneath, which of these lines would you use?

SAGREDO. I would attach a thread to the platform with a lead weight on the end, and would let it hang freely until it touched the floor; and I would say that the length of this thread, being the shortest straight line that could be drawn from this point to the floor, would be the correct height of the platform.

SALVIATI. Excellent; and if, from the point on the floor which was established by this hanging thread (assuming that the floor is level and not sloping) you were to draw two other straight lines, one to determine the length and the other the width of the surface of the floor, at what angle would they be to the thread?

SAGREDO. They would certainly be at right angles, if the thread was perpendicular and the floor was smooth and properly levelled.

SALVIATI. So, if you establish a starting and finishing point for your measurements, and draw a straight line to determine the first measurement, namely length, then the line to determine width must necessarily be at right angles to the first; and the line to determine the third dimension, namely height, if it starts from the same point, must form right angles and not oblique angles with the other two. Thus from these three perpendiculars, being three lines all of which are unique, fixed, and the shortest possible, you will have determined the three dimensions: AB for length, AC for width, and AD for height.

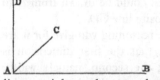

And since it is clear that no other line can run to this point at right angles with these three, and the dimensions must be determined solely by straight lines at right angles to each other, it follows that there cannot be more than three dimensions, and any body which has three dimensions has them all; if it has all three dimensions it is divisible on all sides, in which case it is perfect; and so on.

SIMPLICIO. But who is to say that it isn't possible to draw other lines? Why can I not draw another line coming up to point A from below, which is at a right angle to the others?

SALVIATI. It is certain that you can't make more than three lines converge on a single point which are at right angles to each other.

SAGREDO. Yes, because what signor Simplicio proposes is simply the line DA extended downwards; and indeed, in this way one could draw two more lines, but they would still be the same three original lines, the only difference being that whereas now they simply meet, they would then intersect each other, but they wouldn't introduce any new dimensions.

SIMPLICIO. I won't deny that this proof of yours may be *Geometrical* conclusive, but I will follow Aristotle* in saying that the truths *precision is not* of the natural world don't necessarily have to be demonstrated *to be sought* *in proofs* by mathematical proofs.
regarding the
natural world. SAGREDO. Maybe not, if no such proof is available; but if it exists, why should we not use it? But let us not expend more words on this point, as I think signor Salviati will concede to you and to Aristotle without further proof that the world is a

body, and is perfect, indeed is the height of perfection, as the greatest of God's works.

SALVIATI. Yes indeed. So let us move from the general consideration of the whole to a discussion of the parts, which according to Aristotle's first division are twofold, different and in some respects opposed to each other, namely the heavenly and the elemental. The former is ingenerable, incorruptible, inalterable, impassible, etc.; the latter exposed to continual alteration, mutation, etc. This difference he derives, as its first cause, from the difference in their local motion, and he proceeds to argue as follows.

Aristotle divides the world into two parts, the celestial and the elemental, which are contraries to each other.

Leaving, so to speak, the world of the senses and withdrawing to the world of ideas, Aristotle begins to reason like an architect. Given that nature is the principle of motion, it follows that natural bodies have a local motion. He goes on to say that local motions are of three kinds, namely circular, rectilinear, and a mixture of rectilinear and circular. The first two he calls simple motions, because of all the different kinds of line only a circle and a straight line can be called simple. He then narrows down the argument and specifies that, of these simple motions, circular motion is motion around a centre and rectilinear motion can be either towards or away from a centre—upward if it is moving away from the centre and downward if it is moving towards the centre. From this he deduces that all simple motions can be limited to these three, namely towards, away from, and around a centre; and he finds a certain pleasing correspondence between what was said above about a body, that it too is perfected in three ways, and its motion. Having established these motions he goes on to say that, since natural bodies are either simple or composite (simple bodies being those which have their own natural principle of motion,* such as fire and earth), simple motions belong to simple bodies and mixed motions to composite bodies, with the proviso that composite bodies follow the motion which belongs to the predominant element in their composition.

Three kinds of local motion, rectilinear, circular, and mixed.

Rectilinear and circular motions are simple, because they follow simple lines.

SAGREDO. Please stop there for a moment, signor Salviati: listening to this argument, I feel so many questions pressing in on me from all sides that I must either voice them, if I am to listen attentively to what you say next, or else cease paying attention to what you say in order not to forget my questions.

SALVIATI. I'll gladly stop here, because I too am running the same risk: I am constantly on the point of getting lost, and I feel as if I'm steering a course between rocks and stormy waves which threaten to make me, as they say, lose my bearings. So do explain your difficulties, rather than letting them accumulate any further.

SAGREDO. You began, with Aristotle, by drawing me away from the world of the senses to show me the architecture underlying its construction; and I appreciated your explanation of how natural bodies are by their nature mobile, since it was established elsewhere that nature is the principle of motion. This prompted a doubt in my mind, and it was this: Aristotle's definition stated that nature is the principle of motion and rest; so why did he not say that some natural bodies are by their nature mobile and others immobile? If natural bodies all have a principle of motion, then either it was superfluous to include 'rest' in his definition of nature, or it was not appro-

Aristotle's definition of nature is either faulty, or inappropriate in this context.

priate to cite this definition in this context.

Then, I have no problem with your explanation of what Aristotle means by simple motions, and how he derives them from their movement in space, defining simple motions as those which follow simple lines, namely a straight line and a circle. I won't quibble over the example of a cylindrical screw, since this is the same in all its parts, and so could be included

A cylindrical screw can be defined as a simple line.

among examples of simple lines. But I do object to the way he narrows the discussion, while implying that he is simply repeating the same definition with different words, when he equates circular motion with movement around a centre and rectilinear motion as movement *sursum et deorsum*, that is, up and down. These terms make sense only in the physical world, and presuppose not only that it is a physical world but that it is inhabited by us humans. For if rectilinear motion is simple because a straight line is simple, then it must be applicable to a simple natural body regardless of its direction, whether up or down, forward or back, to the left or to the right, or in any other conceivable direction, as long as it is in a straight line; if not, then Aristotle's supposition is false. He then goes on to state that there is only one circular motion in the world, and therefore only one centre, to which all references to rectilinear motion as being 'upward' or 'downward' relate.

All these points suggest that Aristotle is resorting to sleight of hand, and is adapting his architecture to the building rather than constructing his building according to architectural principles. For, if I were to state that there could be a thousand circular movements, and therefore a thousand centres, in the natural world, then it would follow that there would also be a thousand upward and downward movements. Or again, we have seen that he postulates simple and mixed motions, defining simple motions as those which are either circular or rectilinear, and mixed motions as those which are a composite of these two; he defines natural bodies as either simple (those which have a natural principle of simple motion) or composite, and he attributes simple motions to simple bodies and mixed motions to composite bodies. But in saying this, he no longer takes mixed motion to mean a mixture of straight and circular motions—something which can exist in the real world—but instead introduces a concept of mixed motion which is as impossible as it would be to mix movements in opposite directions in the same straight line, as if you could have motion which is partly upward and partly downward. To mitigate this absurdity and impossibility, he is reduced to saying that composite bodies move according to the simple element which predominates in their composition; and this makes it necessary to say that even motion in the same straight line is sometimes simple and sometimes mixed, so that simplicity of movement no longer depends solely on following a simple line.

Aristotle adapts his architectural principles to the construction of the world, not the construction to his principles.

According to Aristotle, rectilinear motion is sometimes simple and sometimes mixed.

SIMPLICIO. Isn't it a sufficient difference to say that absolutely simple movement is much faster than movement which is derived from the predominant element? After all, how much faster does a piece of solid earth fall than a piece of wood?

SAGREDO. That's a valid point, signor Simplicio. But if simplicity of motion is affected by its velocity, not only will there be a hundred thousand kinds of mixed motion, but how will you be able to identify which motions are simple? What's more, if simplicity of motion depends on greater or lesser velocity, then no simple body will ever move with simple motion, because any natural rectilinear motion constantly increases in velocity, and hence its 'simplicity' is constantly changing—but 'simplicity' implies that it is immutable. More seriously, you will be burdening Aristotle with another error, because in his definition of mixed motion he makes no mention of greater or lesser velocity, and now you are saying that this is a necessary and essential factor. What's more, such a rule would serve no useful purpose, because there will be mixed motions—plenty of them—some of which will be slower and others faster than simple motion: the movement of lead or wood in comparison with earth, for example. So how would you decide which of these motions to call simple, and which mixed?

SIMPLICIO. I would define simple motion as that of a simple body, and mixed motion that of a composite body.

SAGREDO. Exactly. But think what this means, signor Simplicio. You argued earlier that simple and mixed motions serve to distinguish between simple and composite bodies, and now you are saying that simple and composite bodies will enable me to differentiate between simple and mixed motions. That sounds like a recipe for never being able to identify either motions or bodies. What's more, you are now saying that greater velocity is not enough, and you're adding a third condition for your definition of simple motion, which Aristotle was content to define with just one, namely simplicity in space: now, according to your definition, simple motion is motion in a simple line, at a specified velocity, by a simple mobile body. Well, that's up to you, but I think we should go back to Aristotle, who defines mixed motion as being a mixture of rectilinear and circular motion, although he was unable to give any examples of a body which moved naturally in such a way.

SALVIATI. So I shall return to Aristotle. Having begun his exposition clearly and methodically, he now goes off at a tangent, being more concerned to arrive at a conclusion he had already reached in his own mind than to follow through the logic of his argument. He states as something well known and self-evident that upward and downward rectilinear motions belong naturally to fire and earth respectively, and that therefore there must be, in addition to these bodies which are close at hand, another body to which circular movement belongs naturally; and that this body is as much more excellent as circular motion is more perfect than rectilinear motion. He defines the greater perfection of circular movement by reference to the perfection of a circular line compared to a straight line, calling the former perfect and the latter imperfect. He considers a straight line imperfect because, if it is infinite, it has no termination or end, and if it is finite, there is something beyond it to which it can be extended. This is the cornerstone and foundation of the whole Aristotelian structure of the world, on which all its other qualities depend—of having neither weight nor lightness, of not being susceptible to generation or corruption, of being immune to change, of having no location, etc. He attributes all these qualities to the simple body which moves naturally in circular motion, while the opposite conditions—of weight and lightness, susceptibility to corruption, etc.—belong to bodies which move naturally in rectilinear motion.

Aristotle's view that a circular line is perfect and a straight line imperfect, and why.

It follows that if there is found to be any flaw in what has been established up to this point, there is reason to have doubts about everything else which is posited on it. I don't deny that what Aristotle has introduced here in general terms, based on universal first principles, is confirmed later in his exposition with more detailed arguments and experiments, which must all be considered and evaluated in their own right. But given that there are many difficulties—and not insignificant ones—in what has been expounded thus far, and that the first principles and foundations need to be absolutely firmly established, so that we can build on them with confidence, I suggest that it would be a good thing, before we accumulate any more doubts, to see whether we can (as I think) follow another path which might be safer and more direct, and establish our basic foundations on more carefully considered architectural principles. So, leaving aside Aristotle's argument for now, to return to examine it in detail later, I can say that of the points which he has stated so far, I agree with him that the world is a body endowed with all the dimensions, and therefore absolutely perfect; and I would add that it is therefore necessarily perfectly ordered, that is, that its parts are disposed with supreme and perfect order among themselves. I don't imagine that anyone will disagree with this assumption.

The author's supposition that the world is perfectly ordered.

SIMPLICIO. How could anyone deny it—first because it comes from Aristotle, and second because the order which the world perfectly contains is the basis of the name 'ordered world' itself?

SALVIATI. So having established this principle, we can immediately conclude that, if the integral bodies* of the world must in their nature be mobile, their motion cannot possibly be rectilinear, or indeed anything but circular, for a very obvious reason. Anything which moves in a straight line changes its location, and the longer it goes on moving the further away it gets from its starting point and from all the other points it has passed through; and if this is its natural motion, then it was not in its natural place when it started, and therefore the parts of the world were not disposed in perfect order. But we have presupposed that they are in perfect order, in which case it cannot possibly be in their nature to change their location, and therefore to move in a straight line. Moreover, rectilinear motion is by its nature infinite, because a straight line is infinite and has no defined end; and so no mobile body can possibly have a natural principle of rectilinear motion, as this would be motion towards a point which is impossible to reach, since it has no defined end. And Aristotle himself rightly says that nature does not undertake to do anything which is impossible, or to move towards a point which is impossible to reach.

Rectilinear motion is impossible in a perfectly ordered world.

Rectilinear motion is by nature infinite; it is impossible in nature.

Nature does not undertake to do anything which is impossible.

It may be objected that even if a straight line, and hence rectilinear motion, can be extended to infinity, i.e. has no defined end, nonetheless nature has, so to speak, arbitrarily set limits on them and has endowed her natural bodies with a natural instinct to move towards those limits. To this I would reply that we might well speculate that such a situation existed in the primeval chaos, where indistinct matter moved erratically in a confused and random way, and that nature quite properly used rectilinear motion to put it in order. For if rectilinear motion produces disorder when it moves bodies which are rightly disposed, it produces order when it moves

Rectilinear motion perhaps existed in primeval chaos. It is suitable for setting in order bodies which are wrongly ordered.

those which are wrongly disposed. But once everything had been optimally distributed and put in place, no body could possibly still have a natural inclination to move in a straight line, which could only mean moving out of its natural and proper place, and therefore producing disorder. So we can say that rectilinear motion can serve to move materials in the work of construction, but that once the construction is complete, it either remains motionless or, if it is mobile, moves only in a circular motion.

We might, however, follow Plato in saying that even the heavenly bodies, once they had been created and fully established, were temporarily endowed by their Maker with rectilinear motion, and that when they reached their assigned places they began one by one to move in a circle, exchanging rectilinear for circular motion, which they have maintained ever since. This is an inspired idea, well worthy of Plato,* about which I recall our mutual friend the Lincean Academician* speaking. His argument, if I remember rightly, was as follows: any body which for whatever reason is constituted in a state of rest, but which is by nature mobile, will move when it is free to do so, provided it has a natural inclination to some particular place; for if it had no such inclination it would remain in a state of rest, as it would have no more reason to move to one place than to another. The fact that it has such an inclination necessarily means that its motion will continually accelerate; and, starting with the slowest degree of motion, it will only reach a given degree of velocity by passing through all the lower degrees of velocity or, rather, the higher degrees of slowness. For, starting from a state of rest, which is the infinite degree of slowness, there is no reason for it to acquire a given degree of velocity without first passing through a lower degree, and an even lower degree before that. Indeed, it is reasonable to say that it passes first through the degrees which are closest to the one from which it started, and from these to the degrees which are furthest from it—and the degree from which a mobile body begins to move is the highest degree of slowness, namely a state of rest. Such acceleration will not happen unless the mobile body gains something by moving; what it gains is to come closer to its desired place, that is, the

According to Plato, the heavenly bodies were moved initially with rectilinear, and then with circular motion.

A mobile body in a state of rest will not move unless it has an inclination towards a particular place. Its motion accelerates as it moves towards the place to which it is inclined. In changing from a state of rest, a mobile body passes through every degree of slowness. A state of rest is the infinite degree of slowness.

A mobile body accelerates only when it draws closer to its goal.

place to which its natural attraction draws it; and it will take the shortest route to that point, i.e. a straight line. So we can reasonably conclude that nature, to confer a given velocity on a mobile body which has been constituted in a state of rest, does so by making it move, in a given interval of time and space, in a straight line. If this reasoning is correct, we may imagine that God created a heavenly body, e.g. Jupiter, and determined to confer on it a given velocity, which it should then maintain constantly in perpetuity; and, following Plato's argument, we can say that God moved it initially with accelerating rectilinear motion, and that once it had reached that degree of velocity its rectilinear motion became circular, the velocity of which is naturally constant.

To confer on a mobile body a given degree of velocity, nature moves it in rectilinear motion. Constant velocity is a property of circular motion.

SAGREDO. I like this reasoning very much, and I will like it even better when you have resolved a difficulty for me. I don't quite understand how it must necessarily be the case that when a mobile body changes from a state of rest to its natural motion it has to pass through all the intermediate degrees of slowness, of which there must be an infinite number. Why could nature not endow Jupiter with its circular motion, at a given velocity, the moment it was created?

Between a state of rest and any intermediate velocity there are infinite lesser degrees of velocity.

SALVIATI. I didn't say, and I wouldn't presume to say, that it was impossible for God or nature to endow a body with its natural velocity immediately; I say only that in practice nature does not do so, so that this would be something outside the course of nature, and therefore miraculous. [If a body of whatever weight, moving at whatever speed, encounters a body at rest, however weak and incapable of resistance, it will not immediately impart its velocity to that body. Clear evidence of this is the sound of the impact, which would not be heard (or rather, would not exist) if the body at rest were to acquire the same velocity as the moving body at the moment of impact.]*

Nature does not immediately endow bodies with a specific degree of velocity, although it could do so.

SAGREDO. Is it your view, then, that when a stone changes from a state of rest to its natural motion towards the centre of the Earth, it passes through every degree of slowness up to a given velocity?

SALVIATI. I'm quite sure of it, so much so that I can make you sure of it as well.

SAGREDO. If understanding this was the only thing I learnt from today's discussion I would be well satisfied.

SALVIATI. As far as I understand your thinking, your difficulty consists largely in the fact that a mobile body has to pass in a very short time through all the infinite degrees of slowness leading up to whatever speed it reaches in that time. So before going on to anything else, let me try to resolve this doubt; this should not be difficult, if I say that the mobile body passes through all these degrees without pausing in any of them, taking only an instant to do so. And since any interval of time, however small, contains an infinite number of instants, there will always be enough instants for it to pass through every degree, however short the time may be.

When a mobile body changes from a state of rest, it passes through every degree of velocity without pausing in any of them.

SAGREDO. I follow you so far; but I still find it extraordinary that a cannon-ball—for that is how I imagine this falling body—which we see falling so fast that it travels more than two hundred *braccia** in less than ten pulse beats, can still have passed through such a small degree of velocity that, if it had continued to move at that speed without accelerating further, it would not have travelled this distance if it had continued all day.

SALVIATI. Or all year for that matter, or ten years, or a thousand years, as I will try to show you, I hope without your contradicting a few simple points which I shall put to you. So tell me first, do you have any problem in agreeing that the cannon-ball continually gains impetus and speed as it falls?

SAGREDO. Not at all; I'm quite sure that it does.

SALVIATI. And would you agree that the impetus it had gained at any point in its motion was sufficient to take it back to the height from which it started?

SAGREDO. I would certainly agree, provided the whole of the impetus it had gained was directed without any obstacle solely to taking the body, or another equal to it, back to its original height. So, if there were a hole through the centre of the Earth, and the cannon-ball was dropped from a hundred or a thousand *braccia* above its surface, I don't doubt that it would pass the centre of the Earth and rise to the equivalent of the height from which it had fallen. I deduce this from what happens when a pendulum is moved from the perpendicular,

A falling body gains sufficient impetus to take it back to the height from which it has fallen.

which is the state in which it is at rest, and then allowed to fall freely: it falls back to the perpendicular and then goes past it by the same amount, apart from such impetus as it loses through the resistance of the air, the cord, and any other accidental impediments. The same can be deduced from water flowing into a siphon which rises by the same amount as it falls.

SALVIATI. Your reasoning is impeccable. Now, I know you have no doubt that the body gains impetus by moving away from its starting point towards the centre to which its motion is drawn; so can you agree that two equal bodies will, if they are not impeded, gain an equal impetus if they move towards the centre by an equal amount, even if they move along different lines?

SAGREDO. I'm not sure that I understand the question.

SALVIATI. I can explain more clearly with a drawing, like

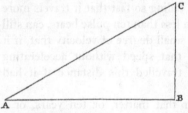

this: The line AB is a horizontal, and BC is a perpendicular above it, with an inclined line joining CA. If we now take the line CA to be a hard, perfectly polished inclined plane, and if a perfectly round, hard ball is rolled down it while another similar ball is dropped down the perpendicular CB, my question is whether you agree that the impetus gained by the ball which has rolled down the plane CA is the same, when it reaches A, as that gained by the other ball when it reaches B having been dropped down the perpendicular CB.

SAGREDO. It seems clear that the answer is yes, because both balls have moved towards the centre by the same amount, *The impetus is* and so, on the basis of what I have already agreed, both will *equal in bodies* have gained an impetus sufficient to take them back to the *moving equally* same height. *towards the* *centre.*

SALVIATI. Now tell me what you think this same ball would do if it were placed on the horizontal plane AB.

Mobile bodies SAGREDO. It would not move, provided the plane was not *remain* inclined in any way. *motionless on a* *horizontal* SALVIATI. But on the inclined plane CA it would move, but *plane.* with a slower motion than if it fell along the perpendicular CB.

SAGREDO. I was about to reply definitely that it would, because movement along the perpendicular CB seems necessarily to be faster than along the incline CA; but in that case how can the body which has reached A along the inclined plane have the same impetus, i.e. the same velocity, as the one which has moved along the perpendicular to B? These two propositions seem to be contradictory.

SALVIATI. In that case, you will certainly not agree with me when I say categorically that the velocity of a falling body along an inclined plane and along a perpendicular are equal. But this is an absolutely true proposition; just as it is also a true proposition that a falling body moves more rapidly along a perpendicular than along an incline.

Velocity along an inclined plane is the same as velocity along a perpendicular, but motion along a perpendicular is faster than along an incline.

SAGREDO. These sound to me like contradictory propositions; what do you think, signor Simplicio?

SIMPLICIO. They seem contradictory to me as well.

SALVIATI. I think you are teasing me by pretending not to understand something which you understand better than I do. Tell me, signor Simplicio: when you imagine one moving body to be faster than another, how do you picture that in your mind?

SIMPLICIO. I picture it as one body covering a greater distance than the other in the same time, or else covering the same distance in a shorter time.

SALVIATI. Fine; and how do you picture two bodies moving at the same speed?

SIMPLICIO. As covering the same distance in the same time.

SALVIATI. Is that all?

SIMPLICIO. This seems to me to be the correct definition of equal motion.

SAGREDO. I think we could add a further statement: if we also say that velocities are equal when the distances covered are in the same proportion to the time taken to cover them, this will be a more universally valid definition.

Velocities can be said to be equal when the distances covered are in proportion to the time taken.

SALVIATI. So it will, because it includes both equal distances covered in an equal time, and unequal distances covered in times which are unequal but proportionate to the distances. Now look again at the figure, and using your concept of faster motion, tell me why you think the velocity of the body falling along CB is greater than that of the body descending along CA.

SIMPLICIO. Because I think that in the time it takes for the falling body to cover the distance CB, the descending body will cover less of CA than the total distance CB.

SALVIATI. Indeed it will, which confirms that a body moves more rapidly along a perpendicular than along an incline. Now consider whether this same figure can be used to confirm the proposition that the bodies moved with equal velocity along both CA and CB.

SIMPLICIO. I can't see how this can be so; on the contrary, what you have just said seems to me to be a contradiction.

SALVIATI. What about you, signor Sagredo? I've no wish to teach you what you already know, since you've just given me a definition of it.

SAGREDO. The definition I gave you was that moving bodies can be said to have equal velocity when the distances they cover are in the same proportion to the time which they take to cover them. For this definition to apply in the present case, the time taken for a body to descend along the line CA would have to be in the same proportion to the time taken for one to fall on the line CB as the lines CA and CB are to each other; but I don't see how that can be the case given that the movement along CB is more rapid than that along CA.

SALVIATI. You must see, surely: both movements continually accelerate, don't they?

SAGREDO. Yes, they do, but the acceleration is greater on the perpendicular than on the incline.

SALVIATI. But is the acceleration on the perpendicular such that if you took two sections of equal length at any point on the two lines, the perpendicular and the incline, the acceleration on that section of the perpendicular would always be greater than on the corresponding section of the incline?

SAGREDO. No; on the contrary, if I took a section of the perpendicular at the end nearest to C and a section of the incline at the opposite end from it, the velocity would be much greater on the incline than on the corresponding section of the perpendicular.

SALVIATI. So you see that the proposition 'movement along a perpendicular is more rapid than on an incline' is only universally true if both movements start from the same state, namely a state of rest. If this condition is not met, then the proposition is so flawed that its opposite could also be true, namely that movement on an incline is more rapid than on a perpendicular, since it's perfectly possible to take a space on an incline which a moving body would cover more rapidly than an equivalent space on a perpendicular. So, if movement on an incline is more rapid than on a perpendicular in some places and less rapid in others, it follows that in some places the proportion between the time taken by a mobile body on an incline and the time taken on a perpendicular will be greater than the proportion between the distance covered in each case, and in other places the reverse will be true, i.e. the proportion of time will be less than the proportion of distance. So, for example, if two bodies move from a state of rest at C, one along the perpendicular CB and the other along the incline CA, in the time it takes for one body to cover the whole distance CB the other will have covered only the shorter distance CT. Therefore the time taken to travel CT will be greater in proportion to the time taken to travel CB (as they are equal) than the distance TC to the distance CB; the same measurement is greater in proportion to a smaller measurement than to a larger one. Conversely, if we took a section of the line CA, extended as necessary, which was equal to CB but could be travelled in a shorter time, then the proportions of time between the perpendicular and the

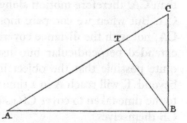

incline would be less than the proportions of distance. So, if it is possible to conceive of an incline and a perpendicular with distances and velocities such that the proportions between distances and times can be both greater and less, then we can reasonably allow that there can also be inclines and perpendiculars with distances where the time taken to cover them is in the same proportion as the distances themselves.

SAGREDO. You've resolved my main doubt, and I can see now that what I originally thought was a contradiction is in fact possible and indeed necessary. But I still don't see how one of these possible or necessary cases applies to what we are trying to establish here, namely that the proportion between the time taken to descend along CA and to fall along CB is the same as between the lines CA and CB themselves, so that it can be said without contradiction that the velocities along the incline CA and the perpendicular CB are the same.

SALVIATI. Let it suffice for now that I've dispelled your disbelief; the scientific explanation will become clear when you read our Academician's proofs concerning local motion.* As you will see, he shows that in the time which it takes for a body to fall the whole of the distance CB, a body descending along CA will reach T, which is the point from which a line at right angles to CA reaches B. To find out the point which the body falling along CB has reached when the other body reaches A, draw a line from A at right angles to CA and extend CB until it meets this line; this will be the point you are seeking. This shows how motion along CB is more rapid than along the incline CA (assuming that in both cases the moving body starts from a state of rest at C), because the distance CB is greater than CT, and the distance from C to the point at which CB, extended, intersects with a line at right angles to A is greater than CA; therefore motion along CB is more rapid than along CA. But when we compare motion along the whole distance CA, not with the distance covered in the same time along the extended perpendicular but just with the distance CB, it is quite possible that the object moving along CA, continuing beyond T, will reach A in a time that is in the same proportion to the time taken to cover CB as between the distances CA and CB themselves.

So to return to our original discussion, which was to establish that a moving body passes through every degree of slowness in moving from a state of rest to whatever velocity it acquires, let us look again at the same figure. We have agreed that a body falling along the perpendicular CB and another descending along the incline CA will acquire the same degree of velocity by the time they reach points B and A. So I imagine

you will have no dif-
ficulty in conceding
that on another plane
with less elevation than
AC, as for example
DA, a moving body
would descend even
more slowly than on

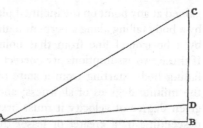

CA. Clearly we could add other planes with such a small
degree of elevation above the horizontal AB that the time
taken by the moving object, the ball we imagined, to reach
point A was greatly prolonged; for on the plane AB itself
that time would be infinite, and the smaller the degree of
inclination the slower motion becomes. So it must be possible
to fix a point such a small distance above B that, if you made a
plane from that point to A, the ball would not traverse it even
in a year. Now you should know that the impetus or degree of
velocity which the ball has gained when it reaches point A is
such that, if it continued to move uniformly with the same
velocity, without accelerating or slowing down, it would cover
twice the distance in the same time as it had covered so far; so,
for instance, if the ball had taken an hour to cover the distance
DA, and if it continued to move uniformly at the velocity it
had gained when it reached A, it would cover twice the distance DA in the next hour. And we have seen that the degree of
velocity gained at points B and A by bodies moving from any
point on the perpendicular CB, one along the inclined plane
and the other along the perpendicular, is always equal; so it
would be possible for a body to fall along the perpendicular
from a point so close to B that the velocity it had gained when
it reached B was not sufficient for it to cover a distance double
the length of the inclined plane in a year, or for that matter in
ten years or a hundred years. So we can conclude as follows.
Let us assume, first, that in the course of nature, and in the
absence of any external accidental impediments, a body moves
along an inclined plane at a degree of slowness determined by
the degree of inclination, and that the degree of slowness
becomes infinite if the plane is no longer inclined but horizontal. Second, let us also assume that the degree of velocity

gained at any point on the inclined plane is equal to that gained by a body falling along a perpendicular at the point intersected by a horizontal line from that point on the inclined plane. If these two assumptions are correct it must be the case that a falling body, starting from a state of rest, passes through all the infinite degrees of slowness; and consequently, to gain a given degree of velocity it must first move in a straight line, descending by a greater or lesser distance depending on the velocity it is to acquire and on the degree of inclination of the plane along which it is moving. So there could be a plane with so little inclination that, to gain a given degree of velocity, a body would have to move a very great distance over a very long period of time; and on a horizontal plane it would never naturally gain any velocity at all, as the body would not move. But motion along a horizontal line which is not inclined either upwards or downwards is circular motion around a centre. Therefore circular motion can never be acquired naturally unless it is preceded by rectilinear motion; but once acquired, it will continue perpetually at a uniform velocity.

Circular motion can never be acquired naturally unless it is preceded by rectilinear motion. Circular motion is perpetually uniform.

I could explain and even demonstrate these truths with other arguments, but it would be a very long digression, and I would rather leave it for another occasion and return to the main theme of our discussion—particularly as it is relevant here not for any necessary demonstration, but as an illustration of a concept of Plato's. And here let me add another observation of our Academician's which is indeed marvellous. Let us imagine that among the decrees of the divine Architect was that he would create the universe with these spheres which we see continually revolving, and that he placed the Sun as the fixed centre of their revolutions. Let us suppose, further, that he created all these spheres in the same place, and that he conferred on them the inclination to move by descending towards the centre until they acquired that degree of velocity which it pleased the divine Mind to give them; and that once they had reached it their motion should become circular, each in its own orbit and maintaining its allotted velocity.* And let us ask what would be the height and distance from the Sun at which these spheres were created, and whether it is possible that they were all created in the same place. For this we need to take the calculations of the most expert astronomers for the size of the planets' orbits and their orbital periods, from which we can work out, for example, how much the velocity of Jupiter is greater than that of Saturn. Once we have established that Jupiter moves more rapidly than Saturn (as indeed it does), it follows that, if they both started from the same height, Jupiter must have fallen further than Saturn, as indeed we know to be the case, since Jupiter's orbit is below Saturn's. But we can go further: given the proportions between the velocities of Jupiter and Saturn, the distance between their orbits and the proportions between the natural acceleration of each, we can discover the height and the distance from the centre of their revolutions of the place from which they started. Having found and established this, we can see whether Mars [also derives its orbit and velocity from] its descent from the same place, and we find that the size of its orbit and its velocity correspond to what this calculation suggests; and the same is true of the Earth, Venus, and Mercury, all of whose orbits and velocities accord so closely with the result of these computations that it is a marvel to behold.

The size of the planets' orbits and their velocity of motion are in proportion to their having all descended from the same place.

SAGREDO. I am delighted with this idea, and if I didn't think that making such calculations accurately would be a long and laborious undertaking and beyond my powers of comprehension, I would ask you to illustrate it for me.

SALVIATI. It would indeed be a long and difficult task, and I'm not sure I could reproduce it here and now; so I suggest we leave it for another time.

[SIMPLICIO. Please make allowance for my lack of expertise in the mathematical sciences if I say that your explanations, based on greater or lesser proportions and other terms which I don't sufficiently understand, haven't resolved my doubt, or rather my incredulity. How can a heavy lead ball weighing a hundred pounds, when it is dropped from a height, pass from a state of rest through every extreme degree of slowness, when it falls more than a hundred *braccia* in four pulse beats? I find it completely incredible that there could be any moment when it was moving with such a degree of slowness that, if it continued to move at that rate, it would take more than a thousand years to travel half an inch. But if this is really the case, I'd be glad to have it explained to me.

SAGREDO. Signor Salviati is such a learned man that he often thinks that terms which are well known and familiar to him are equally familiar to everyone else, so he sometimes forgets that when he's talking to us he needs to help our lack of understanding by using less abstruse terms. Since I don't rise to such heights, with his leave I will try at least to mitigate signor Simplicio's incredulity by referring to the evidence of our senses. So, still staying with our example of the cannon-ball, please tell me, signor Simplicio: do you not agree that in passing from one state to another it is naturally quicker and easier to pass to a state which is closer at hand than to one which is further away?

SIMPLICIO. I understand and agree this, yes; so I don't doubt that, for example, as a piece of hot iron cools it passes from 10 degrees of heat to 9 before it passes from 10 to 6.

SAGREDO. Excellent. So now tell me this: if the cannon-ball is fired straight up in the air, does its motion not continually slow down until it reaches its highest point, when it is in a state of rest? And as its speed decreases—or if you like, as its slowness increases—is it not reasonable that it should pass from 10 degrees to 11 rather than from 10 to 12, or from 1,000 to 1,001 rather than to 1,002? In short, that it passes from one degree to another that is closer to it, rather than to one which is further away?

SIMPLICIO. Yes, that's reasonable.

SAGREDO. But what degree of slowness could be so far removed from any motion that a state of rest, which is an infinite degree of slowness, is not even further removed from it? So we can't doubt that before the cannon-ball reaches a state of rest it passes through greater and greater degrees of slowness, including that degree where it would take more than 1,000 years to travel an inch. This being the case, surely it shouldn't seem improbable to you that as the cannon-ball begins to fall and moves away from a state of rest, it regains its velocity of motion by passing through the same degrees of slowness as it did on the way up, and not by missing out those degrees which are closer to a state of rest and jumping straight to another degree which is further away.

SIMPLICIO. This explanation has helped me to understand far better than all those mathematical subtleties; so now let signor Salviati continue his exposition.]

SALVIATI. So let's return to our main discussion, picking it up at the point where we digressed. If I remember rightly, we were establishing how rectilinear motion cannot serve any useful purpose where the world is properly ordered, and we were going on to say that the same is not true of circular motion, since the motion of a body rotating on its axis keeps it always in the same place, and the motion of a body on the circumference of a circle around a fixed and unmoving centre does not cause any disorder either to itself or to the bodies around it. For such motion, first of all, is finite and defined; *Circular motion which is finite and defined does not disrupt the order of the world.* more than that, there is no point on its circumference which is not both the beginning and the end of its revolution, so it continues in its allotted circumference, leaving every other space both inside and outside its circle free for other bodies, without obstructing or disordering them in any way. Because it *In circular motion any point on the circumference is both beginning and end.* is a motion in which the body is always both departing from and arriving at its end, such motion alone can be uniform; for acceleration occurs when a body is moving towards the place to which it is inclined, and its motion is retarded because of its resistance to moving away from the same place. But because in *Circular motion alone is uniform.* circular motion a body is always both moving away from and returning to its natural place, its resistance and its inclination are always equally balanced, and this equilibrium produces a motion which is neither retarded nor accelerated, but uniform. Being uniform and defined, such motion can continue perpetually, always repeating the same revolution, but this cannot *Circular motion can continue perpetually.* occur naturally in a line which has no defined end or in motion which is continually accelerating and slowing down. I say 'naturally' because when rectilinear motion is slowed down this is due to an external force, which cannot be perpetual; and *Rectilinear motion cannot naturally be perpetual.* when it accelerates it will necessarily reach its goal, if there is one; if there is not, then there can be no motion towards it, because nature does not move towards a point which is impossible to reach. I conclude therefore that the only natural motion in the natural bodies which make up the universe and are in their optimal place is circular; and that rectilinear motion can be said at the most to be used by nature when some body or part of a body has moved away from its proper place and is wrongly disposed, and therefore needs to return by the shortest route to its natural state. Hence it seems reasonable to conclude that the only motion which serves to maintain the parts of the world in perfect order is circular, and that if there are any bodies which do not move in circular motion they must necessarily be fixed, since only circular motion and a state of rest serve to maintain order. And I am surprised that Aristotle, who believed that the earthly globe was located and fixed in the centre of the universe, did not state that some natural bodies were mobile by nature and others immobile, especially since he defined nature as being the principle of motion and rest.

Rectilinear motion occurs in nature to restore bodies to perfect order when they have moved away from it.

Only circular motion and a state of rest serve to maintain order.

SIMPLICIO. Aristotle, for all his great insight, did not expect more from his reason than it could provide; he considered in his philosophy that the experience of the senses should take precedence over any argument constructed by human reason, and he said that anyone who denied the evidence of one of the senses deserved to be deprived of that sense. Who then is so blind that they cannot see that earth and water, being heavy, move naturally downwards, that is, towards the centre of the universe, this being the goal and end fixed by nature for downward rectilinear motion; and in the same way that fire and air naturally move upwards in a straight line towards the concave surface of the moon's sphere, as the natural goal of upward motion? And if this is manifestly true, and given that we know that *eadem est ratio totius et partium* ('the same reasoning applies to the whole and to the parts'),* is it not self-evidently true that the natural motion of the Earth is rectilinear motion towards the centre, and that of fire is rectilinear motion away from the centre?

The experience of the senses should precede human reasoning. Anyone who denies a sense deserves to be deprived of it.

Our senses show that heavy bodies move towards the centre, and light bodies move towards the sphere's concave surface.

SALVIATI. The most I'll concede in response to your argument is that, if parts of the Earth which are separated from the whole, i.e. from the place where they naturally belong and so, ultimately, in violation of their place in the natural order, will spontaneously and therefore naturally return to their place in a straight line, then we can infer that—granted that *eadem sit ratio totius et partium*—if the terrestrial globe itself were to be moved from its natural place by some external force, it would move in a straight line to return there. This, as I say, is as much as can be conceded to you, making all possible allowances. But anyone who examined these arguments rigorously would, first of all, deny that when parts of the Earth move to be reunited with the whole they move in a straight line, rather than with *It is doubtful* circular or mixed motion; and you would have your work cut *whether heavy* out to argue against them, as you will see from the replies to *straight line.* the arguments and experiments adduced by Ptolemy and Aristotle. Secondly, if it were put to you that parts of the Earth move not in order to be at the centre of the world but rather to rejoin the whole of which they are part, and that their natural motion is therefore towards the centre of the terrestrial globe, then what other whole and what other centre could you give *The Earth is a* to which the terrestrial globe itself would seek to return, if it *sphere because* were dislodged from it? Yet this would have to happen, if the *together around* same reasoning is to apply to the whole and to the parts. *its centre.* Finally, neither Aristotle nor you will ever be able to prove that the Earth is in reality at the centre of the universe; in fact, if a centre can be assigned to the universe at all, we shall find it is more likely to be where the Sun is, as you will see in due *It is more* course. *probable that*

the Sun is the
If all the parts of the Earth come together to form the whole *centre of the* because they are drawn together by an equal inclination from *universe than* all sides, and form themselves into a sphere so as to be as *the Earth.* closely united as possible, is it not likely that the Moon, the Sun, and the other heavenly bodies are also spherical for the same reason, simply because all their component parts are *The parts of all* naturally and instinctively drawn together? In which case, if *the heavenly* any part of them were separated from the whole by some *globes have a* external force, is it not reasonable to suppose that they would *inclination* spontaneously and instinctively return to them, and would this *towards their* not mean that rectilinear motion belongs equally to all the *centre.* heavenly bodies?

SIMPLICIO. It's quite clear that you will never be persuaded or shifted from any of your preconceived opinions, because you intend to deny not just the basic principles of science but even the plain evidence of our senses. So I shan't object, not because I am convinced by your arguments but because *contra negantes principia non est disputandum* ('there is no arguing with those who deny first principles').* But staying with what you have just said, since you even cast doubt on whether falling objects move in a straight line, how can you deny that heavy objects, being parts of the Earth, fall towards the centre in a straight line? If you drop a stone from a high tower which has straight, perfectly vertical walls, will it not land exactly where a plumb-line suspended from the point from which you dropped it ends, almost brushing against the wall as it falls? Surely this is clear evidence that this is rectilinear motion towards the centre?

Evidence of the senses that falling objects move in a straight line.

Second, you question whether the parts of the Earth move towards the centre of the universe, as Aristotle affirms, even though he proved it conclusively by means of contrary motion,* as follows: the motion of heavy bodies is the contrary of that of light bodies; light bodies can be seen to move directly upwards, i.e. towards the outer circumference of the universe; therefore heavy bodies move directly towards the centre of the universe, which coincidentally is also the centre of the Earth, since this is where the Earth happens to be placed. As for asking what would happen if part of the lunar or solar globe were separated from the whole, this is a vain question because it is asking what would follow from an impossibility. For Aristotle also shows that the heavenly bodies are impassible, impenetrable, and unbreakable, so such a thing could never happen; and even if it did and the separated part did return to the whole, it would not do so by virtue of being either heavy or light, because Aristotle himself proves that heavenly bodies are neither heavy nor light.

Aristotle's argument to prove that falling bodies move towards the centre of the universe. Falling bodies move coincidentally towards the centre of the Earth.

It is vain to seek what would follow from an impossibility. Aristotle's view that heavenly bodies are neither heavy nor light.

SALVIATI. Whether my doubts about falling objects moving vertically in a straight line are reasonable or not will become clear, as I have said, when I come to discuss this question in detail. But as for the second point, I'm surprised that I have to point out the flaw in Aristotle's logic to you, since it seems self-evident. Surely you realize that Aristotle is presupposing the point which he is trying to prove? Note that . . .

SIMPLICIO. Please, signor Salviati, don't speak so dis-
respectfully of Aristotle. No one is going to believe that the
first and unequalled exponent of syllogistic logic, of proofs, of
the Socratic dialogue, of how to recognize sophisms and faulty
logic, in short of the whole science of logic, should have made
such a basic error as to presuppose as given the point which is
under discussion. At least understand him first, before you try
to refute him.

*Aristotle
cannot make
logical
mistakes, since
he was the
inventor of
logic.*

SALVIATI. Signor Simplicio, by all means point out my
errors and I'll be grateful for the correction, since this is a
discussion among friends and we are trying to establish the
truth; so please tell me if I have failed to grasp Aristotle's
meaning. But in the meantime do allow me to explain the
difficulty I have, and first of all to respond to your last point.
Logic, as you know, is the organ which we use as the instrument on which to practise philosophy. But just as there can be
excellent organ-builders who have no expertise in playing,
so it's possible to be a great logician but have little expertise
in knowing how logic is to be used. In the same way, there
are many who know all the rules of poetics off by heart, but
would struggle to put four verses together; and there are
others who know all Leonardo's rules but are unable to paint a
chair. We learn to play the organ from an organist, not an
organ-builder; we learn poetry by constantly reading the
works of poets, and painting by constant practice in drawing
and painting. And so we learn how to construct a proof
by reading books full of proofs, which are written by
mathematicians, not logicians.

So to return to the question we were discussing, Aristotle's
view of the motion of light bodies is based on his observation
of fire, which moves directly upwards from anywhere on the
surface of the terrestrial globe. This is indeed motion towards
a circumference greater than that of the Earth; Aristotle
himself defines it as motion towards the concave surface of
the Moon's sphere. But we can only say that this is also the
circumference of the universe, or concentric with it, and
therefore that motion towards it is also motion towards the
circumference of the universe, if we presuppose that the centre
of the Earth, from which we see light bodies rising, is the same

*Aristotle's
faulty logic in
his proof that
the Earth is
at the centre of
the universe.*

as the centre of the universe—in other words, that the centre of the Earth is located at the centre of the universe, which is the point Aristotle is trying to prove. How can you deny that this is a clear logical fallacy?

SAGREDO. I found this argument of Aristotle's faulty and inconclusive in another respect, even if it is granted that the circumference towards which fire moves is the sphere which encloses the whole universe. For if a body moves in a straight line starting from any point inside a circle, not just from its centre, it will undoubtedly be moving towards the circumference, and will reach it provided it keeps going long enough; so this is not in doubt. But it does not follow that by moving along the same straight line in the contrary direction it would necessarily reach the centre; this would only be true either if its starting point were the centre itself, or if it were to move along the line which, if extended from the starting point, passed through the centre. So to say 'fire moving in a straight line moves towards the circumference of the universe; therefore those parts of the Earth which move along the same lines in the opposite direction move towards the centre of the *Aristotle's* universe' is only valid if the lines along which fire moves *faulty logic* would, if extended, pass through the centre of the universe. *exposed from* We know for certain that a line which is perpendicular to the *another angle.* surface of the terrestrial globe passes through the centre of the globe; so the statement is only valid if we assume that the centre of the Earth is the same as the centre of the universe, or at least that fire and earth in their upward and downward motion move only along one line which passes through the centre of the universe. This is clearly false and contrary to our experience, which is that fire rises vertically from the surface of the terrestrial globe along any number of lines produced from the centre of the Earth to every part of the universe.

SALVIATI. Signor Sagredo, you cleverly show how Aristotle's argument leads to the same inconsistency by exposing its evident misunderstanding. But another inconsistency follows from this. We see that the Earth is a sphere, and so we can be sure that it has a centre; and we see that all its parts move towards its centre, as this necessarily follows if their motion is always perpendicular to the surface of the globe. We understand this motion towards the centre of the Earth as an instinct drawing them towards the whole of which they are part, and towards their universal mother; and we are so sympathetic to this idea that we are persuaded that their instinct draws them not to the centre of the Earth but to the centre of the universe, even though we do not know where that is, or even if it exists—and if it does exist it is no more than an imaginary point with no physical properties at all.

Proof that it is more reasonable to say that falling objects move towards the centre of the Earth than the centre of the universe.

Signor Simplicio's last point was that it was vain to ask whether parts of the Sun, the Moon, or other celestial bodies which became separated from the whole would naturally return to it, because such a thing could not happen, as Aristotle showed that the heavenly bodies are impassible, impenetrable, indivisible, etc. To this my response is that none of the qualities which Aristotle cites as differentiating the heavenly bodies from the elemental has any substance except in relation to the different motions which he assigns to them. So if we deny that circular motion belongs solely to the heavenly bodies and affirm that, on the contrary, it is equally appropriate to all moving bodies in nature, then we must acknowledge one of two things: either the attributes of being subject or not subject to generation, change, division, etc., apply equally to all bodies, whether heavenly or elemental; or Aristotle was wrong in saying that those which he assigned to the heavenly bodies were the consequence of their circular motion.

The conditions by which Aristotle differentiates the heavenly bodies from the elemental follow from the movements which he assigns to them.

SIMPLICIO. This kind of reasoning subverts the whole of natural philosophy and leaves heaven, Earth, and the whole universe in confusion and disorder. But I believe that the foundations of Peripatetic philosophy are sound enough for there to be no danger that new sciences will be built out of their ruins.

SALVIATI. You needn't worry about the heavens or the Earth being subverted, or philosophy for that matter. As far as the heavens are concerned, there is no reason to fear for them, for you yourself say that they are inalterable and impassible; and as for the Earth, we are trying to perfect it and give it a more noble status, making it more like the celestial bodies and, in a way, giving it a place in the heavens, from which your philosophers have banished it. And philosophy itself can only

benefit from our disputes, because if our conclusions are correct new truths will have been gained, and if they are wrong, the original positions will have been strengthened by refuting them. Think rather of certain philosophers and try to help and sustain them, for science itself cannot fail to advance. So to return to our theme, please tell us the arguments which you think will maintain Aristotle's fundamental distinction between heavenly bodies and the elemental part of the universe, such that the former are ingenerable, incorruptible, unchanging, etc., whereas the latter are corruptible, changeable, etc.

SIMPLICIO. I don't see that Aristotle has any need of support, since he is still standing as firmly as ever, and I don't think he has yet even been attacked, let alone defeated, by you. How are you going to defend yourself against his first assault? Aristotle writes as follows: 'Generation is produced by a contrary in a subject, and similarly corruption in a subject is pro-

duced by one contrary being corrupted into another', so that—note this—corruption and generation occur only in contraries. 'But the movements of contraries are contrary to each other. So if celestial bodies have no contrary, because they move in circular motion which has no contrary motion, it is entirely fitting that nature should have made the heavens, which are not subject to generation or corruption, exempt from contraries.'*

Once this fundamental principle is established, it clearly follows as a consequence that the heavens cannot be increased or altered and are impervious to change, and in short are eternal and a fit habitation for the immortal Gods, as is held by

all those who have any concept of the Gods. Our senses, too, confirm this, for throughout recorded time no change has been observed in the highest heaven or in any part of the heavens. As for there being no motion contrary to circular motion, Aristotle proves this in numerous ways and I won't rehearse them all. But the clearest proof is this: there are only three simple motions, towards, away from, and around the centre; of these the two rectilinear motions, upwards and downwards, are clearly contraries, and each can have only one contrary; therefore no motion remains to be the contrary of circular motion. And there you have Aristotle's brilliantly conclusive proof that the heavens are incorruptible.

SALVIATI. This is indeed exactly the step in Aristotle's argument which I've already alluded to; and if I deny that the motion which you attribute to the heavenly bodies is not also appropriate to the Earth, the conclusion which he draws from it is invalid. So, if I say that the circular motion which you claim belongs exclusively to the heavenly bodies belongs also to the Earth, and if the rest of your argument holds, then one of three things follows. I've said this already but I repeat it now: either the Earth, like the celestial bodies, is unaffected by generation and corruption; or the celestial bodies, like those composed of the elements, are subject to generation, change, etc.; or these different kinds of motion have nothing to do with generation and corruption at all. Aristotle's argument, which you reproduce, contains many propositions which cannot be taken for granted, so to examine it more closely it will be as well to reduce it as far as possible to its essentials—and I ask signor Sagredo's indulgence if I bore him by repeating things that have been said more than once already: perhaps he can imagine that he is hearing the arguments rehearsed in a public debate. So your argument runs as follows: 'Generation and corruption are produced only where there are contraries; contraries are found only among simple natural bodies which move with contrary motion; the only contrary motions are rectilinear motion between contrary goals, of which there are only two, towards the centre and away from the centre. These movements occur only among natural bodies composed of earth, fire, and the other two elements; therefore generation and corruption are produced only among these elements. And since the third simple motion, which is circular motion around the centre, has no contrary (the other two being contrary to each other, and each can have only one contrary), a natural body to which such motion belongs also has no contrary; since it has no contrary, it is not subject to generation and corruption, etc., because generation and corruption, etc., are produced only where there are contraries. But circular motion belongs only to the heavenly bodies; therefore these alone are not subject to generation, corruption, etc.'

My first response to this is to say that I think it is much easier to establish whether the Earth, a vast body which is easy for us to study because it is so close at hand, moves at such a speed that it turns on its axis in twenty-four hours, than it is to understand whether generation and corruption are the product of contraries, or for that matter whether corruption, generation, and contraries exist in nature at all. And if, signor Simplicio, you can explain to me how nature is able to generate thousands of flies in almost no time from a few fumes of grape must, and to show me what are the contraries in this process, what it is that is corrupted and how, then my estimation of you will be even greater than it is, because these are things which I cannot understand at all. I would dearly like to know, too, why these corrupting contraries are so benign in their treatment of crows and so harsh towards doves, or why they are so tolerant of deer and so impatient with horses, for the former are granted more years of life—i.e. of incorruptibility—than the latter are weeks. Or again, peach trees and olive trees both have their roots in the same soil, are exposed to the same cold and heat, the same wind and rain, in short to all the same contraries, and yet peach trees die in a short time whereas olive trees live for centuries. What's more, I have never been able to understand this idea of transmutation of substance (still speaking in purely natural terms) whereby matter is so totally transformed that its original being is completely destroyed so that nothing of it remains, and another completely different body is produced from it. For a body to appear first under one aspect and then under another quite different one a short time later, I don't find it impossible that this can come about through a simple transposition of parts, without corruption or generation of anything new, for we see such metamorphoses happening all the time. So I repeat, you will have an uphill task if you want to persuade me that the Earth cannot move in circular motion because of the principle of corruption and generation, because I will prove the opposite to you with arguments which are no less conclusive, though they may be more difficult.

It is easier to establish whether the Earth moves than whether corruption is the product of contraries.

Bodies can appear under different aspects through a simple transposition of parts.

SAGREDO. Signor Salviati, forgive me if I interrupt your exposition, fascinating as I find it, for I find myself embroiled in exactly the same difficulties. But I fear we could only get to the end of it by setting aside our main subject altogether; so I propose that, in order to continue with our original discussion, we keep this whole debate about generation and corruption as a topic for a separate dialogue in its own right. And if you and signor Simplicio agree, I will keep a note of this and any other specific questions which come up in the course of our discussions, so that we can devote another day to examining them in detail. So returning to the matter in hand, you say that if we deny Aristotle's claim that circular movement does not affect the Earth as it does the other celestial bodies, it follows that whatever happens on Earth as regards being subject to generation, change, etc., also affects the heavens. So, leaving aside whether or not generation and corruption are to be found in nature, let us return to investigating the motion or otherwise of the terrestrial globe.

SIMPLICIO. I can't bring myself to listen to someone denying that generation and corruption exist in nature, when the evidence for them is continually before our eyes, and Aristotle has written two whole books* on the subject. If you start denying the principles of science and casting doubt on things which are perfectly plain to see, then of course you can prove whatever you want and maintain any paradox at all. If you don't see the constant generation and corruption of grasses, plants, and animals all around you, what do you see? How can you fail to see the perpetual clash of contraries, and earth turning into water, water into air, air into fire, and then condensing again into clouds, rain, hail, and tempests?

Any paradox can be maintained by denying the principles of science.

SAGREDO. We do see all these things, of course, and so we grant you Aristotle's argument as far as these examples of generation and corruption produced by contraries are concerned. But if I were to prove to you, on the basis of these same propositions of Aristotle's, that the celestial bodies are subject to generation and corruption just as much as those made up of the elements, what would you reply?

SIMPLICIO. I would reply that you have done the impossible.

SAGREDO. Then tell me, signor Simplicio, are not these qualities contraries?

SIMPLICIO. Which qualities?

SAGREDO. The qualities of being alterable or inalterable, passible or impassible, generable or ingenerable, corruptible or incorruptible.

SIMPLICIO. Of course they are contraries.

SAGREDO. In that case, if the celestial bodies are ingenerable and incorruptible, I will prove to you that they must necessarily be generable and corruptible.

SIMPLICIO. This can only be pure sophistry.

SAGREDO. Listen to the argument, then identify it and resolve it. The celestial bodies, being ingenerable and incorruptible, have contraries in nature, namely bodies which are generable and corruptible; but generation and corruption are found where there are contraries; therefore the celestial bodies are generable and corruptible.

Heavenly bodies are generable and corruptible, because they are ingenerable and incorruptible.

SIMPLICIO. I said this was nothing but sophistry. This is one of those horned arguments known as sorites,* like the example of the Cretan who said that all Cretans are liars. If he was a Cretan he must have been lying when he said that all Cretans were liars; therefore Cretans must tell the truth, and he, as a Cretan, must have been telling the truth when he said that Cretans were liars, including himself; and therefore he must have been lying. This kind of sophism simply goes round in circles without ever reaching any conclusion.

A horned argument, otherwise known as a sorites.

SAGREDO. So you've identified the argument; now resolve it, and show where the fallacy lies.

SIMPLICIO. Resolve it and demonstrate the fallacy? For a start, surely you can see the evident contradiction—'the celestial bodies are ingenerable and are incorruptible; therefore they are generable and corruptible'? In any case, the contraries are not in the celestial bodies but in the elements, which are contraries in that they are light and heavy and that they move upwards and downwards. The heavens, on the other hand, move in a circle, to which there is no contrary motion; therefore they have no contraries; therefore they are incorruptible, etc.

There are no contraries in the celestial bodies.

SAGREDO. Just a moment, signor Simplicio. This contrary which you say causes some simple bodies to be corruptible: does it reside in the body which is corrupted, or in relation to some other body? For example, does the humidity which causes corruption in some part of earth reside in the earth itself, or is it in some other body such as air or water? I think you will say that, just as upward and downward motion, weight and lightness, which you call the primary contraries, cannot exist in the same subject, so it is with the contraries of moist and dry, hot and cold. It follows then that when a body is corrupted this is the result of a quality found in another body which is contrary to its own. Therefore a heavenly body is corruptible if there are bodies in nature which have qualities contrary to it; and that is exactly what the elements are, if it is true that corruptibility is the contrary of incorruptibility.

The contraries which are the cause of corruption are not inherent in the body which is corrupted.

SIMPLICIO. That's not a sufficient proof. The elements change and are corrupted because they touch and mingle with each other, and so exercise their contrary qualities; but the celestial bodies are separate from the elements and are not affected by them, although they do have an influence on the elements. To prove that generation and corruption exist in the celestial bodies you would have to show that contraries reside among them.

Heavenly bodies touch the elements, but are not touched by them.

SAGREDO. I'll demonstrate this as follows. The contrary qualities in the elements derive in the first place, in your view, from their contrary upward and downward motion; therefore the principles from which these motions derive must also be contraries. Now upward motion is the result of lightness, and downward motion the result of weight; therefore these must also be contraries, as too must be the qualities which make one body light and another heavy. But your own school of philosophy maintains that lightness and weight are the consequence of rarity and density; so rarity and density must also be contraries. And these are such pervasive qualities among the heavenly bodies that you hold the stars to be simply the densest parts of their heavenly sphere. If this is correct, then the density of the stars must be almost infinitely greater than that of the rest of the heaven. This is clear from the fact that the heaven is almost wholly transparent and the stars are almost wholly opaque, and varying degrees of density and rarity are the only qualities found in the heavens which could be the cause of these varying degrees of transparency.

Weight and lightness, and rarity and density, are contrary qualities.

The stars are infinitely more dense than the substance of the rest of the heaven.

Therefore, since these contraries exist among the celestial bodies, the celestial bodies must also be subject to generation and corruption, in the same way as elemental bodies; or else corruptibility, etc., are not caused by contraries.

SIMPLICIO. Neither of these conclusions follows. Density and rarity in the celestial bodies are not contraries as they are in elemental bodies, because they do not depend on the prime qualities of heat and cold, which are contraries, but on the greater or lesser amount of matter in proportion to their quantity. The opposition between more and less is what is known as a relative opposition, which is the least significant kind, and has nothing to do with generation and corruption.

*Rarity and density in the celestial bodies are not the same as they are in the elements (Cremonino).**

SAGREDO. So you are saying that the density and rarity which are the source of weight and lightness in the elements, which cause contrary upward and downward motion, which in turn produce the contraries underlying generation and corruption, [are different from the kind of density and rarity which exist in the heavens. The latter are the result of] varying amounts of matter contained in the same quantity or mass of a body, but only density and rarity deriving from the primary qualities of heat and cold produce the required effect in the elements; otherwise they would have no effect. But if this is the case, Aristotle has deceived us: he should have said straight away that generation and corruption occur in simple bodies which are subject to simple upward and downward motion, caused by lightness or weight, as a result of rarity or density produced by a greater or lesser amount of matter depending on heat or cold; instead of which he just refers to upward and downward motion, without any other qualification. Because I can assure you that any kind of density or rarity is enough to make bodies heavy or light, and therefore subject to contrary motion, whether as a result of heat and cold or anything else. Heat and cold have nothing to do with this effect, as is clear from the fact that a piece of iron which has been in the fire, and so can be called hot, has the same weight and moves in the same way as one that is cold. But leaving this aside, how do you know that density and rarity in the heavens are not dependent on heat or cold?

Aristotle's account of the causes of generation and corruption in the elements is incomplete.

SIMPLICIO. Because these qualities do not occur among the celestial bodies, which are neither hot nor cold.

SALVIATI. I can see that we're adrift again in a boundless sea from which there is no escape, since we are navigating without compass, stars, oars, or rudder, so that all we can do is go from rock to rock, or run aground, or else sail aimlessly for ever. So if, as you suggest, we are to press ahead with our main topic, we must leave aside for now this general discussion of whether rectilinear motion is necessary in nature and belongs to some bodies and not others, and come to consider the specific demonstrations, observations, and experiments which have been put forward, first by Aristotle, Ptolemy, and others to prove that the Earth is fixed, and try to resolve them; and then those which have convinced others that the Earth, no less than the Moon or any other planet, is to be numbered among the natural bodies endowed with circular motion.

SAGREDO. I gladly agree to this, all the more so because I find your general architectural exposition much more satisfactory than Aristotle's: yours resolves all my difficulties, whereas with Aristotle's I find obstacles in my path at every step. In fact I don't see how signor Simplicio can fail to be convinced by your argument proving that rectilinear motion can have no place in nature if we assume that the parts of the universe are optimally disposed in perfect order.

SALVIATI. Allow me to interrupt you, signor Sagredo, since I have just thought of a way to convince even signor Simplicio, as long as he isn't so wedded to every word of Aristotle's that he considers any departure from him to be sacrilege. There is no doubt that maintaining the optimal disposition of every part of the universe in perfect order, in terms of their location, requires circular motion and a state of rest; whereas I cannot see that rectilinear motion serves any function apart from restoring to its natural place some fragment of an integral body which has been separated from it, as we have seen. So now let us consider what is needed for the terrestrial globe, given that it, like all the other bodies in the universe, must be kept in its optimal natural disposition. There are three possibilities: that it rests and remains immobile in its proper place in perpetuity; or that it turns on its axis while remaining always in the same place; or that it moves around a centre, following the circumference of a circle.

Of these, Aristotle, Ptolemy, and their followers say that its natural disposition has always been the first, and that it will remain so for all eternity, i.e. that it is perpetually at rest in the same place. So why, I ask, do we not say that the Earth's natural condition is to remain motionless, instead of saying that it naturally moves downwards, since it never has moved downwards and there is no reason to think that it ever will? And as for rectilinear motion, let us allow that nature uses it to restore to their proper place all those fragments of earth, water, air, fire, or any other integral body which for any reason have become separated from the whole, and so removed from their properly ordered place—unless even such a restoration might not be achieved more effectively by a circular motion of some kind. So it seems to me to fit much better with all the other consequences, even in Aristotle's own terms, to say that the Earth's natural condition is to be in a state of rest, instead of making rectilinear motion the intrinsic natural principle of all the elements. This is clearly the case: the Peripatetics believe that the heavenly bodies are incorruptible and eternal; but if I were to ask one of them if he believes that the terrestrial globe is corruptible and mortal, and therefore that there will come a time when the Sun, the Moon, and the other stars continue in their courses but the Earth no longer exists, having been corrupted and dissolved along with all the other elements, I am sure that he would say no. So corruption and generation must belong to the parts, not to the whole, and what's more to very small and superficial parts, so small that they are almost imperceptible in comparison to the whole. Therefore, since Aristotle attributes generation and corruption to the contraries of rectilinear motion, let us leave these motions to the parts, as it is only they that undergo change and corruption, and conclude that the globe or sphere as a whole either moves with circular motion or is permanently motionless in the same place, since only these two contribute to the maintenance of perfect order. What we have concluded about the Earth applies equally to fire and to the greater part of the air. The Peripatetics end up saying that the intrinsic

Aristotle and Ptolemy say that the terrestrial globe is motionless. The natural condition of the terrestrial globe should be defined as a state of rest rather than downward rectilinear motion.

There are more grounds for attributing rectilinear motion to parts of the elements than to the elements as a whole.

natural motion of these two elements is one with which they never have moved and are never likely to, while they define as contrary to nature the motion with which they do move, as they always have and always will. They say that air and fire naturally move upwards, even though neither of these elements has ever moved upwards but only fragments of them, simply so as to return to their properly ordered position from which they had been displaced; and they say that for these elements the circular motion with which they move constantly is an aberration, quite forgetting that Aristotle said on several occasions that no violent disturbance can last for long.

The Peripatetics groundlessly say that the natural motions of the elements are ones with which they never move, and that the motions with which they move all the time are aberrations.

SIMPLICIO. We have conclusive replies to all these points, but I shall leave these for the moment to concentrate on more specific arguments and the experience of the senses, which Aristotle rightly says should take precedence over what can be worked out by human reason.

The experience of the senses should precede human reasoning.

SAGREDO. Good; so let everything that has been said so far serve to set out two general accounts and to help us consider which appears the more probable. Aristotle's view, based on the different kinds of simple motion, would persuade us that the nature of sublunary bodies, being generable, corruptible, etc., is entirely different from the nature of the celestial bodies which are impassible, ingenerable, incorruptible, etc. Signor Salviati, on the other hand, assumes that the integral parts of the universe are optimally constituted, and consequently that rectilinear motion cannot belong intrinsically to simple natural bodies, because it would serve no purpose in nature; so he considers the Earth to be itself one of the celestial bodies, endowed with the same prerogatives as all the others. I must admit that this latter seems to me so far to be much the more satisfactory of the two. So let me now invite signor Simplicio to put forward the detailed arguments, experiments, and observations, both natural and astronomical, to persuade us that the Earth is different from the celestial bodies, and that it is fixed immobile at the centre of the universe, together with any other qualities which exclude its being mobile like Jupiter, the Moon, or any of the other planets; and I'll ask signor Salviati to be so good as to reply to each point in turn.

SIMPLICIO. To start with, here are two powerful demonstrations to prove that the Earth is completely different
from the celestial bodies. First, bodies which are generable,
corruptible, mutable, etc., are completely different from those
which are ingenerable, incorruptible, immutable, etc. The
Earth is generable, corruptible, mutable, etc., and the celestial
bodies are ingenerable, incorruptible, immutable, etc.; therefore the Earth is completely different from the celestial bodies.

SAGREDO. Your first argument is simply reintroducing the
one we've just disposed of.

SIMPLICIO. Wait a minute; listen to what follows and you'll
see how different it is. The previous argument was to prove the
minor premise *a priori*; now I shall prove it *a posteriori*,* and
you'll see that it is not the same at all. So this is the proof of
the minor premise, since the major premise is self-evident.
The experience of our senses shows us that the processes of
generation, corruption, change, etc., are continually going on
here on Earth, whereas we have never seen them in the
heavens, and none of the traditions or writings of our forebears

*The heavens
are immutable
because no
change has ever
been seen in
them.*

has any record of them; therefore the heavens are unchangeable, etc., and the Earth is changeable, etc., and hence different
from the heavens. The second argument is one which is based
on a principal and essential quality, and it is this: a body which
is naturally dark and devoid of light is different from bodies

*Bodies which
are naturally
sources of light
are different
from those
which are dark.*

which shine with their own light. The Earth is dark and
devoid of light; the heavenly bodies shine brilliantly with their
own light; therefore, etc. Perhaps you would reply to these
arguments before I go on to others, so as not to accumulate too
many at a time.

SALVIATI. As regards the first argument, which you derive
from experience, I would like you to tell me more specifically
what are the changes which you see taking place on Earth and
not in the heavens, on the basis of which you say the Earth is
mutable and the heavens are not.

SIMPLICIO. On the Earth I see a continuous process of
generation and corruption in grasses, plants, and animals; I see
wind, rain, storms, and tempests arising; in short, this Earth
which we see is in a constant state of metamorphosis. And I see
none of these changes in the celestial bodies, whose position

and configuration remain exactly as they have always been recorded, without any new phenomenon being generated or any former one having decayed.

SALVIATI. But if your belief rests on these visible phenomena —or rather, ones which you have seen—then you must consider China and America to be celestial bodies, since you have certainly not seen in them the changes which you see here in Italy; so in your view they must be unchangeable.

SIMPLICIO. I may not have physically seen these changes in those places, but there are reliable accounts of them; and in any case, *cum eadem sit ratio totius et partium*—since the same reasoning applies to the parts and to the whole—these countries are parts of the Earth the same as ours, and so they must be subject to change just as our country is.

SALVIATI. And why have you not seen and observed them with your own eyes, so that you have to rely on accounts provided by other people?

SIMPLICIO. Because these countries are not accessible to our eyes, and in any case they are so far away that even if we could see them our eyes would not be able to perceive changes of this kind.

SALVIATI. See how you have now yourself exposed the fallacy in your argument. If you say that the changes which we see before our eyes here on Earth would not be visible to you in America because it is so far away, how much less would you be able to see them in the Moon, which is hundreds of times further away? And if you believe in changes in Mexico on the basis of reports which have come from there, what reports have you received from the Moon to tell you that there are no changes there? So while you quite rightly argue for the existence of changes on Earth because we can see and recognize them, you can't argue that no changes take place in the heavens because you don't see them, since if there were any you wouldn't be able to see them because they are so far away, or because you have no reports of them, since there is no way for such reports to reach us.

SIMPLICIO. I can give you examples of changes on Earth such that, if anything comparable were to happen on the Moon, it could easily be seen from here on Earth. There is a very ancient memory that Abyla and Calpe* were once joined together at what is now the strait of Gibraltar, and that together with other smaller mountains they held back the western ocean; then, for whatever reason, the mountains separated, opening the way for the sea-water to rush in so that it formed the whole of the Mediterranean sea. Now if we consider the size of the Mediterranean, and the difference in appearance that there must be between the surface of the sea and the land when seen from a great distance, we can be sure that this change could easily have been visible if anyone had been looking at it from the Moon; so a similar change on the Moon's surface would be visible to us from Earth. But there is no record that any such thing has ever been seen; so there are no grounds for saying that the celestial bodies are mutable, etc.

The Mediterranean formed by the separation of Abyla and Calpe.

SALVIATI. Well, I don't presume to know whether changes of such magnitude have ever happened on the Moon or not, but nor can I say categorically that they could not have happened. And since in any case all we would see of such a change would be some variation in the lighter and darker areas on the Moon's surface, I don't know whether anyone on Earth has mapped the Moon accurately enough over a sufficiently long period of years for us to be certain that there never has been any such change on its surface. The best we can do in describing it is that some people have said it is like a human face, others that it resembles a lion's muzzle, and others that it is Cain carrying a bundle of sticks on his back. So it doesn't prove anything to say 'The heavens are unchangeable because there are no alterations in the Moon or the other celestial bodies which can be seen from the Earth'.

SAGREDO. This first argument of signor Simplicio's has raised another doubt in my mind which I would like to have resolved. So let me ask him this: was the Earth generable and corruptible before the flood which formed the Mediterranean, or did it only become so afterwards?

SIMPLICIO. There's no doubt that it was already generable and corruptible before that; but the flood was such a huge mutation that it could have been seen even from the Moon.

SAGREDO. But if the Earth was generable and corruptible even before this flood, why can the same not be true of the

Moon even if there has never been such a mutation there? Why should that be a necessary condition on the Moon if it had no effect at all on the Earth?

SALVIATI. A very telling example. But I wonder if signor Simplicio hasn't slightly changed the meaning of what Aristotle and the other Peripatetic philosophers wrote. They say that they consider the heavens to be immutable because no one has ever seen the generation or corruption of a star, which is a very small part of the heavens, possibly less than a city would be on Earth. And yet innumerable cities have been destroyed so that not a trace of them remains.

SAGREDO. I understood what signor Simplicio said differently; I thought that he was deliberately obscuring the meaning of this text so as to spare the Master and his disciples from an even greater absurdity. What kind of empty argument is it to say 'The heavenly part of the universe is unchangeable because stars are not subject to generation and corruption'? Has anyone ever seen a terrestrial globe decompose and a new one generated from it? And yet don't all philosophers accept that there are very few stars in the heavens which are smaller than the Earth, and that many of them are very many times bigger? So for a star to become corrupted in the heavens would be at least the equivalent of the whole terrestrial globe being destroyed. Well, if introducing the principle of generation and corruption into the universe requires bodies as vast as a star to be corrupted and regenerated, we might as well forget the principle altogether, because I can assure you that seeing the terrestrial globe or any other integral body in the universe dissolve so that no trace of it is left, after it has been seen to be there for centuries, is simply not going to happen.

It is as impossible for a star to decompose as it would be for the whole terrestrial globe.

SALVIATI. But to be generous to signor Simplicio and rescue him from error, if we can, we should acknowledge that there have been such new discoveries and observations in our own time that I'm quite sure Aristotle, if he were alive today, would change his opinion. This is clear from his own method of reasoning: since he says that he considers the heavens to be immutable, etc., because no one has seen any new bodies being generated or old ones dissolving there, he implies that if

Aristotle would change his opinion if he saw the discoveries that have been made in our time.

he ever did see such a thing he would believe the contrary. He would quite rightly have rated the evidence of his senses above that of human reason, since it was the lack of any sense evidence of change in the heavens that convinced him of their immutability.

SIMPLICIO. Aristotle based his conclusions primarily on arguing *a priori*, and he demonstrated the necessity of immutability in the heavens by means of clear natural principles; and then he confirmed this *a posteriori*, drawing on the evidence of the senses and the traditions of the ancients.

SALVIATI. He did indeed write his works in this way, but I don't think that was how he arrived at his conclusions. I'm quite sure that he tried first to assure himself as far as he could of the conclusion on the basis of observation and the evidence of the senses, and that he then tried to find the means to demonstrate it. This is the usual way of proceeding in the demonstrative sciences, because if the conclusion is correct, then the analytical method will readily lead to some proposition which has already been proved or to some generally accepted principle; whereas if the conclusion is false, the argument can proceed ad infinitum without ever reaching any recognized truth, if indeed it does not lead to some impossibility or manifest absurdity. There can be no doubt that Pythagoras was already confident in his own mind that the square on the hypotenuse of a right-angled triangle is equal to the sum of the squares on the other two sides, long before he arrived at the proof for which he sacrificed a hundred oxen; for in the demonstrative sciences, certainty about the conclusion is a great help in finding a proof. But whichever way Aristotle reasoned, whether the argument *a priori* came before the sense evidence *a posteriori* or vice versa, the important thing is that, as we have already said more than once, Aristotle rated the evidence of the senses more highly than argument of any kind.

Being certain of the conclusion helps to find proof by the analytical method.

Pythagoras sacrificed a hundred oxen* when he discovered his geometrical proof.

So to return to our subject, I say that the things which have been discovered in the heavens in our own time are and have been enough to satisfy all philosophers. For both in particular bodies and in the great expanse of the heavens as a whole we have seen, and continue to see, occurrences similar to what we describe as generation and corruption here on Earth. Eminent astronomers have observed numerous comets come into being and disintegrate in regions beyond the orbit of the Moon, as well as the two new stars* which appeared, undoubtedly far beyond all the planets, in the years 1572 and 1604. And now, thanks to the telescope, we can see dense dark spots forming and dissolving on the face of the Sun itself, very similar in appearance to the clouds around the Earth, many of them so vast that they are not just bigger than the Mediterranean sea but the whole of Africa and Asia as well. What do you think Aristotle would say and do, signor Simplicio, if he were to see these things?

New stars have appeared in the heavens. Spots which form and dissolve on the face of the Sun. Sunspots bigger than the whole of Africa and Asia.

SIMPLICIO. Aristotle was the master of the sciences, and what he would do or say I don't know, but I do know what his followers do and say—quite rightly, if they are not to be left without any kind of guide or leader in philosophy. As regards the comets, surely those modern astronomers who claimed that they were celestial bodies have been proved wrong by the *Anti-Tycho*?* Proved wrong, what's more, by their own arguments, by means of parallax* and a hundred and one different calculations, all pointing to the conclusion that Aristotle was right to say that the comets are all elemental bodies. And once that claim has been disproved, since it was the foundation of all their other innovations, I don't see that they've got a leg to stand on.

Astronomers refuted by the Anti-Tycho.

SALVIATI. Just a minute, signor Simplicio. What does this modern author say about the new stars of 1572 and 1604, and about the sunspots? Because as far as the comets are concerned, I'm quite ready to accept that they are generated either below or above the sphere of the Moon, and I've never set much store by Tycho's verbosity. And I don't have any problem in believing that they are made of elemental material, or that they can ascend freely in the heavens without encountering any obstacle in passing through the moving spheres, which I consider to be much thinner, finer, and more yielding than our atmosphere. As for calculations of parallax, I'm equally sceptical of both sets of opinions, first because I doubt whether comets are subject to such accidental factors, and secondly because the observations on which their computations are based are so inconsistent—especially since the

The Anti-Tycho *adapts astronomical observations to its own purposes.*

Anti-Tycho seems to me to adapt them at will, or to dismiss as erroneous any which don't fit into its scheme.

SIMPLICIO. The *Anti-Tycho* deals with the new stars in a few words: the author says there is no certainty that these recently discovered stars are celestial bodies at all, and that if his adversaries want to prove that change and generation are found in the heavens they will have to demonstrate changes in the stars which are indisputably celestial, having been recorded there over many years; and this they will never be able to do. As for the materials which some say form and disintegrate on the surface of the Sun, he makes no mention of them; which I take to mean that he regards them as a myth, or as illusions produced by the telescope, or at most as minor disturbances in the atmosphere—anything, in short, but celestial matter.

SALVIATI. And you yourself, signor Simplicio, what response have you come up with to these importunate sunspots which have come along to disrupt the heavens, and even more, the Peripatetic philosophy? You must surely have found some reply and solution, intrepid defender of Aristotle as you are, so please give us the benefit of your thoughts.

SIMPLICIO. I have heard a range of opinions on this par-
*Various opinions on the sunspots.**
ticular matter. Some say that they are stars which revolve in their own orbits around the Sun in the same way as Venus and Mercury; that they appear dark to us as they pass between us and the Sun; and that because they are so numerous it often happens that some of them cluster together and then separate. Others believe that they are impressions in the atmosphere; others that they are optical illusions produced by the lenses; and others offer different explanations. For myself, I am inclined to believe—in fact I am convinced—that they are a conglomeration of many different opaque bodies which come together almost at random, so that when we look at a sunspot we can often count ten or more of these minute, irregularly shaped bodies, which appear to us like snowflakes or flecks of wool or flying insects. Their relative positions change, and they join together and separate again, especially below the sphere of the Sun, around which they revolve. But it does not follow from this that they are subject to generation and

corruption; rather, they are sometimes hidden behind the body of the Sun, and at other times they are invisible because of their proximity to the dazzling light of the Sun, although they are not on the Sun's surface. For the Sun's eccentric sphere* has layers like an onion, one inside the other, each of which moves and each of which is marked by a number of small spots; and although their motions appear initially to be inconstant and irregular, nonetheless I understand that recent observations have shown that the same spots recur within a fixed time period. This seems to me to be the best explanation that has been put forward so far to account for these appearances, while at the same time maintaining the incorruptibility and ingencrability of the heavens. If this explanation should be found wanting, there will be other more elevated minds who will be able to find better ones.

SALVIATI. If we were discussing some point of law or some other humanistic discipline, where there is no final truth or falsehood, then we could indeed have confidence in a writer's subtlety of mind and greater skill and experience in speaking, and hope that someone who excelled in these abilities would also be able to make their own argument prevail. But conclusions in the natural sciences are necessarily true, and have nothing to do with human choice; so we must be careful not to commit ourselves to defending what is false, because then all the skill of Demosthenes* and Aristotle would be left standing by any run-of-the-mill thinker who happened to have latched on to the truth. So, signor Simplicio, you should stop thinking and hoping that there might be men so much more learned and erudite than us that they could defy nature and make what is false become true. You have concluded that, of all the opinions put forward so far to account for the nature of these sunspots, the one you have just expounded is true; so it follows that, if you are correct, all the others must be false. Now in order to show you that this opinion, too, is completely fanciful and mistaken, leaving aside all its other improbabilities, I will cite just two pieces of evidence against it.

The art of oratory is ineffective in the natural sciences.

The first is that many of these spots appear in the middle of the Sun's disc, and many similarly disintegrate and disappear when they are a long way from its circumference. This must

*Argument
proving that
the sunspots
must form and
disintegrate.*
necessarily show that they form and disintegrate, for if they were simply the product of local motion we would see them all enter and leave the Sun's disc at its circumference. The second observation concerns the apparent change in their shape and

*Conclusive
demonstration
that the
sunspots are on
the surface
of the Sun.*
the velocity of their motion, and conclusively shows, to anyone who is not completely ignorant of perspective, that the sunspots must be contiguous with the body of the Sun, and that they are in contact with its surface and move with it or above it; they cannot possibly revolve in an orbit separate from it. This is shown by their motion, since they appear to move very slowly as they approach the circumference of the solar disc and faster towards the middle; and it is shown by their shape, since at the circumference they appear much narrower

*The motion of
the sunspots
appears slower
towards
the Sun's
circumference.
The sunspots
are narrow in
shape when
they approach
the Sun's
circumference,
and why.*
than when they are in the middle. This is because in the middle they appear in all their glory as they really are, whereas near the circumference they appear foreshortened because the surface of the Sun's globe is moving away from us. What is more, it is clear to those who have been able to observe and calculate their motions systematically that both these changes, in shape and in motion, correspond exactly to what we would expect if they were contiguous with the Sun, and are completely incompatible with their moving in circles separated from the solar body, even by only a small distance. This has all been demonstrated at length by our friend in his *Letters on the Sunspots* to signor Mark Welser.*

The fact that they change shape in this way also shows that none of them are stars or any other kind of spherical body, for the sphere is the only shape which never appears foreshortened, or anything but perfectly round. So if any one of these spots was a spherical body, as we assume all stars to be,

*The sunspots
are not
spherical in
shape, but
stretched out
like fine flakes.*
then its shape would always appear equally round, whether it was in the middle of the Sun's disc or at the edge; whereas the fact that they are so foreshortened at the edge, and by contrast so expansive and wide at the middle, clearly shows that they are like flakes, with very little depth or thickness in relation to their length and breadth. Finally, signor Simplicio, you should not believe anyone who claims that recent observations have shown the sunspots reappearing unchanged after a fixed period of time; whoever told you that is deceiving you. Why

else would they have said nothing to you about the spots which form and disintegrate on the face of the Sun, well away from its circumference; or about their apparent foreshortening, which necessarily proves that they are contiguous with the Sun's surface? As for the spots reappearing unchanged, all this means is that some of them may occasionally last longer than it takes for the Sun to rotate once on its axis, which is less than a month, before they disintegrate. Our friend has made this clear in the *Letters* we have already mentioned.

SIMPLICIO. To tell the truth, I haven't observed them for long enough or systematically enough to have mastered the *quod est* of this matter; but I certainly shall do so, and try to see for myself whether I can harmonize what we learn from experience with what is demonstrated by Aristotle, because it's clear that two truths cannot contradict each other.

SALVIATI. As long as you aim to harmonize what you perceive with your senses with the soundest teachings of Aristotle, you will have no trouble at all. After all, doesn't Aristotle say that we cannot have complete certainty in matters concerning the heavens, because of their great distance from us?

SIMPLICIO. He does indeed.

SALVIATI. And doesn't he also say, with great emphasis and without reservation, that what we learn from the experience of our senses should take precedence over any reasoned argument, however well founded it may seem?

SIMPLICIO. Yes, he does.

SALVIATI. So of these two propositions, both of them Aristotle's, the second, saying that the senses should be given precedence over reason, is much more firmly established than the one which says that the heavens are immutable. Hence it is more faithful to Aristotle to say 'The heavens can change, because this is what the evidence of my senses shows', than to say 'The heavens are immutable, because this was what reason led Aristotle to conclude'. What's more, we are in a much better position to speak about the heavens than Aristotle was. He admitted that such knowledge was difficult for him because the heavens were so far removed from his senses, and he conceded that someone whose senses gave them a better representation of the heavens would have a more

We cannot speak with certainty about the heavens because of their great distance, according to Aristotle.

The senses prevail over reason, according to Aristotle.

It is more in conformity with Aristotle to say that the heavens can change than to say that they are immutable.

*Thanks to the
telescope, we
are better able
to speak about
the heavens
than Aristotle.*

secure basis for speculating about them. We, thanks to the
telescope, have made the heavens thirty or forty times closer to
us than they were to Aristotle, so that we can see very many
things there which he could not, including the sunspots, which
were completely invisible to him. So we have a more secure
basis for speculating about the Sun and the heavens than
Aristotle.

SAGREDO. I sympathize with signor Simplicio: I can see
that he is moved by the strength of these all too conclusive
arguments, and yet on the other hand he sees the universal
authority in which Aristotle is held, all the famous commentators who have laboured to expound his meaning, and the other
sciences, so necessary to the common good, which base such
a large part of their reputation on the standing of Aristotle,
and he is confused and alarmed.

*Simplicio's
declamation.*

It is as if I hear him say: 'Who are we to turn to for a resolution of our disputes if Aristotle is unseated from his throne?
What other author should we follow in schools, academies, and
universities? What other philosopher has written on every part
of natural philosophy, in such a systematic way that not a
single conclusion is omitted? Are we to lay waste the building
which gives shelter to so many travellers? Must we destroy that
refuge, the Prytaneum,* which enables so many scholars to
shelter in comfort and, without being exposed to the outside
air, to acquire knowledge of every aspect of nature simply by
turning a few pages? Are we going to demolish that rampart
within which we are safe from every enemy attack?' I feel
sorry for him, as I would for a landowner who has spent years
building a magnificent palace, sparing no expense and employ-
ing hundreds of craftsmen, only to find it starting to crumble
because its foundations are unsound. Rather than have the
grief of seeing the walls decorated with so many fine paintings
collapse, the pillars supporting the splendid loggias fall down,
the gilded balconies, the doorways, the cornices, and the
marble mouldings all in ruins, he would do all he could with
chains, props, buttresses, embankments, and supports to stave
off collapse.

SALVIATI. Oh, I don't think signor Simplicio need fear any
such collapse; I would undertake to insure him against loss for

far less expense than that. There's no danger of such a large number of wise and prudent philosophers being overcome by one or two individuals creating a bit of commotion. They won't even need to sharpen their pens against them; simply meeting them with silence will expose them to general scorn and derision. It's vain to think that anyone could introduce a new philosophy simply by criticizing this or that author; they would need first to remake men's minds and make them able to distinguish true from false, and that's something that only God can do.

The Peripatetic philosophy is immutable.

But how did we get on to this? I'll need the help of your memory to get me back onto the right track.

SIMPLICIO. I remember exactly where we were. We were discussing the response of the *Anti-Tycho* to the arguments against the immutability of the heavens, and you added the question of the sunspots, which the *Anti-Tycho* doesn't mention. I think you were going to consider his response to the example of the new stars.

SALVIATI. Yes, now I remember the rest. So to carry on where we left off, there are several points in the *Anti-Tycho*'s response which seem to me to deserve criticism. First of all, he says that the two new stars, which he has no choice but to place in the highest part of heaven, and which lasted a long time before they finally disappeared, don't undermine his belief in the immutability of the heavens, because it is not certain that they are celestial bodies at all, and in any case they are not changes in the stars which have been observed since antiquity. But in that case, why does he go to such trouble over the comets, to exclude them at all costs from the celestial regions? Surely he could just have said the same thing about them as he did about the new stars—that it is not certain that they are part of the heavens and they do not involve changes to any of the stars, and so they don't affect either the heavens or the teaching of Aristotle. Secondly, I can't follow his line of thought when he admits that any changes in the stars would destroy the prerogatives of the heavens, of being incorruptible, etc., because everyone agrees that the stars are clearly celestial bodies; and yet it doesn't seem to trouble him that the same changes might affect the rest of the expanse of the heavens, away from the stars themselves. So does he not consider the heavens to be part of the celestial regions? I always thought that the stars were called celestial bodies by virtue of their being in the heavens or made of celestial material, and that therefore the heavens were more celestial than the stars—in the same way as there can be nothing more terrestrial than the Earth, and nothing more fiery than fire itself. And as for his not mentioning the sunspots, which have been shown conclusively to form and disintegrate, to be close to the body of the Sun, and to rotate with it or around it, this seems to me to show that this author was writing to please others rather than for his own satisfaction. I say this because he clearly shows that he understands mathematics, and so he cannot possibly fail to be convinced by the demonstrations that the sunspots must be contiguous with the body of the Sun, and that they are examples of generation and corruption on a far larger scale than any that ever occur on Earth. And if they are found so frequently and on such a scale in the sphere of the Sun itself, which must surely be considered among the noblest parts of the heavens, then what grounds remain for denying that other examples can occur in the other spheres?

SAGREDO. I am astonished when I hear it said that this quality of impassibility, immutability, unchangeableness, etc., is a source of great nobility and perfection in the integral natural bodies of the universe, whereas being changeable, generable, mutable, etc., is considered a great imperfection. In fact I find such an idea repugnant, for I consider the Earth to be noble and marvellous because of the many different changes, mutations, generations, etc., which constantly occur in it. If it were not subject to any kind of change but was just a vast solitude of sand or a mass of jasper-hard rock, or if the water which covered the Earth at the time of the Flood had frozen to form an immense sphere of crystal where there was never any change or mutation of any kind, I would consider that to be a vile body which served no useful purpose but was simply superfluous, as if it had no existence in nature. It would be like the difference between a living body and a dead one; and the same goes for the Moon, for Jupiter, and all the other spheres in the universe.

To be subject to generation and change is a greater perfection in the bodies of the universe than the opposite. The Earth is noble because of the many mutations which occur in it.

The Earth would be vain and useless if it did not undergo change.

In fact, the more I reflect on the vacuous way the common people reason, the more shallow and insubstantial it appears. What could be more foolish than to say that gold, silver, and gems are precious, and that earth and mud are merely vile? Does it not occur to them that if earth was as scarce as jewels or precious metals, there would not be a prince who would not willingly give a sackful of diamonds and rubies and four cartloads of gold simply to have enough earth to plant a jasmine in a little pot, or to grow an orange tree from seed and see it sprout, grow, and produce such lovely foliage, such fragrant blossom, and such noble fruit? So it is scarcity and abundance which make the common people esteem or despise things. They will say that a diamond is beautiful because it is like pure water, and yet they would not exchange it for ten barrels of water. I think that those who exalt incorruptibility, immutability, etc., so highly say these things because they are so anxious to have a long life and are so terrified of death; they don't realize that if men were immortal they would never have been brought into the world. It would serve them right if they were to encounter a Gorgon's head which would turn them into jasper or diamond statues, so that they could become more perfect than they are.

Earth is more noble than gold or jewels.

Things are held in esteem or scorn because of their scarcity or abundance.

The common people celebrate incorruptibility because of their fear of death.

Those who despise corruptibility deserve to be turned into statues.

SALVIATI. It might even be a change for the better, because I think not talking at all is preferable to talking nonsense.

SIMPLICIO. There's no doubt that the Earth is far more perfect as it is, mutable, subject to change, etc., than it would be if it were a mass of rock, even if it were a single diamond, hard and impassible. But these qualities which confer nobility on the Earth would make the celestial bodies more imperfect, because they would be superfluous for them: the celestial bodies—the Sun, the Moon, and the other stars—are ordered solely for the benefit of the Earth, and to achieve this end they need only two things, motion and light.

The celestial bodies are ordered so as to serve the Earth, for which they need only motion and light.

SAGREDO. Are you saying, then, that nature has produced and ordained all these vast, perfect, and noble celestial bodies, impassible, immortal, and divine, for no other purpose than to serve the needs of the Earth, which is changeable, transient, and mortal? Purely for the benefit of what you call the dregs of the universe, the bilge where all the filth accumulates? What would be the point of making the celestial bodies immortal, etc., simply to serve one which is transient, etc.? The whole vast array of celestial bodies would be completely useless and superfluous if it were not for serving the needs of the Earth, given that they are all immutable and impassible and so can't possibly have any reciprocal effect on each other. *The celestial bodies have no reciprocal effect on each other.* If the Moon, for example, is impassible, what effect can the Sun or any other star have on it? Surely it would be less than that of someone who tried to liquefy a large mass of gold by looking at it or thinking about it. Indeed, it seems to me that if the celestial bodies work together to produce generation and change on Earth, they must themselves be subject to change. Otherwise, expecting the Moon or the Sun to produce generation on Earth would be like putting a marble statue alongside a bride and expecting the union to produce offspring.

SIMPLICIO. Corruptibility, change, mutability, etc., don't apply to the entire terrestrial globe, which as a whole is no less eternal than the Sun or the Moon, but its external parts are *Mutability applies not to the whole terrestrial globe, but to some parts of it.* subject to generation and corruption. In these parts, however, generation and corruption are perpetual, and as such they need the operation of eternal effects from the heavens; hence it is necessary for the celestial bodies to be eternal.

SAGREDO. So far so good. But if the eternity of the terrestrial globe as a whole is not undermined by the corruptibility of its external parts, and if indeed this susceptibility to generation, corruption, change, etc., is an adornment to it and enhances its perfection, could you not—indeed, should you not—apply the *Celestial bodies are subject to change in their external parts.* same argument to the celestial spheres? Why not allow that change, generation, etc., in their external parts are an adornment to them, without in any way diminishing their perfection or limiting their effects? Indeed, their effects would be increased, because as well as acting on the Earth they would also produce reciprocal effects on each other, and the Earth on them.

SIMPLICIO. That's not possible, because the generation, mutation, etc., that was produced, for example, on the Moon, would be vain and would serve no useful purpose, and nature does nothing in vain: *natura nihil frustra facit.**

SAGREDO. Why would they be vain and serve no useful purpose?

SIMPLICIO. Because it's plain to see that all the generations, mutations, etc., that are produced on Earth are designed, directly or indirectly, to serve the use, convenience, and benefit of man. Horses are born for the convenience of men; the Earth produces hay to feed the horses, and the clouds water it; grass, grain, fruits, animals, birds, fish are all produced for the convenience and nourishment of men. In short, if we carefully examine and resolve all these things we shall find that the end to which they are all directed is the need, the utility, the convenience, and the delight of men. Now, what use to the human race would be any generation that was produced on the Moon or another planet? Assuming, that is, that you don't claim that there are also men on the Moon to enjoy its fruits, an idea that is either fantastic or impious.

The generations and mutations on Earth are all for the benefit of man.

SAGREDO. I have no way of knowing whether herbs, plants, or animals similar to ours are generated on the Moon or any other planet, or whether they have rain, wind, or thunderstorms like those around the Earth. I don't believe so, and even less do I believe that the Moon is inhabited by men. But I don't see how it follows that just because they do not have generated species similar to ours, there cannot be change or alteration of any kind, or other things which mutate, generate, and disintegrate which are not just different from ours, but beyond our imagination and wholly unknowable to us. I'm quite sure that a man who was born and brought up in a vast forest, among wild animals and birds, who had no knowledge of the element of water, could never imagine that there was another natural world quite different from dry land, full of creatures which move swiftly without legs or wings. Such creatures move not only on its surface as animals do on land, but anywhere in its depths as well; and they can also remain motionless wherever they wish, something which is impossible for birds in the air. Men, too, live in this element and build palaces and cities there, and can travel so easily that they can transport their whole families, households, and entire cities to distant countries. So if such a man, however fertile his imagination, could never conceive of fish, the ocean, ships,

The Moon does not have generated species similar to ours; it is uninhabited by men.

The Moon may have generated species which are different from ours.

Anyone who had no knowledge of the element of water would be unable to imagine ships or fish.

fleets, and navies, how much less can we conceive what substances and effects there might be on the Moon, which is such a great distance away from us and which could, for all we know, be made of material quite different from the Earth? They might well be not just remote from our experience but completely beyond our imagination, having no resemblance to anything we know and hence being inconceivable to us. For the products of our imagination are necessarily either things we have already seen or a combination of things or parts of things we have seen before; this explains such things as sphinxes, mermaids, chimeras, centaurs, etc.

SALVIATI. I have often exercised my imagination about these matters, and I think I am now able to identify some of the things which are not and cannot be on the Moon, but not any of the things which I believe are or could be there, except in the most general terms. Of the latter, I can imagine only that they are an adornment to the Moon, that they act, move, and live and, in ways that may be completely different from ours,

The Moon may have substances which are different from ours.

contemplate and marvel at the greatness and beauty of the universe and of its Creator and Ruler, continually singing His glory. In short, I imagine them doing what is so frequently affirmed in Holy Scripture, namely that all creatures are perpetually occupied in praising God.

SAGREDO. If these are, in the most general terms, what might be found on the Moon, I would be glad to hear what you think are the things which are not and cannot be there. You must be able to identify these more specifically.

SALVIATI. Signor Sagredo, this will be the third time that our conversation has led us away from the main topic which we set ourselves, without our realizing it. If we keep on digressing we'll never get to the end of our discussion, so I think it would be a good idea if we set this to one side, together with the other matters which we've agreed to come back to on a separate occasion.

SAGREDO. Please, since we've reached the Moon, let's deal with the questions concerning it, so as not to have to make such a long journey again.

SALVIATI. Very well, if you wish. Starting, then, at the most general level, I consider that the sphere of the Moon is very different from that of the Earth, although we can also see some similarities; I'll describe the similarities first, and then the differences. We can be sure that the Moon is similar to the Earth in shape, as there is no doubt that it is spherical: this is proved by the fact that we see it as a perfectly circular disc, and by the way in which it receives light from the Sun. If its surface were flat, then it would all be bathed in light at the same moment, and similarly it would all simultaneously become dark; whereas in fact the parts facing the Sun are illuminated first, followed by the rest, so that it is only at full moon, when it is in opposition to the Sun, that the whole disc appears light. Conversely, the opposite would happen if its visible surface were concave; in that case the parts turned away from the Sun would be illuminated first.

First similarity between the Moon and the Earth: its shape, as is proved by the way it is illuminated by the Sun.

Secondly, the Moon is dark in itself and opaque, like the Earth; its opacity means that it can receive and reflect the light of the Sun, as it could not do otherwise. Third, I consider the matter of which it is made to be dense and solid, no less than the Earth; I find clear evidence of this in the fact that the greater part of its surface is uneven, with many peaks and hollows which can be seen with the aid of a telescope. There are very many such peaks, in every respect like the steepest and most rugged of our mountains, some of which are in ranges which stretch for hundreds of miles; others are in more compact groups, and there are also many isolated rocks which are very steep and precipitous. But the features which occur most often are high banks (this is the best word I can find to describe them) which surround and encircle plains of different sizes. These plains form various shapes, but the majority of them are circular; many of them have a mountain which stands out prominently in the middle, and a few are filled with darkish matter, similar to the spots which we can see with the naked eye. These are the largest of these flat areas; added to which there are a great number of smaller ones, nearly all of them circular.

The second similarity is that the Moon is dark, like the Earth. Third, the Moon is made of dense matter, and is mountainous, like the Earth.

Fourth, in the same way as the surface of our globe is divided into two large areas, the land and the sea, so too on the surface of the Moon we can see a marked difference between some large areas which shine brightly and others less. I think

Fourth, the Moon has distinct light and dark areas, like the sea and the land surface on Earth.

The surface of the sea, seen from a distance, would appear darker than that of the land.

Fifth, the Earth changes shape in the same way as the Moon, and with the same periodicity.

that anyone who could observe the Earth from the Moon or some other point a similar distance away would see the Sun illuminate it in the same way, with the seas appearing as darker areas and the land lighter. Fifth, we see the Moon from the Earth sometimes wholly illuminated, sometimes half, sometimes more or less; sometimes we see it as a crescent, and sometimes it is completely invisible to us, as happens when it is directly below the Sun's rays, so that the part of it facing the Earth is in darkness. The Sun's light on the face of the Earth would appear exactly the same from the Moon, and the Earth's shape would change in just the same way in the same time period. Sixth . . .

SAGREDO. Just a moment, signor Salviati. I can quite well see that the Earth would appear illuminated in the same changing shapes to someone observing it from the Moon as the Moon does to us. But I don't see how this could follow the same time period, since the Sun shines on the surface of the Moon over the course of a month in the same way as it does on the Earth in twenty-four hours.

SALVIATI. It's true that when the Sun shines on these two bodies and seeks out every part of their surface with its light, it illuminates the whole of the Earth in a natural day and the Moon in a month; but this is not the only factor affecting the different shapes in which the bright part of the Earth would appear if it were observed from the Moon. This depends also on the varying positions which the Moon has in relation to the Sun. If, for instance, the Moon's movements exactly followed the Sun's, and if it happened always to be directly interposed between the Sun and the Earth—what we mean when we say it is in conjunction with the Sun—then the hemisphere of the Earth facing the Sun would also be facing the Moon, and it would always appear fully illuminated. If, on the other hand, the Moon were always in opposition to the Sun, then an observer on the Moon would never see the Earth, because the Earth's dark side would always be facing the Moon and so would be invisible. But when the Moon is at its first or last quarter, then, of the hemisphere of the Earth visible from the Moon, the half which is facing the Sun would be shining and the half which is facing away from the Sun would be dark;

and so the illuminated part of the Earth would appear to the Moon as a semicircle.

SAGREDO. That's all quite clear to me now. I can see how none of the illuminated part of the Earth's surface is visible from the Moon when it is in opposition to the Sun, but then as it moves day by day towards the Sun it gradually begins to see a small part of the Earth, which appears as a fine crescent because the Earth is round. Then, as the Moon's motion brings it daily closer to the Sun, more and more of the illuminated hemisphere of the Earth is revealed, so that when the Moon reaches first quarter it sees exactly half of it, just as we see half of the Moon. As it continues towards its conjunction with the Sun, more and more of the Earth's bright surface appears, until finally at conjunction the whole hemisphere is shining brightly. And I can see how the experience of Earth-dwellers in watching the phases of the Moon would be replicated for anyone observing the Earth from the Moon, but in reverse: when we see the full Moon when it is in opposition to the Sun, for someone on the Moon the Earth would be in conjunction with the Sun and so would be wholly dark and invisible; and what for us is the conjunction of the Moon with the Sun, when the Moon appears absent and invisible, for them the Earth would be in opposition to the Sun and it would be, as it were, 'full Earth', i.e. fully illuminated. And finally, at any given time whatever proportion of the Moon's surface appears illuminated to us, a corresponding proportion of the Earth's surface would appear dark to the Moon, and however much of the Moon appears dark to us, the equivalent part of the Earth would appear light to the Moon. Only at the first and last quarter, when we see a semicircle of the Moon illuminated, would a Moon-dweller see the Earth in the same way. There's just one respect in which, as I see it, these reciprocal effects differ: assuming for the sake of argument that there was someone on the Moon in a position to observe the Earth, they would see the whole of the Earth's surface every day, because of the Moon's motion around the Earth every twenty-four or twenty-five hours; whereas we only ever see half of the Moon, because it does not turn on its axis, as it would have to do for us to be able to see it all.

SALVIATI. Unless the opposite is true—that it's because the Moon turns on its axis that we never see the other side, as would have to be the case if the Moon had an epicycle.* But aren't you overlooking another difference, which compensates for the one you've already noted?

SAGREDO. What difference is that? For the moment I can't think of any others.

SALVIATI. It's this: that while, as you've rightly pointed out, only half of the Moon is visible from the Earth, but the whole of the Earth is visible from the Moon, it's also the case that the Moon is visible from the whole of the Earth, but the Earth is only visible from half of the Moon. The inhabitants, so to speak, of the upper hemisphere of the Moon, which is invisible to us, cannot see the Earth: perhaps these are the Antichthons.*

The whole of the Earth can see only half of the Moon, but only half of the Moon can see the whole of the Earth.

But I have just remembered a particular detail which has recently been observed in the Moon by our friend the Academician, from which two consequences follow: one is that we actually see rather more than half of the Moon, and the other is that the Moon's motion has a precise relation to the centre of the Earth. His observation is as follows. If it is indeed the case that the Moon has a natural correspondence with and attraction to the Earth, and that a specific part of it always faces towards the Earth, then it follows that a straight line joining the centres of the two bodies must always pass through the same point on the surface of the Moon. This means that if anyone could look at the Moon from the centre of the Earth, the disc of the Moon's surface that they would see would always be the same, always bounded by the same circumference. But when someone standing on the surface of the Earth looks at the Moon, the ray from their eye* to the centre of the Moon would only pass through the same point on the Moon's surface as the line joining the centres of the two bodies if the Moon were directly overhead. If the Moon is either in the east or in the west, the observer's line of vision will strike the Moon's surface at a higher point than the line joining the two centres, and therefore part of the hemisphere above the circumference will appear, and a corresponding part below the circumference will be hidden—in relation, that is, to

More than half of the Moon's globe is visible from the Earth.

the hemisphere which would be visible from the true centre of the Earth. And since the upper part of the Moon's circumference when it rises is the lower part when it sets, there should be a quite noticeable variation in the appearance of these parts, which should be visible by the presence or absence of their spots or other distinctive features. The same should be true of the northern and southern extremities of the Moon's disc, depending on the Moon's location in relation to the ecliptic: when it is in the north, part of the area towards the northern circumference should be hidden, and part of the southern area should be revealed, and vice versa. Now the telescope enables us to confirm that these consequences do in fact follow. There are two distinctive spots on the Moon, one which is towards the north-west when the Moon is on the meridian, and another diametrically opposed to it; the first is visible even without a telescope, but not the second. The one to the north-west is an oval shape, and is separate from the other great spots; the one opposite it is smaller, and is also separate from the great spots, and is surrounded by a light-coloured area. Both of them clearly reveal the variations mentioned above; they appear opposite each other, at one time close to the edge of the Moon's disc and at another time further away. In the case of the north-western spot, its distance from the circumference more than doubles; the other is closer to the circumference so the variation is even greater, and is more than three times as much at one time than another. So it is clear that one side of the Moon is constantly facing the terrestrial globe, as if held there by magnetic force, and never diverges from it.

Two spots on the Moon which show that its motion is related to the centre of the Earth.

SAGREDO. Is there no end to the new observations and discoveries which can be made with this wonderful instrument?

SALVIATI. If they progress in the same way as discoveries made by means of other great inventions, we can hope that as time passes we shall be able to see things which now we cannot even imagine. But to come back to our earlier discussion, the sixth similarity between the Moon and the Earth is that, just as the Moon makes up for our lack of sunlight much of the time by reflecting the Sun's rays, making our nights relatively light, so the Earth repays the debt when the Moon needs it most, reflecting the Sun's rays to give it a very strong light, which I

Sixth, the Earth and the Moon illuminate each other reciprocally.

reckon must exceed the light which the Earth receives from the Moon by the same amount as the Earth's surface is larger than the Moon's.

SAGREDO. Don't say any more, signor Salviati; allow me the pleasure of showing you how what you have just said has enabled me to penetrate the reason for a phenomenon which I have thought about a hundred times without ever being able to fathom it. You're saying that the faint light which is sometimes visible in the Moon, especially at new Moon, is the light of the Sun reflected from the surface of the Earth and the sea; and the thinner the crescent Moon, the brighter this light shines, because that's when the illuminated area of the Earth's surface visible from the Moon is largest. This follows the principle you stated just now, that the illuminated part of the Earth which faces the Moon is as large as the dark area of the Moon facing the Earth; so when the crescent Moon is thinnest, and hence its dark area is largest, the illuminated area of the Earth seen from the Moon is largest, and its reflection of the Sun's rays is at its most powerful.

Light from the Earth is reflected in the Moon.

SALVIATI. That was exactly my meaning. What a pleasure it is to talk to people of sound judgement who learn quickly, especially when we are exploring and discussing truths in this way! I've lost count of the number of times I have encountered such obtuse minds that, however many times I repeated to them what you have just worked out for yourself, they have never been able to grasp it.

SIMPLICIO. I'm very surprised if you mean that you've never managed to make them understand it, because I find your explanation perfectly clear, so if they can't understand when you explain it to them I'm sure they will never understand it from anyone else. But if you mean that you haven't been able to persuade them to believe it, that doesn't surprise me at all; I must confess that I myself am one of those who understand your reasoning but am not persuaded by it. In fact, I have many difficulties with this and with some of the other six similarities you have listed; and I'll put these forward when you have finished your exposition of them all.

SALVIATI. I'll deal with the rest very briefly, because I'm always eager to discover new truths, and the objections voiced

by an intelligent person like yourself can be a great help. The
seventh similarity, then, is that the reciprocity between the
Earth and the Moon extends to offences as well as favours: so
as the Moon, when it's at its brightest, is often deprived of
light and eclipsed by the Earth coming between it and the Sun,
so it avenges itself by coming between the Earth and the Sun
and casting its shadow on the Earth. It's true that the revenge
is not equal to the offence, because the Moon is often totally
immersed in darkness by the Earth's shadow for quite a considerable time, whereas it never happens that the whole Earth
is overshadowed by the Moon, nor is it eclipsed for long
periods of time. Nonetheless, considering how much smaller
the Moon's body is than the Earth's, it can't be denied that the
Moon puts up a spirited defence. So much for the similarities
between the two. I could go on to discuss their differences; but
signor Simplicio is going to favour us with his doubts about
the similarities, so I think we should hear and consider these
before we go on.

*Seventh, the
Earth and the
Moon eclipse
each other
reciprocally.*

SAGREDO. Indeed, because I dare say he won't object to the
disparities and differences between the Earth and the Moon,
since he considers their substances to be utterly different.

SIMPLICIO. Of the resemblances you've listed in establishing your parallel between the Earth and the Moon, the only
ones I can accept without reservation are the first and two
others. I accept the first, namely its spherical shape; although
even here I don't entirely agree with you, because I consider
the Moon to be absolutely clear and smooth, like a mirror,
whereas we can see that the Earth is very broken up and
rugged. But the unevenness of its surface comes under one
of the other similarities you have mentioned, so I shall leave
what I have to say about this until we come to that point.
On your second resemblance, that the Moon is opaque and
dark in itself like the Earth, I accept only the first attribute, its
opacity. Solar eclipses are proof enough of this: if the Moon
were transparent, the air would not be as dark as it is when
the Sun is eclipsed, because some refracted light would still
pass through the Moon's transparent body, as it does through
very thick clouds. As for its being dark, I don't believe the
Moon is completely lacking in its own light, as the Earth is;

Secondary light considered to be the Moon's own light.

in fact I think that the faint light visible in the rest of the Moon's disc when a thin crescent is illuminated by the Sun, is the Moon's own natural light, not light reflected from the Earth. I consider the Earth to be incapable of reflecting the

The Earth is incapable of reflecting the Sun's rays.

Sun's rays because it is so uneven and dark. As for your third parallel, I agree with you on one count, and I dispute the other: I agree that the Moon is solid and hard, like the Earth— indeed much more so, because we learn from Aristotle that the

The substance of the heavens is impenetrable, according to Aristotle.

heaven itself is impenetrably hard, and the stars are the densest part of their heavens, so they must be truly solid and impenetrable.

SAGREDO. What wonderful material the heavens would be for building a palace, if you could get hold of any—so hard and so transparent!

SALVIATI. On the contrary, it would be quite useless: because it's so transparent as to be invisible, you couldn't walk through its rooms without risking colliding with the door-jambs and breaking your head.

SAGREDO. There wouldn't be any risk of that if, as some of

Heavenly matter is intangible.

the Peripatetics say, heavenly matter is intangible; because if you can't touch it, you certainly couldn't collide with it.

SALVIATI. That wouldn't help at all. It may be true that you can't touch heavenly matter because it lacks the quality of tangibility, but it can still touch elemental bodies. And it would hurt just as much, if not more, for it to collide with us as for us to collide with it. But enough of these palaces, or rather castles in the air; we mustn't stand in the way of signor Simplicio.

SIMPLICIO. The question which you've raised in passing is one of the most difficult in philosophy, and I've heard some wonderful thoughts about it from a distinguished professor in Padua;* but there isn't time for us to go into this now. So to return to our subject, I reply that I consider the Moon to be even more solid than the Earth; but I deduce this, not as you do from the ruggedness and unevenness of its surface, but rather from the opposite. I see it as being, like the hardest gemstones, capable of being polished to a lustre greater than

The Moon's surface is smoother than a mirror.

that of the most highly polished mirror, as its surface must be for it to reflect the Sun's rays so brightly. All the appearances which you describe—mountains, rocks, banks, valleys, etc.—

are illusions. I've heard the case made convincingly in public debates against those who put forward these novelties, that these appearances derive simply from varying degrees of opacity and clarity in the body of the Moon and on its surface. We can often observe this in crystals, amber, and other highly polished precious stones, in which some parts are opaque and others transparent, so that they appear to have various concave and protruding parts.

Peaks and hollows on the Moon are illusions produced by opacity and clarity.

As regards your fourth similarity, I accept that the surface of the terrestrial globe, seen from a distance, would present two different appearances, some parts being brighter and others darker, but I think that these would be the opposite way round from what you say: I believe that the surface of the water would shine more brightly because it is smooth and transparent, and the dry land would be darker because it is opaque and uneven, and so less capable of reflecting the Sun's rays. The fifth comparison I accept entirely, and I can quite well see that if the Earth were to shine like the Moon it would appear, to someone observing it from above, in the same shapes as those we see in the Moon. I can see, too, that its illumination would pass through its phases in the space of a month, even though the Sun seeks out every part of it in twenty-four hours; and finally I have no difficulty in agreeing that only half of the Moon can see the whole of the Earth, and that the whole of the Earth can see only half of the Moon. On the sixth point, I regard as totally false the idea that the Moon can receive light from the Earth, since the Earth is dark, opaque, and quite unsuited to reflecting the rays of the Sun, as the Moon reflects them to us. And as I have said, I consider that the light which can be seen in the rest of the Moon's surface when its crescent is brightly illuminated by the Sun is the Moon's own natural light, and it will take a great deal to persuade me otherwise. The seventh similarity, the reciprocal eclipses, is acceptable, although what you choose to call an eclipse of the Earth is normally more properly called an eclipse of the Sun. These, I think, are all the points I have to make in refuting your seven similarities, and if you have anything to say in reply I shall be glad to hear it.

SALVIATI. If I've understood your response correctly, we differ on some of the features which I said were common to the Moon and the Earth, as follows. You consider the Moon to be smooth and polished like a mirror, and therefore able to reflect the Sun's rays, whereas you don't believe that the Earth is capable of reflecting them because its surface is so rough. You agree that the Moon is hard and solid, but you deduce this from your belief that it is clear and smooth, not from its mountainous surface; and you attribute its mountainous appearance to its having different parts which are more and less opaque and clear. Finally, you consider that the Moon's secondary light is intrinsic to it and not reflected from the Earth, although you do agree that the sea could reflect light because of its smooth surface. I don't hold out much hope of disabusing you of your erroneous belief that the Moon reflects light like a mirror, since it's clear that you've learnt nothing from what our mutual friend wrote about this in the *Assayer* and his *Letters on the Sunspots**—assuming that you have read what he says about this carefully.

SIMPLICIO. I've skimmed it rather superficially, as much as the time I could spare from more weighty studies allowed; so if you think you can resolve my difficulties by reproducing some of his arguments, or by introducing new ones, I'll give them my best attention.

SALVIATI. I'll give you my thoughts as they occur to me now, so they may be a mixture of my own ideas and those I read in these two books. I remember being entirely convinced by them, although when I first read them I found them highly paradoxical. So, signor Simplicio, we're trying to establish whether, in order to produce reflected light such as we receive from the Moon, the reflecting surface needs to be smooth and polished like a mirror, or whether a surface which is uneven and unpolished would be more effective. Suppose, then, we had light reflected from two different surfaces, one light brighter than the other, which surface do you think would shine more brightly and which would be darker?

SIMPLICIO. I have no doubt that the one which reflected the light more vividly would appear brighter, and the other would be darker.

SALVIATI. Be so good as to take that mirror that's attached to the wall, and let's go out into the courtyard; you come too, signor Sagredo. Now let's fix the mirror to the wall here where it's in the Sun, and we shall retire over here into the shade. So there we have two surfaces both exposed to the Sun, the wall and the mirror. Now tell me: which of the two shines more brightly, the wall or the mirror? What, no answer?

Lengthy proof that the Moon has a rough surface.

SAGREDO. I'll let signor Simplicio reply, since he's the one who has the difficulty: for myself, just this small experiment is enough to convince me that the Moon's surface must be very unpolished.

SALVIATI. Signor Simplicio, tell me whether, if you had to paint a picture of that wall with the mirror on it, where would you use the darker colours: to paint the wall or the mirror?

SIMPLICIO. The paint for the mirror would be much darker.

SALVIATI. Then if it's true that the surface which reflects the light more strongly is the one which appears brighter, the wall must reflect the Sun's rays more strongly than the mirror.

SIMPLICIO. Really, is that the best experiment you can come up with? You've made us stand away from the reflection of the mirror, but if you come a little this way—no, just come over here.

SAGREDO. Are you looking for the place where the mirror casts its reflection?

SIMPLICIO. That's right.

SAGREDO. It's there on the opposite wall, exactly the same size as the mirror, and almost as bright as if the Sun were shining directly on it.

SIMPLICIO. Come here then, and look at the surface of the mirror over there, and then tell me whether it's darker than the wall.

SAGREDO. You can look at it if you want to; I don't want to be blinded, and I know very well without needing to look that it's as bright and clear as the Sun, or only slightly less so.

SIMPLICIO. How can you say, then, that a mirror reflects light less strongly than a wall? Looking at this wall opposite, onto which light is reflected both from the sunlit wall and from the mirror, the reflection of the mirror is much the brighter of the two; and when I look at the mirror itself it, too, is much brighter than the wall.

SALVIATI. You've cleverly anticipated what I was going to say next: I was going to use precisely this observation to point out what follows. You see the difference, then, between the reflections produced by the two different surfaces, the wall and the mirror, when they are both identically exposed to the Sun's rays. You can see that the reflection from the wall is evenly diffused across the whole of the surface opposite, whereas the mirror is reflected in just one place which is no bigger than the mirror itself. You can see, too, how the surface of the wall has the same brightness from wherever you look at it, and is much brighter than the mirror except from that one small place where the mirror's rays fall; and from there the mirror appears much brighter than the wall. Now I think it's quite easy to work out from this tangible evidence of our senses whether the light of the Moon comes to us as from a mirror or as from a wall, in other words from a smooth or a rough surface.

SAGREDO. I don't think I could better grasp the roughness of the Moon's surface if I was there and could reach out and touch it than I have from listening to your explanation. Whatever the Moon's position in relation to the Sun and to us, the part of its surface illuminated by the Sun always appears equally bright. This is exactly the same as the wall, which appears equally bright from wherever you look at it, and is quite different from the mirror, which only appears bright from one position and is dark from everywhere else. And also, the light reflected off the wall is weak and not dazzling at all, compared to the reflection of the mirror which is so strong that it hurts the eyes almost as much as looking at the Sun itself. In the same way, we can easily look at the face of the Moon, but if it were a mirror its brightness would be absolutely intolerable and it would be like looking at another Sun, especially as its closeness to us would make it appear as big as the Sun itself.

SALVIATI. Please, signor Sagredo, don't make my demonstration carry more weight than it can bear. Let me put a point to you which may not be very easy to resolve. You make it a point of great difference between the Moon and the mirror that the Moon sheds its light equally in all directions, like the wall, whereas the mirror is reflected only in a single specific place; and hence you conclude that the Moon is like the wall and not like the mirror. But the reason the mirror is reflected in only one place is that its surface is flat; and since the angle of reflection of the rays must be equal to their angle of incidence, the rays reflected off a flat surface must necessarily all go to the same place. But the Moon's surface is spherical, not flat; and rays striking a spherical surface are reflected at the same angle as their angle of incidence in all directions, because the surface of a sphere has an infinite number of inclinations. Hence the Moon can shed its reflected light in all directions, and the light doesn't all have to be reflected in the same place, as it does from the mirror which is flat.

Plane mirrors reflect in only one direction, but spherical mirrors in every direction.

SIMPLICIO. This is precisely one of the objections I was going to raise.

SAGREDO. If this was one of your objections, that means you must have others. Let's hear them, because I think that this first point may count against you rather than in your favour.

SIMPLICIO. You stated it as a self-evident fact that the light reflected off that wall was as bright as the light that comes to us from the Moon, but I consider it incomparably less. For in this matter of illumination it is important to distinguish the sphere of activity; and no one doubts that the celestial bodies have a greater sphere of activity than our elemental, transient, and mortal bodies. And what is that wall, in the last resort, but a small piece of earth, dark and quite incapable of shedding light?

The celestial bodies have a greater sphere of activity than elemental bodies.

SAGREDO. Here too I think you are much mistaken. But let me come to signor Salviati's first point. For an object to appear illuminated to us, it's not enough for rays from the source of light to fall on it; they must also be reflected to our eyes. This is clear from the example of the mirror, which is clearly exposed to the sun's rays even though it didn't appear illuminated to us unless we were looking from the precise spot where the reflected rays fell. Now let's consider what would happen if the mirror were spherical: we would undoubtedly find that only a tiny part of the light reflected from its illuminated surface would fall on our eyes at any particular point, because the reflection at that point would come from only a tiny part of the spherical surface. So only a very small part of the sphere's surface would appear bright to us, and all the rest would remain dark. If, then, the Moon were smooth like a mirror, our eyes would only perceive the light from a very small part of its surface, even if its whole hemisphere was exposed to the Sun; the rest of it would not appear illuminated and so would be invisible. The final result would be to make the Moon itself invisible, since the part reflecting the light would be so small and so far away that it would be lost to our sight; and if it was invisible to our eyes, we would also receive no illumination from it, because it's impossible for a luminous body to shed light in the darkness without being visible.

If the Moon were like a spherical mirror, it would be invisible.

SALVIATI. Stop there, signor Sagredo, because I can see from signor Simplicio's expression that either he hasn't grasped what you've so evidently and correctly said, or he isn't convinced by it; and I've just had an idea for another experiment which will remove all his doubts. I noticed a big spherical mirror in one of the rooms upstairs; let's have it brought down here, and while we're waiting for it to be brought, perhaps signor Simplicio would like to look again at the brightness of the light reflected from the plane mirror on the wall here under the loggia.

SIMPLICIO. I can see that it's scarcely less bright than if it were in direct sunlight.

SALVIATI. It is indeed. Now tell me, if we were to take away the small plane mirror and put the big spherical one in its place, what effect do you think its reflection would have on the same wall?

SIMPLICIO. I think the light on it would be much greater and more widely spread.

SALVIATI. What would you say if there proved to be no light at all, or so little you hardly noticed it?

SIMPLICIO. I'll wait until I see the effect, and then I'll consider how to reply.

SALVIATI. Here's the mirror; I'd like it to be placed alongside the other one. But first let's look closely at the reflection from the plane mirror, and note its brightness: look how bright it is here where it strikes the wall, and how clearly you can make out every detail of the wall.

SIMPLICIO. I've seen and noted it. Now have the other mirror placed alongside the first one.

SALVIATI. It's there already. It was put there as soon as you began looking at the detail of the wall, and you didn't notice it; that's the measure of how much it increased the light on the rest of the wall. Now let's take away the plane mirror. You see how all the reflection has gone, even though the big convex mirror is still there. You can take it away and put it back as many times as you like, but you won't see any change in the light on the wall. So there you have the evidence of your senses to show that sunlight reflected in a convex spherical mirror does not cast any appreciable light on its surroundings. Now how do you respond to this experiment?

SIMPLICIO. I fear you may have introduced some sleight of hand. But when I look at that mirror I see it shedding a brilliant light, so bright it almost blinds me; and more important, I see it from wherever I look, and the light comes from a different place on the mirror's surface depending on where I am standing to look at it. This must mean that the light is reflected brightly in every direction, and so it must shine as strongly on the whole of that wall as it does to my eyes.

SALVIATI. That shows how cautious you must be in assenting to a conclusion which is arrived at solely by means of argument. There's no doubt that what you say seems very plausible; and yet you can see that the evidence of your senses tells you otherwise.

SIMPLICIO. So how do we make any progress in resolving this question?

SALVIATI. I'll tell you what I think, though I don't know how far you will be convinced by it. To start with, that brilliance which you see when you look at the mirror, and which seems to you to occupy quite a large part of it, is not actually as large as it looks; in fact it comes only from a very small part. But its brightness has an effect on your eyes, because it is reflected in the moisture at the edge of your eyelids covering the surface of the pupil, which looks like a kind of secondary irradiation, rather like the illusion of a halo around a candle seen from a distance. Or you could compare it to the secondary brightness which you see around a star. If you compare the size, for example, of the Dog Star when you see it through a telescope in the daytime, when there is no irradiation, with its appearance at night seen with the naked eye, you will be left in no doubt that the irradiation makes it appear more than a thousand times larger than the bare body of the star itself. The image of the Sun that you see in that mirror is enlarged in the same way—in fact more, because the sun's light is so much stronger than that of a star, as is clear from the fact that looking at a star is much less dazzling to the eyes than looking at this reflection in the mirror. Therefore, the reflected light which has to be shared over the whole surface of this wall comes from only a small part of that mirror, whereas the reflection from the whole of the plane mirror was restricted to only a small part of the wall. So it's not surprising that the first reflection appeared to shine very brightly, while the second is barely perceptible.

Irradiation around the body of a star makes it appear many times larger than it really is.

SIMPLICIO. I'm more confused than ever, and now I have another difficulty: how can it be that that wall, which is made of such dark material and has such an unpolished surface, can reflect light more strongly and vividly than a smooth, polished mirror?

SALVIATI. Not more vividly, but diffused more widely. As for being vivid, you can see that the reflection from that plane mirror where it strikes the wall there under the loggia is very bright, but the rest of the wall, which receives reflected light from the wall with the mirror on it, is far less brightly lit than the small area where the mirror is reflected. If you want to understand the principle behind this, think of the rough surface of that wall as having innumerable tiny surfaces all at different angles, of which many will necessarily reflect rays in one direction and many in another. So there will not be anywhere which does not receive many rays reflected from many tiny surfaces scattered over the whole of the rough surface which is exposed to the light; and hence any part of any surface facing the one on which the primary rays fall will receive some reflected rays, and consequently will be illuminated. It follows, too, that the object on which the illuminating rays fall will

Light reflected from objects with a rough surface is diffused more widely than from those with a smooth surface, and why.

appear wholly illuminated and bright, from whatever point one looks at it; and that is why the Moon, since its surface is rough not smooth, reflects the Sun's light in all directions, and appears equally bright to anyone who looks at it. But since it is a sphere, if its surface was smooth like a mirror it would be invisible, because the very small part of its surface which could reflect the Sun's image to any one observer would be invisible from such a great distance, as we have said.

If the Moon were smooth and polished, it would be invisible.

SIMPLICIO. I understand your explanation perfectly; but I think I can answer it very easily, and maintain that the Moon is round and polished and that it reflects the Sun's light to us like a mirror. Nor does this mean that we should be able to see the reflected image of the Sun in the middle of the Moon; for 'we should not expect to see the small image of the Sun in the form of the Sun itself at such a great distance; rather we should understand the illumination of the whole lunar body as being the light produced by the Sun. We can observe a similar phenomenon in a well-burnished gilded plate, which when it is struck by rays from a luminous body, appears when seen from a distance to be resplendent from its whole surface; only when it is seen close at hand is it possible to discern in the middle the small image of the luminous body.'*

SALVIATI. I must confess my obtuseness, and admit that I don't understand anything of what you have said apart from the bit about the gilded plate. In fact, if you'll allow me to speak freely, I have an idea that you don't understand it either, but have simply learnt by heart what someone has written in order to contradict and appear more intelligent than his opponent. But this only impresses those who, wanting to appear intelligent themselves, applaud what they don't understand, and whose opinion is highest of those they understand least. That's always assuming that the writer himself is not one of those many who write things which they don't understand, so that no one else can understand what they write either.

There are some who write what they do not understand, and therefore what they write cannot be understood.

So leaving the rest aside and responding to what you say about the gilded plate, it can indeed appear as if its whole surface shines when a strong light strikes it, provided it is flat and not very big, and provided you are looking at it directly from the point to which the light is reflected. And it will appear more fiery than if it were made, for example, of silver, because of its colour and because the greater density of the metal makes it more receptive to being highly burnished. If its surface were not only highly polished but also not entirely flat, but had various inclined surfaces, then its splendour would be visible from even more places, as many as caught the reflections from its various surfaces; this is why diamonds are worked with many facets, so that their delightful brilliance can be seen from many different points. If the plate were very large, however, it would not appear from a distance as if its whole surface were shining, even if it were completely flat. To show what I mean more clearly, imagine a very large flat gilded plate which is exposed to sunlight: when looked at from a distance the image of the Sun will appear to occupy only part of the plate—the part from which the incident rays of the Sun are reflected—but the brilliance of the light will make the image appear to be surrounded with many rays, so that it will look as if it occupies a larger part of the plate than it really does. To confirm this, note the particular point on the plate where the reflection comes from, and also estimate how large the bright area appears to be; then cover up most of this area, leaving just a small space around the centre exposed. You will see that, when looked at from a distance, the bright area will not appear diminished in the least; in fact it will spread out over the cloth or whatever you used to cover it up. So if anyone who sees a small gilded plate appear from a distance as if its whole surface shone, and then imagines that the same thing would happen with a plate as big as the Moon, they are as much mistaken as if they imagined the Moon to be no bigger than the bottom of a barrel. If the surface of the plate were spherical, then the strong reflection would be seen in only one small part of it, but its vividness would make it appear to be surrounded by many very bright rays. The rest of the sphere would appear as coloured, as long as it was not highly polished; if it was highly burnished, it would appear dark.

We can see an everyday example of this with silver vessels which, if they are simply boiled in bleach, have a white frosting on them, like snow, and they don't give any reflection at all; but if any part of them is polished, that part immediately appears

Diamonds are worked into many facets; the reason for this.

Burnished silver appears darker than silver which has not been burnished; the reason for this.

dark, and reflects images like a mirror. The reason for its appearing dark is simply that polishing it has smoothed off a very thin patina which made the surface of the silver uneven, and therefore reflect light in all directions, so that it appeared illuminated from wherever you looked at it. Burnishing it then smoothed away these tiny irregularities so that the incident rays were all reflected to the same point. This meant that when seen from that point the burnished part appeared much brighter and clearer than the rest, which was just bleached, but it appeared dark when seen from anywhere else. It's well *Burnished steel* known that because the appearance of burnished surfaces *appears very* varies so much when seen from different viewpoints, if a *from some* painter wants to represent something like a piece of burnished *angles, and very bright* armour in a picture, he has to use black and white alongside *from others.* each other for places where the armour is in fact equally lit.

SAGREDO. So supposing these philosophers were willing to concede that the surface of the Moon, Venus, and the other planets is not as polished and smooth as a mirror, but just marginally less—like a silver plate which has been bleached but not burnished—would that be enough to make them visible and able to reflect the Sun's rays to us?

SALVIATI. It would be partly enough, but the Moon's light would be less bright than it is thanks to its mountainous surface, with its high peaks and deep hollows. But these philosophers will never concede that the Moon is less polished than a mirror; in fact they insist that it is far more so, if such a thing can be imagined. They believe that perfect bodies must have a perfect shape, and therefore the celestial bodies must be absolutely perfect spheres; and in any case, if they conceded even the tiniest irregularity I would have no scruples about claiming much greater ones, because perfection is indivisible and so if it falls short by a hair's breadth it might as well fall short by the height of a mountain.

SAGREDO. This raises two questions in my mind: first, how it is that a more irregular surface produces a more powerful reflection of light; and second, why these Peripatetics are so insistent on this perfect shape.

SALVIATI. I'll reply to the first question, and I'll leave signor Simplicio to respond to the second. You need to know, then, that the same surface can be illuminated to a greater or lesser extent by the same rays, depending on whether the rays strike it more or less obliquely; the greatest illumination is produced when they fall vertically. Here's a demonstration of what I mean. I'll fold this piece of paper so that one part of it is at an angle to the rest. If I hold it up to the light reflected from that wall over there, you can see how this surface of the paper, which the rays strike obliquely, is less bright than the other where they strike it vertically; and see how the illumination diminishes as I turn the paper to receive the light more and more obliquely.

A rough surface produces a greater reflection of light than one which is less rough. Rays which strike vertically give more illumination than those which strike obliquely; the reason for this.

SAGREDO. I can see the effect, but I don't understand the cause.

SALVIATI. I'm sure you would if you thought about it for a moment; but to save time, here is a drawing to demonstrate it.

SAGREDO. Yes, I see now just by looking at the drawing— but carry on.

SIMPLICIO. Please explain the rest to me; I'm not so quick to understand.

The reason why oblique rays illuminate less.

SALVIATI. Let these parallel lines starting from points A, B be rays of light which strike the line CD at right angles. Now

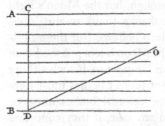

incline the line CD so that it is in a new position, DO: don't you see how many of the rays which struck the line CD pass over DO without touching it? If DO is illuminated by fewer rays, it follows that the light it receives must be weaker. Now consider the Moon: as it is a sphere, if its surface were as smooth as this sheet of paper then the parts near the edge of the hemisphere illuminated by the Sun would receive very much less light than those in the centre, since in the centre the Sun's rays would strike the surface at right angles and at the edges at a very oblique angle. This would mean that at full moon, when almost the full hemisphere is illuminated, the central parts ought to appear much brighter to us than those at the circumference; and yet this doesn't happen. But if you now picture the Moon's surface with very high mountains, you can see how their slopes and ridges, standing out from the perfectly spherical surface, are exposed to the Sun and in a position to receive its rays at a much less oblique angle, and therefore to appear as brightly illuminated as the rest.

SAGREDO. I see all that. But if there are such mountains near the circumference, it's certainly true that the Sun's rays will fall more directly on them than on the inclination of a polished surface; but it's also true that the valleys between these mountains would all be dark, because of the very long shadows the mountains would cast at such a time. But the parts near the centre, even if they were also full of mountains and valleys, would not have shadows because the Sun would be much higher in the sky. So the parts in the centre should shine much more brightly than those at the edge which would have as much shadow as light; and yet no such difference is apparent.

SIMPLICIO. I was just thinking of the same difficulty.

SALVIATI. Isn't it remarkable how much more readily signor Simplicio grasps the difficulties which support Aristotle's view than he does their solutions? But I think he sometimes deliberately keeps quiet about them; and in the present case, since he was able to see the objection by himself—which is quite an ingenious one—I can't believe he hasn't also spotted the reply. So let me try to winkle it out of him, so to speak. Tell me then, signor Simplicio: do you think it is possible for there to be shadow in a place exposed to the Sun's rays?

SIMPLICIO. I think not, in fact I'm sure of it: the Sun is the greatest luminary which scatters the darkness with its rays, so anywhere touched by the Sun's rays cannot possibly remain dark. Besides, we have the definition that *tenebrae sunt privatio luminis*: darkness is the absence of light.*

SALVIATI. Therefore, when the Sun beholds the Earth, the Moon or any other opaque body, it never sees any of their parts that are in shadow, since it has no eyes to see apart from its light-giving rays; and hence any observer who was on the Sun would never see anything that was in shadow, because his visual rays* would always accompany the Sun's rays of light.

SIMPLICIO. That's very true, beyond contradiction.

SALVIATI. But when the Moon is in opposition to the Sun, what difference is there between the line of your visual rays and that of the Sun's rays?

SIMPLICIO. Now I understand; you mean that since the rays of our vision always follow the same lines as the Sun's rays, we can't see any of the valleys of the Moon that are in shadow. Please don't think that I am pretending or dissembling with you; I give you my word as a gentleman that this reply had not occurred to me, and I doubt whether I would have found it without your help, or at least without thinking about it at length.

SAGREDO. The solution to this latest difficulty which you have arrived at between you has convinced me as well. But at the same time, this notion of visual rays accompanying the Sun's rays has raised another doubt in my mind on the other side of the argument. I don't know if I'll be able to explain it, because it has only just occurred to me and I still haven't clarified my thoughts about it, but let's see whether we can resolve it between us. We've established that the areas near the circumference of a hemisphere which is smooth but not burnished, when the Sun shines on it, are struck by the Sun's rays obliquely and therefore receive fewer of them than the areas in the middle, which the rays strike directly. So it could be that a band which is, say, twenty degrees wide near the edge of the hemisphere receives no more rays than another band near the centre which is only four degrees wide; and therefore the band at the edge will appear much darker than the one at the centre, to anyone looking straight at them—seeing them in all their glory, as it were. But if the observer's eye were so placed that the dark, twenty-degree band appeared no wider than the four-degree band at the centre, then it's quite possible that the two might appear to shine equally brightly: they would both appear as a band four degrees wide, and an equal number of rays would reach the eye from them both—both the band in the middle which really was four degrees wide, and the other which was twenty degrees wide but appeared to be only four degrees because it was being looked at obliquely. And we do indeed observe them from just such a position, because we are between the illuminated hemisphere and the source from which the light is coming, and therefore our vision and the rays of light follow the same line. Hence, it seems not impossible that the Moon's surface is really quite even, and that it nonetheless appears at full moon to shine as brightly at the edges as it does in the middle.

SALVIATI. Your doubt is an ingenious one, and it deserves to be taken seriously. And as it has only just occurred to you, I too will answer with the thoughts that come to mind immediately; maybe when I have thought more about it I will be able to give a better response. First, though, before I introduce any new arguments, it would be a good idea to see experimentally whether your theoretical objection is borne out in practice. So let's take this piece of paper again, and fold it so that a small part of it is at an angle to the rest; and then let's hold it in the light so that the rays fall directly on the smaller part and obliquely on the rest. You see, this clearly shows that the smaller part is noticeably brighter. Now, to see whether your objection is conclusive, we need to look at it from a lower viewpoint, so that we look obliquely at the larger, less bright part until it appears no wider than the other, more brightly illuminated part, so that we see the same arc for both. If you're right, then the light from the wider part will increase so that it appears to be as bright as the other. I'm looking at it now, so obliquely that the wider part appears narrower than the other, but it still remains as dark as ever. Now you look and see if you find the same thing.

SAGREDO. I've looked, and however much I lower my eyes I don't see that surface as any brighter or more illuminated; if anything it seems to get darker.

SALVIATI. We've established, then, that the objection is invalid. As for the explanation, I think that because this paper is not completely smooth, only a few rays are reflected back in the direction of the incident rays compared with the many which are reflected in opposite directions; and of these few, more are lost the more closely our visual rays are aligned with the incident light rays. And since the brightness of the object comes not from the incident rays but from those which are reflected to our eyes, it follows that in lowering our eyes we lose more rays than we gain, and this is borne out by your observation that the paper appeared to become darker.

SAGREDO. I'm convinced, both by the experiment and by your argument. So it remains now for signor Simplicio to reply to my other question, and to tell me what moves the Peripatetics to seek such exactness in the rotundity of the celestial bodies.

SIMPLICIO. The fact that the celestial bodies are ingenerable, incorruptible, unchangeable, impassible, immortal, etc., means that their perfection is absolute; and a necessary consequence of this perfection is that they should be perfect in every respect, and hence that their shape should also be perfect, that is, spherical, and absolutely and perfectly spherical, not rough or irregular.

Why the Peripatetics assume a perfectly spherical shape for the celestial bodies.

SALVIATI. And from what do you derive this quality of incorruptibility?

SIMPLICIO. Directly, from their lack of contraries; indirectly, from their constant circular motion.

SALVIATI. So, it appears from what you say that if the essential quality of the celestial bodies is that they are incorruptible, immutable, etc., then rotundity is not a cause or necessary condition; if it were, and if this were the cause of immutability, then we could confer incorruptibility on a piece of wood, or wax, or any other elemental material, simply by giving it a spherical shape.

SIMPLICIO. But isn't it clear that a ball made of wood will last longer and be better preserved than a spire or some other angular shape made from a piece of the same wood?

Shape is a cause not of incorruptibility, but of greater durability.

SALVIATI. That's very true, but this doesn't mean that it will no longer be corruptible and become incorruptible: rather, it will still be corruptible, but more durable. The point to note is that there can be degrees of corruptibility, so we can say 'This is less corruptible than that'—for instance, that jasper is less corruptible than sandstone. But incorruptibility doesn't admit of different degrees; it's not possible to say 'This is more incorruptible than that', if both are incorruptible and eternal. Differences of shape, therefore, can only affect materials which are capable of being more or less durable; but materials which are eternal must all be equally eternal, and so shape no longer has any effect on them. So, since celestial material is not incorruptible on account of its shape but for

The corruptible can vary in degree, but the incorruptible cannot.

Perfection of shape affects corruptible bodies, but not those which are eternal.

some other reason, there is no need to be so anxious about its
being perfectly spherical; if it is incorruptible it will remain so,
whatever shape it may have.

SAGREDO. I have been thinking about this a little further. If
it were granted that a spherical shape had the capacity to confer incorruptibility, then all bodies, of whatever shape, would
be eternal and incorruptible. For if a spherical body was incorruptible, then corruptibility would be confined to those parts
of the body which make it less than perfectly round. A dice, for
instance, contains within it a perfectly round ball, which as
such is incorruptible, and so the corruptible parts must be the
corners which cover and conceal its rotundity; so at the most it
would be these corners and, so to speak, excrescences which
would be subject to corruption. But if we take the argument a
stage further, those parts which form the corners have within
them other smaller balls of the same material, and so they
too, being round, are incorruptible; similarly in the residue
which remains around each of these eight smaller spheres
it is possible to imagine yet others; so that finally the entire
dice resolves itself into innumerable balls, and we have to
admit that it is incorruptible. And you could pursue the same
argument to a similar conclusion with any other shape.

SALVIATI. The argument works very well. So if, for
example, a crystal sphere is incorruptible, i.e. able to resist any
alteration either internally or externally, because of its shape,
is it not clear that adding more crystal to it in order to make
it, say, into a cube, would change it both internally and
externally? It would then be less able to resist the new material
surrounding it, even though it was the same, than it was before
when it was surrounded by a different material. And yet
Aristotle says that corruption is produced by contraries; and
what could you surround that ball of crystal with that would
be less contrary to it than crystal itself?

But we're losing track of the time, and we shall be very
late concluding our discussions if we pursue every detail at
such length. What's more, my memory is so confused by all
these different topics that I have difficulty in recalling all the
propositions which signor Simplicio set out so logically for us
to consider.

If spherical shape conferred eternity, all bodies would be eternal.

SIMPLICIO. I remember where we were. On this question of the mountainous surface of the Moon, you haven't yet dealt with the explanation which I put forward which I think saves this appearance very well: that it is an illusion deriving from the fact that different parts of the Moon are more and less opaque and clear.

SAGREDO. When signor Simplicio was speaking earlier about the apparent unevenness in the Moon's surface and, following the opinion of a Peripatetic philosopher friend of his, attributed it to the varying degrees of opacity and clarity in its different parts, he compared this to the similar illusions which we see in crystals and various kinds of precious stones. This reminded me of another material which is much more suitable to replicate this kind of effect, so much so that I think this *Mother-of-* philosopher would pay any price for it: mother-of-pearl, *pearl can* which can be worked into various shapes, and even when it is *replicate the* polished to a high degree of smoothness, still gives the optical *apparent* illusion of having such a variety of hollows and protrusions *the Moon's* that even touching it is hardly enough to convince you that it's *surface.* smooth.

Mother-of-pearl can replicate the apparent unevenness in the Moon's surface.

SALVIATI. That's a charming idea; no doubt there will be more which haven't been thought of yet, and if there are any other gems and crystals which have nothing to do with the illusions produced by mother-of-pearl, I dare say they will be excellent ideas as well. Meanwhile, as I don't want to stand in anyone's way, I shall suppress the answer which I could give to this, and just concentrate on answering signor Simplicio's objections.

I say, then, that your argument is too general, and since you haven't applied it to all the varying appearances which can be observed in the Moon, which have prompted me and others to believe it to be mountainous, I don't think you will find anyone who is convinced by it. Indeed, I don't think that either you *The apparent* or the author himself will find it any more satisfactory than *unevenness in* any other explanation that has nothing to do with the case. If *the Moon's* you constructed any kind of ball with a smooth surface and *surface cannot* parts which had varying degrees of opacity and clarity, you *be replicated by* wouldn't be able to replicate a single one of the many different *means of parts* *which are more* appearances which are observed in the course of a lunar *and less opaque* *and clear.*

The apparent unevenness in the Moon's surface cannot be replicated by means of parts which are more and less opaque and clear.

month. But you could take any solid and non-transparent material you like and make it into balls which, solely by means of peaks and hollows and varying kinds of illumination, would replicate exactly all the views and mutations which can be observed hour by hour in the Moon. You would see the ridges of the high peaks which catch the sunlight shining brightly, and the dark shadows which they cast behind them; you would see them appear larger or smaller depending on their distance from the boundary which divides the bright part of the Moon from its dark part. You would see that this boundary or dividing line is not evenly spread, as it would be if the ball's surface was smooth, but jagged and crenellated; you would see numerous peaks catching the light on the dark side of this boundary, detached from the rest of the area which was already illuminated; you would see the shadows mentioned above diminishing and, as the light grew higher, eventually disappearing altogether, until when the whole hemisphere was illuminated none would remain. Then, as the light moved towards the Moon's other hemisphere, you would recognize the same peaks which you had observed before, and this time you would see their shadows on the other side, and growing. And I repeat that you would not be able to replicate a single one of these things with your variations in opacity and clarity.

The changing appearance of the Moon can be replicated with any opaque material. Various appearances from which can be deduced the mountainous surface of the Moon.

SAGREDO. Actually, you could replicate one of them, which would be the full Moon, when none of these shadows or other variations produced by the peaks and hollows are visible because the whole surface is illuminated. But please, signor Salviati, don't waste any more time on this point, because anyone who had had the patience to observe the Moon over one or two months and still could not grasp these plainly evident truths, would have to be considered lacking in any kind of judgement. And there is no point in wasting time and words with such people.

SIMPLICIO. To tell the truth, I haven't made such observations, because I haven't had sufficient curiosity, or a suitable instrument for the purpose, but I would very much like to do so. So let's leave this question pending and move on to our next point. Please give me your reasons for believing that the Earth can reflect the Sun's light no less powerfully than the

Moon; because it seems to me to be so dark and opaque that I find such an effect quite impossible.

SALVIATI. Signor Simplicio, your reason for believing the Earth incapable of shedding light is equally impossible. Wouldn't it be a fine thing if I were able to understand your argument better than you can yourself?

SIMPLICIO. Whether my argument is sound or not is something that I dare say you can judge better than me; but sound or unsound, I'll never believe that you can understand my argument better than I can.

SALVIATI. I think I can make you believe it right now. Tell me: when the Moon is nearly full, so that it can be seen both by day and in the middle of the night, when do you think it shines more brightly, by day or by night?

The Moon appears brighter by night than by day. The Moon seen in the daytime is like a cloud. SIMPLICIO. By night; there's no comparison. I think the Moon is like the pillar of cloud and fire which accompanied the children of Israel, which was like a cloud when the Sun was shining but was brilliantly bright by night.* I've sometimes observed the Moon among clouds in the daytime, and it appeared white just like one of the clouds, but then at night it shines resplendently.

SALVIATI. So if you had only ever seen the Moon in the daytime, you would have thought it was no brighter than a cloud.

SIMPLICIO. I believe so, yes.

SALVIATI. So now tell me this: do you think the Moon is really brighter at night than it is by day, or is there some other factor that makes it appear so?

SIMPLICIO. I think its brightness is intrinsically the same both by day and by night, but its light appears greater at night because we see it against the dark background of the sky. In the daytime, because everything around it is light, it hardly stands out at all, and so it appears much less bright to us.

SALVIATI. Tell me now, have you ever seen the terrestrial globe illuminated by the Sun in the middle of the night?

SIMPLICIO. That's either a trick question, or one which you only ask someone who is known to have no sense at all.

SALVIATI. No, no: I know you to be a man of very good sense, and I'm asking the question quite seriously. So don't hesitate to answer; if then you think I'm talking nonsense, I'm happy to be considered the one who has no sense. After all, if someone asks a stupid question it's they who are stupid, not the person they are asking.

SIMPLICIO. Very well: if you don't take me for a complete simpleton, assume that I have answered and said it is impossible for someone who is on the Earth, as we are, to see at night that part of the Earth where it is day, i.e. where the Sun is shining.

SALVIATI. So you have only ever seen the Earth illuminated in the daytime, whereas you see the Moon shining in the sky even on the darkest night; and that, signor Simplicio, is the reason why you do not believe that the Earth shines like the Moon. If you were able to see the Earth with the Sun shining on it when you were in darkness similar to our night, you would see it shining more brightly than the Moon. Now, to make a valid comparison, you must compare the light of the Earth with the light of the Moon when you see it in the daytime, not with the Moon at night, because we never see the Earth illuminated except by day. Isn't that the case?

SIMPLICIO. Yes, it must be.

SALVIATI. And since you have said yourself that you have seen the Moon by day surrounded by white clouds and similar in appearance to one of them, you already acknowledge that the clouds, although they are elemental matter, are capable of receiving illumination from the Sun just as much as the Moon is—indeed more, if you think of times when you have seen very large clouds which are as white as snow. We can be sure that if such a cloud could shine as brightly as that in the middle of the night, it would light up the area around it more than a hundred Moons. If, then, we could be certain that the Earth receives light from the Sun as readily as one of these clouds, then we could no longer doubt that it shines no less brightly than the Moon. And we can be sure of this, because we can see these same clouds at night, in the absence of the Sun, appear as dark as the Earth; more than this, we have all had the experience of seeing clouds low on the horizon, and not being sure whether they were clouds or mountains. This

Clouds, no less than the Moon, are capable of being illuminated by the Sun.

clearly shows that the mountains receive light no less than the clouds.

SAGREDO. That's enough arguments! Look there at the Moon, which is more than half full, and over there at that high wall with the Sun shining on it. Come over here so you can see the Moon and the wall alongside each other, and look: which of them is brighter? Isn't it clear that, if anything, the wall is brighter? The Sun's rays are striking it; from there, they are reflected onto the walls of this room, and from them into that other small room over there. So they are reflected three times to reach that small room, and yet I'm sure it has more light than it would if it were lit directly by moonlight.

A wall illuminated by the Sun, compared to the Moon, reflects no less light.

The third reflection of light from a wall is stronger than the first reflection from the Moon.

SIMPLICIO. No, I don't agree about that; moonlight is very bright, especially when the Moon is full.

SAGREDO. It seems very bright because the area all around it is dark and in shadow, but in absolute terms it's not; in fact it's less than the light at twilight half an hour after sunset. You can see that, because only then do objects lit by the Moon cast a shadow on the Earth. If you want to prove whether the third reflection in that little room gives more light than the first reflection from the Moon, you can go in there now and read a book, and then try this evening to read by moonlight, and see which is easier to read by; I'm sure that it will be harder by moonlight.

The light of the Moon is weaker than at twilight.

SALVIATI. So now, signor Simplicio, if you're convinced, you can see how you really knew all along that the Earth reflects light no less than the Moon; to convince you of it I had only to remind you of things you knew already, without needing me to teach them to you. I didn't teach you that the Moon shines more brightly by night than in the daytime; you knew it for yourself, just as you knew that a cloud can appear as bright as the Moon. You knew, too, that the Earth's illumination can't be seen at night; in short, you knew everything without knowing that you knew it. So it shouldn't be hard for you to accept the logic that the dark part of the Moon can be illuminated by the reflection of the Earth, with no less light than the Moon gives when it lights up the night—in fact more, for the Earth's light exceeds the Moon's by as much as the Earth is bigger than the Moon, which is forty times more.*

SIMPLICIO. But I did believe that the Moon's secondary light was intrinsic to the Moon itself.

SALVIATI. And yet you knew this too, without realizing it. You knew for yourself, surely, that the Moon shines much more brightly by night than in the daytime, in contrast to the darkness of the area around it. So did you not know in consequence the general principle that any body reflecting light appears brighter when its surroundings are dark?

Illuminated bodies appear brighter in dark surroundings.

SIMPLICIO. Yes, I know that, certainly.

SALVIATI. When the Moon is new and its secondary light is brightest, is it not always close to the Sun, and hence in twilight?

SIMPLICIO. Yes, it is; and I've often wished that the sky would get darker so that I could see this light more clearly, but the Moon has set before it got completely dark.

SALVIATI. So you must have known, then, that the secondary light would appear more clearly when it was completely dark.

SIMPLICIO. Yes, of course; and it would appear more brightly still if it were not for the brightness of the crescent which is illuminated by the Sun, which detracts from the secondary light.

SALVIATI. But doesn't it sometimes happen that you can see the whole of the Moon's face without the Sun shining on it at all?

SIMPLICIO. As far as I know, the only time that happens is when there is a total eclipse of the Moon.

SALVIATI. Then at such a time its light ought to shine very brightly indeed, since its surroundings are completely dark and there is no illuminated crescent to detract from it. But how bright have you seen it under those conditions?

SIMPLICIO. I've sometimes seen it appear copper-coloured with a touch of whiteness, but at other times it has been so faint that I've lost sight of it altogether.

SALVIATI. In that case how can that be its own intrinsic light, since you can see it quite brightly at twilight despite the much greater brightness of the crescent alongside it, but when the night is completely dark and all other light is absent, it's not visible at all?

SIMPLICIO. I understand some people have said that the light is shed on it from the other stars, and particularly from Venus, which is nearest to it.

SALVIATI. This is another vain idea, because in that case it ought to shine more brightly than ever when it is totally eclipsed, since no one supposes that the Earth's shadow conceals it from Venus or the other stars. But in fact it's completely devoid of light at an eclipse, because the hemisphere of the Earth which is facing the Moon at that time is in darkness, so there is a complete absence of light from the Sun. And if you observe carefully, you will see clearly that when the Moon is new it gives very little light to the Earth, but as the area illuminated by the Sun grows so the light which it reflects to us becomes brighter. At the same time, the new Moon appears very bright to us, because it is between the Sun and the Earth and therefore is exposed to a large part of the Earth's sunlit hemisphere; as it approaches first quarter and moves away from the Sun, its brightness diminishes, and once it has passed its first quarter it becomes quite dim, because its exposure to the illuminated part of the Earth is continually decreasing. But if the Moon's light were intrinsic to it or derived from the stars, the opposite would happen, because then we would be able to see it in the darkest night and in completely dark surroundings.

SIMPLICIO. Please stop there. I've just recalled reading a modern book of conclusions* containing many new ideas, which explained this secondary light as follows. It is neither produced by the stars nor intrinsic to the Moon itself; even less is it reflected to the Moon from the Earth. Rather, it comes

Some say that the Moon's secondary light is caused by the Sun.

from the light of the Sun itself, which penetrates through the body of the Moon, the substance of which is slightly transparent, although it shines most brightly on the surface of the Moon's hemisphere which is exposed directly to its rays. Below the surface the Moon soaks up, as it were, the light like a cloud or a crystal, transmits it and so is visibly illuminated. The author proves this, if I remember rightly, by means of authority, experience, and reasoned argument: he cites Cleomedes, Vitellio, Macrobius,* and some other modern author,* and he adds that it can be observed by the fact that the

Moon shines very brightly when it is close to its conjunction with the Sun, when it is new, and also because it shines most brightly around the edge. He also says that at a solar eclipse, when the Moon is in front of the Sun's disc, light shines through it, especially at the outer edge. His arguments, I think, are that since this light cannot derive from the Earth, the stars, or the Moon itself, it must necessarily come from the Sun, and moreover that this assumption gives a coherent explanation for all the observed phenomena. The fact that this secondary light appears brighter around the outer edge is explained by the fact that the Sun's rays only have to penetrate a shorter distance, since the longest line through a circle is the one which passes through its centre, and those further away from this line are always shorter than those close to it. The same principle explains why this light does not diminish very much. Finally, it explains why it is that in a solar eclipse this bright ring around the outer edge of the Moon is visible around the part which is directly in front of the Sun's disc but not around the rest; the reason for this is that the Sun's rays pass through the part of the Moon which is in front of it directly to our eyes, but those which are outside the area of overlap fall outside our field of vision.

SALVIATI. If this philosopher had been the first author to put forward this view, it would have been understandable for him to be so attached to it that he embraced it as true. But since he learnt it from others, there is no excuse for his failure to see the fallacies in it, especially if he had heard the true cause of this effect and had the opportunity to confirm it by any number of experiments and manifest proofs: it is reflected light from the Earth, and nothing else. Of those authors who find this explanation unconvincing, I can excuse the ancients who did not hear it expounded and who did not think of it themselves; I'm sure that if they heard it now they would have no hesitation in accepting it. To speak quite frankly, I can't believe that this modern author doesn't assent to it privately, but I suspect the fact that he can't claim to be its originator prompts him to try to suppress it or discredit it at least among the ignorant, of whom as we know there are a great number. For there are many who have much more pleasure in the applause of the common crowd than in the approval of the select few.

SAGREDO. I don't think you're really getting to the heart of the matter, signor Salviati. These people are always out to set traps for the unwary, and they're quite capable of making themselves authors of other people's ideas, as long as the ideas haven't been common currency for so long that everybody knows them.

SALVIATI. Oh, I'm more cynical than you are. Never mind about ideas being common currency; does it make any difference whether the ideas and discoveries are new to men, or the men are new to the discoveries? If you wanted nothing more than the esteem of those who are beginners in science, you could claim to be the inventor of the alphabet and they would all wonder at you; and even if in time your deception was discovered, that wouldn't do you any serious damage, because there are always newcomers to keep up the number of your admirers. But let's return to our discussion, and show signor Simplicio how his modern author's argument is undermined by various errors, non sequiturs, and untenable opinions. First of all, it's not correct to say that this secondary light is brighter around the Moon's outer edge than in the middle, so as to form a brighter ring or circle around the rest of its surface. It is true that it appears at first sight to be surrounded by such a circle when the Moon is observed at twilight, but this is an illusion caused by the different borders around its surface when it shines with this secondary light. The side which is towards the Sun is bordered by the Moon's brilliantly bright crescent, whereas on the other side it borders on the darkness of twilight; the contrast makes this side of the Moon's disc appear brighter, while the other appears darker against the greater brightness of the crescent. Your modern author could have seen this if he had tried blocking out the crescent by looking at it with a rooftop or some other obstacle in the way; then he would have realized that the rest of the Moon's surface, apart from the crescent, is in fact equally illuminated.

SIMPLICIO. I seem to remember he does say that he used some such device to obscure the bright crescent.

It makes no difference whether ideas are new to men, or the men new to the ideas.

The Moon's secondary light appears in the form of a ring, bright at the outer circumference and not in the middle; the reason for this.

How to observe the Moon's secondary light.

SALVIATI. Well, in that case what I thought was just care-lessness on his part turns out to be a lie, and a barefaced lie at that, since anyone can easily put it to the test for themselves. As for the Moon being visible in a solar eclipse, I doubt very much whether this can be for any reason other than the lack of light shining on it—especially when it is not a total eclipse, as must be the case with any which this author has observed. But even if it did appear to shed some light, that would support our argument rather than undermining it, since in a solar eclipse *In a solar* the whole of the Earth's sunlit hemisphere is facing the Moon, *eclipse the* and even if the Moon itself obscures part of it this is only a *visible only* very small part compared to what remains illuminated. The *through lack of* author goes on to say that in an eclipse the edge of the Moon *light.* shines brightly around the part which overlaps with the Sun, but not around the rest, and that this is because the Sun's rays pass through the overlapping part of the Moon directly to our eyes, but not through the rest. This is one of those fantasies that show up the other fictitious ideas he is propounding; for does the poor fellow not realize that if the Moon's secondary light is only visible to us when the Sun's rays pass directly through it to our eyes, then the only time we would ever see this secondary light would be in a solar eclipse? And if a part of the Moon which is less than half a degree away from the Sun is enough to deflect the Sun's rays away from our sight, what will happen when the Moon is twenty or thirty degrees away from the Sun, as it is when it first rises? And in any case, how will the Sun's rays pass through the body of the Moon to reach our eyes?

This man proceeds by imagining facts as his theory requires them to be, rather than accommodating his theories to fit the facts as they are. So, in order for the Sun's rays to penetrate the *The author of* substance of the Moon he says that the Moon is partly trans- *the book of* parent, in the same way as, for example, a cloud or a crystal; *conclusions* but I don't know how he conceives of a transparency which *adapts the facts* allows the Sun's rays to penetrate a cloud more than two *not his ideas* thousand miles thick. Let's assume that he confidently replies *to the facts.* that this is perfectly possible in celestial bodies, which are quite different from the elemental bodies we know, impure and tainted as they are; and let's refute his error by means which don't allow any riposte, or rather subterfuge. So if he wants to maintain that the Moon's substance is transparent, he will have to explain how its transparency allows the Sun's rays to pass right through it, a distance of more than two thousand miles, but that when they are blocked by a thickness of only a mile or less they are no more able to penetrate it than they could a mountain here on Earth.

SAGREDO. This reminds me of a man who wanted to sell me a secret device* which would enable me to talk to someone two or three thousand miles away by means of some kind of attraction of magnetic needles. I said I would certainly like to buy it, but first I would like to see it working, and that it would be enough for him to go into another room and talk to me from there. He replied that its operation couldn't be seen properly over such a short distance. So I sent him away, and said that I wasn't inclined to go to Cairo or Moscow to try the experiment, but he was welcome to go and then he could talk to me from there while I stayed here in Venice. But let's hear what follows from this author's argument, and why he must admit that the Moon is permeable to the Sun's rays to a depth of two thousand miles, but more opaque than one of our mountains at a thickness of only one mile.

Riposte to a man who wanted to sell a secret device for talking to someone a thousand miles away.

SALVIATI. The mountains on the Moon themselves are evidence of this. When the Sun shines on them on one side, they cast dark shadows on the other, sharper and more clearly defined than the shadows of our mountains; whereas if they were transparent, we would never have discerned any irregularity on the Moon's surface, or seen those illuminated peaks which stand out detached from the boundary between its light and dark areas. In fact, this boundary itself would be much less distinct if the Sun's light really did penetrate through the body of the Moon. It follows from the author's own words that the transition and dividing line between the part which was exposed to the Sun's rays and the part which was not would have to appear blurred and a mixture of light and dark; for a material which lets the Sun's rays pass through a depth of two thousand miles must be so transparent that they would hardly be obstructed at all by a thickness of only a hundredth part of that or less. But in fact, the dividing line between the light and dark areas is as sharp and distinct as the contrast between black and white, especially where it crosses the part of the Moon which is naturally brighter and more uneven; where it cuts across the ancient spots,* which are plains, it is not so clear-cut because the light is fainter, as the inclination of the Moon's spherical surface makes the Sun's rays strike them more obliquely. Finally, what he says about the secondary light not diminishing or becoming fainter as the Moon becomes more full is simply wrong; in fact, this light is hardly seen at all when it is at its first or last quarter, when it would be expected to be most clearly visible because then it is surrounded by complete darkness, well away from the twilight.

We conclude, therefore, that the Earth's reflection has a powerful effect on the Moon. Even more important, another marvellous congruity can be deduced from this: that if it is true that the planets affect the Earth with their light and motion, the Earth may be no less able to affect the planets reciprocally, with its light and possibly with its motion as well. In fact, even if the Earth did not move it could still affect them in the same way, for as we have seen, the action of the light is the same in either case, as it is the reflected light of the Sun, and motion simply produces variations in its aspect which follow in the same way whether the Earth moves and the Sun is fixed or the other way round.

The Earth can reciprocally affect the celestial bodies with light.

SIMPLICIO. No philosopher has said that these inferior bodies affect the celestial bodies, and Aristotle clearly states the contrary.

SALVIATI. This is excusable in Aristotle and the others who did not know that the Earth and the Moon reflect light to each other; but there would be no excuse for philosophers whom we have shown how the Earth gives light to the Moon if they insisted on denying that the Earth can affect the Moon, while expecting us to agree that the Moon's light affects the Earth.

SIMPLICIO. I have to say that I am extremely reluctant to accept this relationship between the Earth and the Moon that you want to persuade me of, placing the Earth, so to speak, among the stars. Apart from anything else, it seems to me that the great distance separating the Earth from the celestial bodies must necessarily mean that there is a huge difference between them.

SALVIATI. That just goes to show, signor Simplicio, how powerful a long-standing attachment and a deeply rooted opinion can be: it is so tenacious that it makes you cite arguments in support of your belief which actually undermine it. If you are convinced that distance and separation are a mark of great divergence in nature, then the converse must also be true, *Affinity* and proximity and contiguity must indicate similarity; and *between the* how much closer is the Moon to the Earth than to any of the *Earth and the* other celestial globes? So by your own admission you must *Moon on* concede that there is a great affinity between the Earth and the *account of their proximity.* Moon—and you will have other philosophers for company if you do. Now let's proceed: tell me if any of the other objections which you raised against the congruity between these two bodies remain to be discussed.

SIMPLICIO. I still have some uncertainty about the solidity of the Moon, which I deduced from its being supremely polished and smooth, and you from its being mountainous. And I also had a difficulty arising from my belief that the sea, because of the smoothness of its surface, must reflect light more strongly than dry land, which is so uneven and opaque.

SALVIATI. I will respond to your first doubt by reference to the Earth, all of whose parts are drawn by their natural gravity to come together as close to its centre as they can, but some remain further away than others, the mountains being further from the centre than the plains. We attribute this to their hardness and solidity, since if they were made of more fluid matter *Solidity of the* they would find their own level. In the same way, the fact that *Moon's sphere* some parts of the Moon remain elevated above the spherical *deduced from* surface of its plains is evidence of their hardness, since it is *its being* reasonable to suppose that lunar matter too has formed itself *mountainous.* into a sphere because its parts are all drawn together to the same centre.

Light is As regards your other doubt, I think it is clear from our *reflected more* investigation of what happens with mirrors that the light *weakly from* reflected from the sea is much less than from dry land, at least *the sea than* as far as its overall reflection is concerned. There's no doubt *from dry land.*

that a calm sea reflects light very strongly to a particular place, and anyone observing from that spot would see a brilliant reflection from the water, but from anywhere else the surface of the water would appear darker than that of the land. We can demonstrate this experimentally if we go into the main room here and pour a little water onto the floor. There: doesn't that brick which is wet look much darker than the dry ones around it? Clearly it does, and it will appear the same from everywhere except one place, which is where it reflects the light coming from that window. Just move back slowly . . .

Experiment to demonstrate that water reflects less brightly than dry ground.

SIMPLICIO. From here the wet area appears lighter than the rest of the floor, and I can see that this is because the light coming in through the window is reflected towards me.

SALVIATI. The water has simply filled up all the little cavities in the brick so that it has a perfectly smooth surface, so that the reflected rays all converge on the same place; whereas the rest of the floor which is dry still has all its irregularities, and hence an innumerable variety of tiny inclined surfaces, so that the light is reflected in all directions, but more weakly than if the rays all converged together. This means that its appearance hardly changes at all from wherever you look at it; it looks the same from any angle, but much less bright than the reflection from the part which is wet. I conclude from this that since the surface of the sea, seen from the Moon, would appear completely smooth—apart from any islands or rocks—it would appear darker than the land, which is mountainous and uneven. And at the risk of appearing to be pushing my luck, so to speak, I will add that I have observed the Moon's secondary light, which I say is reflected from the Earth, to be noticeably brighter two or three days before the Moon's conjunction with the Sun than after it—when, that is, we see this light before dawn in the east, rather than after sunset in the west. This difference is caused by the fact that when the Moon is facing the Earth's eastern hemisphere it is exposed to much more land, since it includes the whole of Asia, and very little sea, whereas when it is facing west it looks down on a great expanse of sea, all the way across the Atlantic Ocean to the Americas. This makes it highly probable that the surface of the sea reflects light less brightly than the land.

The Moon's secondary light is brighter before its conjunction with the Sun than after.

SIMPLICIO. [So in your view the Earth would have a similar appearance to the two main areas that we see in the Moon.] Do you think, then, that the large dark spots that are visible on the Moon are seas, and that the other lighter areas are land or something similar?

SALVIATI. This is the first of the differences that I believe exist between the Earth and the Moon—and it's time we dealt with these quickly, as I think we've lingered too long on the Moon already. I say, then, that if nature had only one way of making two surfaces illuminated by the Sun appear one darker than the other, namely by making one of dry land and the other of water, then we would necessarily have to conclude that the Moon's surface was partly dry land and partly water. But since there are various ways known to us of producing the same effect, and there may well be others which are unknown to us, I do not presume to say that one rather than another occurs on the Moon. We have already seen how a touch with a burnishing-brush changes a bleached silver plate from white to dark; and the parts of the Earth which are moist are darker than those which are arid. On a mountainside, the areas which are forested appear much darker than those which are bare and barren; this is because there is a great deal of shade among the trees, and the open spaces are illuminated by the Sun. We can see the same effect of light and shade in a piece of patterned velvet, where the cut silk appears much darker than where it is uncut, because of the shadow between one thread and another; similarly plain velvet appears much darker than a smooth lining made from the same silk. So, if there were anything on the Moon resembling our great forests, this could appear to us as the dark spots that we can see; the same effect would be apparent if they were seas; and finally, it is not impossible that these spots could in reality be a darker colour than the rest, rather as snow makes mountains appear lighter. What is quite clear on the Moon is that the darker areas are all plains, with a small number of rocks and banks contained in them; the rest, which is brighter, is full of rocks, mountains, and small banks, some round and some in other shapes; in particular, the dark spots are surrounded by very long ranges of mountains. We can see that the surface of the spots is smooth because when

The darker parts of the Moon are plains, and the lighter parts mountainous. The Moon's spots are surrounded by long ranges of mountains.

the boundary line dividing the light and dark parts of the Moon runs across them, it intersects them evenly, whereas in the lighter areas it appears jagged and crenellated. I don't know, however, whether the smoothness of the surface is sufficient on its own to explain the darkness of these areas; I rather think not.

I also consider that the Moon differs greatly from the Earth in other respects. I imagine it is not completely dead and sterile, but nor do I affirm that movement or life are to be found there. Even less do I believe that plants, animals, and other things similar to ours are generated there; if there is any generation there it will be of things totally unlike what we know and beyond anything we can imagine. My first reason for believing this is that I do not think the Moon is made up of earth and water, and this alone means that it cannot have generation and change as we know them; but even if water and earth did exist there it would still not produce plants and animals similar to ours, for two main reasons. First, generation on Earth would not be possible without the changing aspects of the Sun, which follow a quite different rhythm on Earth and on the Moon. For the greater part of the Earth, the alternating cycle of night and day is completed every twenty-four hours, but on the Moon this is spread over a month. The annual rise and fall in the Sun's altitude which brings us the changing seasons and the varying length of day and night is also completed in a month on the Moon. And whereas for us the variation in the Sun's altitude is approximately forty-seven degrees, this being the distance from one tropic to the other, on the Moon the variation is only ten degrees or a little more, this being the maximum extent of the Sun's movement in relation to the ecliptic. Consider now the effect which the Sun would have in the tropics if its rays beat down continually for fifteen days, and it will be clear that all the plants, grasses, and animals would be scattered; so even if there were generation there, it would be of grasses, plants, and animals completely different from ours. Second, I am firmly convinced that there is no rain on the Moon, because if clouds ever formed over any part of its surface as they do around the Earth, they would block out some of the things which we can see

There is no generation on the Moon of things similar to ours; if there is generation, it is of things which are completely different.

The Moon is not made up of earth and water.

The aspects of the Sun, which are necessary for generation on Earth, are not the same on the Moon. Each natural day on the Moon lasts for a month.

The Sun's altitude rises and falls by 10 degrees on the Moon, and by 47 degrees on Earth.

There is no rain on the Moon.

through a telescope, and we would see some change in its visual appearance; but in all my long and diligent observations I have never seen any such change, but I have always seen the Moon under a perfectly clear sky.

SAGREDO. To this one could reply either that the Moon could have very heavy dew, or that it rained during their night, when it is not illuminated by the Sun.

SALVIATI. If other considerations gave us grounds for believing that there might be generation similar to ours on the Moon, and all that was lacking was evidence of rain, then we might find some mitigating phenomenon to fulfil the same function, like the flooding of the Nile in Egypt. But since we have not found any conditions similar to ours of the many that it would take to produce similar effects, there's no point in going to great lengths to introduce just one, especially since there is no positive evidence of it but simply because it is not inconceivable. And in any case, if I were asked what my initial understanding and pure natural reason tell me about the likelihood of the Moon producing things which are similar to or different from ours, I would always say, utterly different and completely beyond our imagination; for this is what the riches of nature and the omnipotence of its Creator and Ruler seem to me to require.

SAGREDO. It has always struck me as the height of temerity to make the limits of human understanding the measure of what nature can do, when there is not a single effect in nature, however small, which even the most penetrating human intelligence can fully understand. This vain presumption of understanding everything can only come from never having understood anything; for anyone who has tried just once to reach a perfect understanding of just one thing, and has truly tasted what knowledge is made of, will know that there is an infinity of other truths of which they understand nothing.

Never having fully understood anything makes some people believe that they understand everything.

SALVIATI. Your argument is irrefutable; and it is confirmed by the experience of those who have reached some understanding, who, the wiser they become, the more they recognize and freely admit how little they know. And the wisest of the Greeks, who was judged to be so by the oracles, openly confessed that he knew nothing.

SIMPLICIO. Then we will have to say that either the oracle or Socrates was lying, when the oracle said he was the wisest of men, and he said that he knew he was the most ignorant.

SALVIATI. Neither of these follows, since both statements can be true. The oracle judges Socrates the wisest of men, and men's wisdom is limited; Socrates knows that he knows nothing in comparison to absolute wisdom, which is infinite. And since much is no more in relation to infinity than little or nothing—for to arrive at the number infinity it makes no difference whether you multiply thousands, tens, or zeros—Socrates knew that the finite wisdom which he possessed was nothing compared to the infinity which he lacked. Since, however, there is some wisdom among men, and it is not equally shared among all, it was possible for Socrates to have a greater share than anyone else, and so for the oracle's statement to be true.

The oracle spoke the truth in declaring Socrates the wisest of men.

SAGREDO. I think I understand the point here very well. In the same way, signor Simplicio, power exists among men, but not all share it equally; the power of an emperor is clearly much greater than that of a private person, but both are as nothing compared to the omnipotence of God. Or among men, there are some who understand agriculture better than others; but what is knowing how to plant a vine-shoot in a trench, compared to being able to make it take root, draw nourishment from the earth, and select the appropriate parts to form leaves, tendrils, bunches, grapes, and pips, all of which are the work of nature in her great wisdom? This is just one small example of the innumerable works of nature, and this alone shows infinite wisdom, so we can conclude that divine wisdom is infinitely infinite.

Divine wisdom is infinitely infinite.

SALVIATI. Here's another example. Would we not all agree that Michelangelo's* ability to sculpt a beautiful statue in a block of marble has raised his genius to a sublime level far above that of ordinary men? And yet he has only imitated just one external, superficial arrangement of the limbs of a motionless man. What is this in comparison with a man made by nature, made up of so many external and internal members, and muscles, tendons, nerves, and bones, allowing so many and such diverse movements? Not to speak of man's senses, the

The sublime genius of Michelangelo.

capacity of his will, and finally his understanding. Surely we can justifiably say that the creation of a statue falls infinitely short of the creation of a living man, or even of the vilest worm?

SAGREDO. Or what difference do we think there was between the pigeon of Archytas* and one produced by nature?

SIMPLICIO. Either I am not one of those men endowed with understanding, or there is a plain contradiction in your argument. You identify as one of the most praiseworthy attributes of man made by nature, indeed the greatest of all his attributes, this quality of understanding; and yet a moment ago you agreed with Socrates that his understanding was as nothing. So you must conclude that not even nature has understood how to make an intellect which can understand.

SALVIATI. A very telling objection; and to answer it we must make a philosophical distinction between intensive and extensive understanding. Extensively—that is, in relation to things that can be understood, of which there is an infinite number—human understanding is as nothing, even if we understand a thousand propositions, because a thousand is the same as zero in relation to infinity. But if we speak of understanding intensively—meaning that we understand a proposition perfectly—then it is possible for the human intellect to understand some propositions as perfectly, and with as much certainty, as nature itself. The pure mathematical sciences, geometry and arithmetic, are of this kind; and while the divine intellect knows infinitely more propositions in these sciences, nonetheless I believe that in those few which the human mind can understand, we can achieve the same objective certainty as the divine mind itself. For we can understand that they are necessarily true, and there can be no greater certainty than this.

Man understands much intensively, but little extensively.

SIMPLICIO. This strikes me as a very bold and daring statement.

SALVIATI. These are generally accepted propositions which are not bold or presumptuous at all, nor do they detract in any way from the majesty of divine wisdom, any more than it diminishes God's omnipotence to say that He cannot undo what has been done. But I suspect, signor Simplicio, that you may have taken my words amiss because you have misunderstood them; so let me try to explain myself better. The truth which we arrive at by means of mathematical proofs is the same truth that is known to divine wisdom; but I readily concede that the way in which God knows the infinite number of truths of which we know only a few, is vastly superior to ours. We proceed by arguing logically from one conclusion to another, whereas God's knowledge is a simple intuition. So, for example, we gain an understanding of some of the properties of the circle, of which there are an infinite number, by starting from one of the simplest and, taking this as our definition, argue step by step to the second, the third, the fourth, and so on; but the divine mind comprehends them all instantly, in their infinity, by a simple act of understanding. These properties are in any case present implicitly in the definitions of all things, and perhaps, being infinite, they are ultimately all one in their essence and in the divine mind. Indeed, this is not entirely unknown to the human mind, but we perceive it through a deep and dense fog, which may be partly dispersed and penetrated when we have mastered some firmly established conclusions which we grasp so readily that we take them for granted. So, for example, to say that the square on the hypotenuse of a right-angled triangle is equal to the sum of the squares on the other two sides is the same as saying that parallelograms on a common base and between parallel lines are equal; and this is ultimately the same as saying that two plane figures which coincide exactly when they are superimposed are equal. These deductions which our intellect makes step by step in time and space, the divine intellect makes instantaneously, with the speed of light, which is the same as saying that they are all constantly present in the divine mind. I conclude, therefore, that both in its manner of understanding and in the number of things it understands, our intellect is infinitely surpassed by the divine; but I do not therefore despise human understanding or consider it to be worthless. Indeed, when I consider how many and how marvellous are the things which men have been able to do and understand, I perceive all too clearly that the human mind is among the most excellent works of God.

God's way of understanding is different from that of men. Human understanding is gained by logical argument.

Definitions contain implicitly all the qualities of the things defined. The infinite number of qualities may perhaps be only one.

Deductions which human reason makes in time, the divine intellect makes instantaneously, which is to say that they are always present in the divine intellect.

Human genius is wonderful in its acuteness.

SAGREDO. As I have reflected on these matters, I have often thought how acute the human mind is; and as I consider all the marvellous discoveries which men have made, both in the arts and in letters, and then think of what I myself know, and how far I am from even learning everything that has been discovered, much less discovering anything new, I am filled with bewilderment and despair, and count myself little more than a wretch. I look at one of these excellent statues, and I say to myself, 'When could you ever remove the excess from a block of marble to reveal such a beautiful figure concealed within it? Or mix and paint different colours on a canvas or a wall, and with them depict every kind of visible object, like a Michelangelo, a Raphael, or a Titian?' If I consider what composers have discovered about the distribution of musical intervals, and how they have established precepts and rules for arranging them to produce delightful sounds for the ear, how can I fail to be amazed? And what am I to say about the number and diversity of musical instruments? Can anyone read the most excellent poets attentively and not wonder at the inventiveness of their conceits and how they are developed? What are we to say of architecture, or the art of navigation?

The invention of writing is the most stupendous of all.

But perhaps the most stupendous inventor of all was the one who dreamt of finding a way to communicate his innermost thoughts to all, however far away they might be in time and place; of speaking to those who are in the Indies, or to those who are not yet born and who will not be born for another thousand or ten thousand years. And how? By means of varied arrangements of twenty little characters on a sheet of paper. So let this set the seal on all humankind's marvellous inventions, and let it be the conclusion to our discussions for today. The hottest hours of the day have now passed, and I think signor Salviati may like to enjoy the cool of the evening by boat; and I shall expect you both to continue our conversation tomorrow.

[. . .]

SIMPLICIO. You take it as self-evident and generally understood that when a projectile separates from its projector, it continues to move because of the force applied to it by that same projector. This applied force is as unacceptable in Peripatetic philosophy as any other example of an accidental quality passing from one subject to another. The Aristotelian view, as I think you know, is that the projectile is carried forward by the medium, in this case the air; so if a stone dropped from the top of a ship's mast were to follow the motion of the ship, this effect would have to be attributed to the air, not to any force which was applied to it. But you assume that the air does not follow the ship's motion, but is at rest. Moreover, whoever drops the stone should not throw it or give it any kind of impetus with their arm, but simply open their hand and let it go. So since neither the force imparted to it by the projector nor the benefit of the air can cause the stone to follow the ship's motion, it must be left behind.

Aristotle's view that a projectile is moved not by applied force but by the medium.

SALVIATI. You seem to be saying that, as the person dropping the stone does not throw it, it is not really a projection at all.

SIMPLICIO. Its motion cannot properly be called projection, no.

SALVIATI. In that case, what Aristotle says about motion, a moving body and the motive force of projectiles has nothing to do with our discussion; so why did you bring it up at all?

SIMPLICIO. Because you introduced this idea of applied force, which doesn't exist and so can have no effect whatever, because *non entium nullae sunt operationes*: entities which do not exist have no effect. The cause of motion, not just of projectiles but anything other than natural motion, should be attributed to the medium, which we have not considered as we should; so everything that has been said so far is beside the point.

SALVIATI. The point will become clear in due course. Tell me: since your whole objection is based on the non-existence of applied force, would you accept that this force exists if I demonstrated to you that the medium has nothing to do with the continued motion of projectiles once they are separated from their projector, or would you just move to attack it from another angle?

SIMPLICIO. If the effect of the medium were removed, I don't see that there could be any other explanation apart from the property imparted by the mover.

SALVIATI. To remove as far as possible any grounds for endless arguments, it would be useful if you could explain, as fully as you can, the effect of the medium in maintaining the motion of a projectile.

The effect of the medium in maintaining the motion of a projectile.

SIMPLICIO. The thrower, or projector, has the stone in his hand; he moves his arm with velocity and force, and its motion is imparted both to the stone and to the surrounding air. Hence when the stone leaves the thrower's hand, it is in air which already has the impetus to move, and is carried along by it. If the air did not affect it in this way, the stone would fall to the thrower's feet.

SALVIATI. Have you really been so credulous as to be persuaded by these vain ideas, when your own senses should enable you to refute them and perceive the true explanation? You said just now that if a heavy stone and a cannon-ball were placed on a table, they would not be moved by the wind, however strong it was. Tell me now: if they had been balls of cork or cotton wool, do you think they would have been moved by the wind?

Many experiences and arguments against the cause of the motion of projectiles as posited by Aristotle.

SIMPLICIO. I have no doubt that they would have been blown away, and the lighter the material the more quickly the wind would have blown it away. This explains why we see clouds being borne along by the wind at the same speed as the wind itself.

SALVIATI. And what is the wind?

SIMPLICIO. Wind is defined simply as air in motion.

SALVIATI. So air which is in motion carries light materials much further and more rapidly than heavy materials?

SIMPLICIO. Certainly.

SALVIATI. If you were to throw a stone and then a ball of cotton wool, which of them would travel further or more rapidly?

SIMPLICIO. The stone, by a long way; the cotton wool would just fall at my feet.

SALVIATI. But if it is just the air moved by the thrower's arm which carries the projectile after it has left his hand, and if air which is in motion carries light materials more easily than heavy ones, why doesn't the cotton-wool projectile travel further and more rapidly than the stone one? The stone must retain something, apart from the motion of the air. Take another example: suppose two cords of equal length were suspended from that beam, and a lead ball was attached to the end of one and a ball of cotton wool to the other. If both were moved an equal distance from the perpendicular and then allowed to fall freely, we know that they would both move back towards the perpendicular and then, carried by their own impetus, move a certain distance beyond the perpendicular, and then back again. But which of the two pendulums do you think would remain in motion longer, before coming to rest at the perpendicular?

SIMPLICIO. The lead ball would move back and forth for a long time, but the cotton wool only two or three times at most.

SALVIATI. So the impetus and motion, whatever causes it, is conserved for longer in heavy materials than in light ones. Now I come to another point, so let me ask you: why does the air not carry away that lemon on the table there?

SIMPLICIO. Because the air itself is not moving.

SALVIATI. In that case the projector must impart its motion to the air, which then moves the projectile. But if this kind of force cannot be imparted, since an accident cannot pass from one subject to another, how does it pass from the thrower's arm to the air? Surely the air is a subject distinct from the arm?

SIMPLICIO. The answer is that the air is in its proper region and so is neither heavy nor light, and therefore can easily receive and conserve any impulse.

SALVIATI. But the pendulums have just shown us that the less weight a moving body has, the less it is able to conserve motion. So how is it that only the air, which when it is in its proper place has no weight, can conserve the motion which is imparted to it? I believe, and I know that for now you share this belief, that as soon as the thrower's arm stops moving, the air around it stops moving as well. So let's go into this small room and stir the air up as much as we can by shaking a towel; then, as soon as we stop, let a small lighted candle be brought in, or let someone drop a piece of gold leaf. You will see that both move quite steadily, showing that the air has immediately returned to a state of rest. I could suggest any number of other experiments, but if one of these does not suffice then our case is desperate indeed.

SAGREDO. Isn't it incredible that when someone shoots an arrow into the wind, the narrow slice of air that was displaced by the bowstring should accompany the arrow in the face of the storm blowing against it! But there's another detail of what Aristotle says that I'd like signor Simplicio to explain to me. If someone were to shoot two arrows from the same bow, with the string drawn back to the same extent, one pointing forwards in the usual way and the other crosswise, in other words with the length of the arrow laid along the bowstring, which one would travel further? Do please forgive me, and answer my question even though it may seem ridiculous, because as you can see, I'm not very bright and my ability to speculate doesn't rise very high.

SIMPLICIO. I've never seen anyone shoot an arrow crosswise, but I think that if they did it wouldn't travel so much as a twentieth of the distance of one pointing forwards.

SAGREDO. I thought the same thing, which was what prompted me to wonder if there is an inconsistency between what Aristotle says and what we see from experience. Experience suggests that, if I were to put two arrows on this table when a strong wind was blowing, one aligned with the direction of the wind and the other crosswise to it, the wind would quickly blow away the latter and leave the first one where it is. If Aristotle is correct, the same should happen with the two arrows shot from a bow: the one which was shot crosswise would be propelled by a large amount of air set in motion by the bowstring, extending the full length of the arrow, whereas the one which was shot pointing forwards would have only the propulsion of the tiny circle of air corresponding to the arrow's thickness. I can't account for this difference, and I'd very much like to understand it.

SIMPLICIO. The reason seems quite clear to me: the arrow which is fired pointing forwards has to penetrate only a small amount of air, whereas the other has to divide the air along its whole length.

SAGREDO. So when an arrow is shot it has to penetrate the air? But how can it, if the air travels with it, indeed if the air is the medium which carries it along? If this were the case, surely the arrow should move with a greater velocity than the air—and where does it get this greater velocity from? Do you mean to say that the air gives it a greater velocity than its own? So you see, signor Simplicio, the reality is exactly the opposite of what Aristotle says: not only is it wrong to say that the medium confers motion on the projectile; the medium is actually the one thing that impedes it. Once you have grasped this, you won't have any difficulty in understanding that when the air really does move, it will much more easily carry the arrow which is crosswise to it than the one which is aligned with it, because in the former position there is a large volume of air pressing on the arrow, but in the latter very little. When they are shot from a bow, on the other hand, the air is stationary, and the arrow fired crosswise encounters a large volume of air and so meets quite strong resistance; but the arrow fired pointing straight ahead easily overcomes the resistance of the very small amount of air which it encounters.

The medium does not confer motion on projectiles, but rather impedes them.

SALVIATI. How many times have I seen propositions in Aristotle—in natural philosophy, that is—which are not just wrong, but wrong in a way which is diametrically opposed to the truth, as in this case! But to pursue our argument, I think signor Simplicio is convinced now that seeing a stone fall always in the same place does not provide any basis for conjecturing whether the ship is moving or not. If he is not persuaded by what has been said so far, the experiment itself should remove his doubts. The most he might find experimentally would be that the falling object would be left behind if it was very light and if the air was not following the ship's motion; but assuming that the air was moving at the same speed, no conceivable difference could be found in this or any other experiment, as I shall now explain to you. Now, given that no difference is apparent in this case, what do you claim to deduce from a stone falling from the top of a tower, where the stone's circular movement is not fortuitous or accidental but natural and eternal, and where the air duly follows the movement of the tower as the tower does that of the terrestrial globe? Have you anything more to reply on this point, signor Simplicio?

SIMPLICIO. Only that I have still to see a proof of the motion of the Earth.

SALVIATI. I haven't claimed to have proved it, only to show that the experiments brought forward by its opponents to prove that the Earth is at rest are inconclusive; and I think I can show the same with the others.

SAGREDO. Signor Salviati, before we go on, please allow me to put forward a difficulty which was going round in my mind while you were so patiently examining the experiment with the ship to signor Simplicio.

SALVIATI. Of course; we're here to discuss, and it's good that we should each express the difficulties which occur to us; that's how we shall come to understand the truth. So carry on.

SAGREDO. If it's true that the impetus of the ship's motion remains indelibly imparted to the stone after it has been dropped from the mast, and if, moreover, this motion does not impede or retard in any way the stone's natural motion directly downwards, then a marvellous effect in nature must necessarily follow. Suppose the ship is stationary, and the time taken for the stone to fall from the top of the mast is two pulse beats. Suppose, then, that the ship begins to move, and the same stone is dropped from the same place: from what has been said, it will still take two pulse beats to reach the foot of the mast. But in that time, the ship will have moved forwards, say, twenty *braccia*: this means that the stone's true motion will have been in an oblique line, notably longer than the original perpendicular line corresponding to the height of the mast; and yet it will have travelled this distance in the same time. Imagine, then, the ship's motion accelerating further, so that the stone has to move in an oblique line even longer than

A marvellous phenomenon in the motion of projectiles.

before; and in short, the more the ship's speed increases, the longer the oblique lines the falling stone will have to travel, and yet it will always do so in the same space of two pulse beats. By the same principle, if a long-range cannon were fired horizontally from the top of a tower, then regardless of the amount of gunpowder that was used, so that the ball fell to the ground a thousand, four thousand, six thousand, or ten thousand *braccia* away, in each case the shot would take exactly the same length of time. And it would always be equal to the time it would take for the ball to fall to the ground, if without any other propulsion it were simply dropped vertically from the top of the tower. It seems extraordinary that in the same short interval as it takes to fall vertically from a height of, say, a hundred *braccia*, a ball fired from a gun can travel four hundred, a thousand, four thousand, or ten thousand *braccia*, so that any ball fired horizontally from a cannon remains in the air for the same length of time.

SALVIATI. That's a very striking and original thought, and if the effect is true—as I don't doubt that it is—it is indeed marvellous. In fact, I'm certain that if it were not for the resistance of the air, another cannon-ball which was dropped from the same height at the same moment as the cannon was fired would hit the ground at the same time as the one fired from the cannon, even if one ball had travelled ten thousand *braccia* and the other only a hundred—always assuming that the Earth's surface was level, as it would be if you fired the shot over a lake. The effect of air resistance would be to retard the very rapid motion of the ball shot from the cannon.

But let us now turn to resolving the other arguments, since I think I'm right in saying that signor Simplicio now accepts that this first objection based on falling objects has no validity.

SIMPLICIO. I can't say that all my doubts have been resolved, though I dare say the fault for that is mine, as I don't grasp things as quickly or easily as signor Sagredo. If it's true, as you say, that the motion shared by a stone when it is at the top of a ship's mast remains indelibly imparted to it even when it is no longer in contact with the ship, then the same thing should happen if someone galloping on horseback should drop a ball to the ground: the ball ought to continue moving in the same direction, and follow the horse's movement without being left behind. But I don't think we see such a thing happen, unless the rider throws the ball forwards; otherwise I think the ball would remain at the point where it hit the ground.

SALVIATI. I think you're much mistaken; I'm sure that if you tried the experiment the opposite would be the case, and when the ball hit the ground it would still move forward together with the horse. It would only be left behind if it were obstructed by the roughness and unevenness of the road. The reason for this seems quite clear: if you stood still and threw the ball along the ground, surely it would still move forward after it had left your hand, and it would travel further if the ground's surface were smoother? On ice, for instance, would it not go a very long way?

SIMPLICIO. Yes, that's not in doubt, assuming I gave it an impetus with my arm. But in the case of the rider on horseback, I was assuming that he would simply drop the ball.

SALVIATI. That was my assumption as well. But when you throw a ball, the only motion it has once it has left your hand is the motion produced by your arm, which is conserved in the ball and continues to propel it forwards. Now, what difference does it make whether this impetus which is imparted to the ball comes more from your arm or from the horse? When you are on horseback doesn't your hand, and therefore the ball, travel at the same speed as the horse itself? Of course it does. So, if you simply open your hand, the ball is launched with the impetus derived from the horse, which is imparted to you, to your arm, to your hand, and finally to the ball. In fact, I will go further and say that if the rider on the horse throws the ball behind him, it sometimes happens that when the ball hits the ground it will still follow the horse's movement, even though it was thrown in the opposite direction. Sometimes it will just fall to the ground and stay there; and it will only travel in the opposite direction to the horse if the motion it receives from the thrower's arm has a greater velocity than the motion of the horse. And it's quite mistaken to say, as some do, that a horseman can throw a spear in the direction the horse is travelling, and then draw level with it and catch it again; because the only way you can throw a projectile and catch it again is by throwing it straight up in air, just as you would if you were standing still. However fast the horse was running, as long as its motion was uniform and the projectile was not something very light, it would always come back down to the thrower's hand, however high it was thrown.

SAGREDO. This explanation reminds me of some curious problems concerning this matter of projectiles. The first of these will strike signor Simplicio as very strange, and it's this: it can happen that when someone moving rapidly in any way simply drops a ball, the ball will not just follow his motion when it hits the ground but will actually run ahead of it. This is connected to another curious fact, that an object thrown horizontally can gain an additional velocity which is much greater than that which it received from the thrower. I have often wondered at this when I have watched people playing with hoops,* which move through the air at a given speed when they leave the thrower's hand, but their speed then increases significantly when they hit the ground. Then, if as they roll they strike an obstacle which makes them bounce up into the air, the hoops move very slowly when they are in the air and then revert to moving fast when they touch the ground again. And I have noticed the strangest thing of all, that not only do they always move more rapidly on the ground than in the air, but when they roll along the ground more than once their motion is sometimes more rapid over the second stretch of ground than over the first. I wonder what signor Simplicio makes of this?

Various curious problems concerning the motion of projectiles.

SIMPLICIO. My first reaction is that I have never observed such a thing; second, that I am reluctant to believe it; and third, that if you were able to convince me of this and teach it to me with demonstrative proofs, I would say that you have demonic powers.

SAGREDO. In that case I would be like Socrates' demon,* not one from hell. But you keep talking about us teaching you, to which I reply that if someone doesn't know the truth for themselves, no one else can teach it to them. I could teach you all kinds of things which are neither true nor false, but truths—things which are necessarily true and cannot possibly be otherwise—are things which anyone of average understanding either knows for themselves or else will never be able to know. I'm sure signor Salviati takes the same view. So in the case of these problems, I say that you know the explanations for them, though you may not have noticed them.

SIMPLICIO. Well, let's not argue about that now: allow me to say that I don't know or understand these matters we are talking about, and see if you can make me understand them.

SAGREDO. This first problem depends on another: how is it that when a hoop is set in motion with a string, it travels much further, and therefore with greater force, than if it is just started off by hand?

SIMPLICIO. Aristotle has a number of problems concerning this matter of projectiles.

SALVIATI. Indeed he does, especially the one explaining how it is that round hoops run more smoothly than square ones.*

SAGREDO. Signor Simplicio, is that a problem that you would be bold enough to explain on your own, without needing anyone to teach you?

SIMPLICIO. Of course, and I think this joke has gone on long enough.

SAGREDO. In that case you know the answer to the other question. So tell me: you know that when a moving object is impeded it comes to a stop?

SIMPLICIO. Yes, provided the impediment is strong enough.

SAGREDO. And you know that moving along the ground is a greater impediment to an object than moving through the air, since the earth is hard and uneven, and the air is soft and unresisting?

SIMPLICIO. Yes, and for this reason I know that the hoop will move more rapidly through the air than along the ground, which is exactly the opposite of what you thought.

SAGREDO. Just a moment. You know that a mobile body turning on its axis has parts moving in all directions—that some parts are rising, others falling, some moving forwards, others backwards?

SIMPLICIO. Yes, I know that because it was taught to me by Aristotle.

SAGREDO. Do please tell me how he demonstrated it.

SIMPLICIO. By means of the senses.

SAGREDO. You mean that Aristotle made you able to see something which you wouldn't have seen on your own? Did he lend you his eyes? What you mean is that Aristotle told you, pointed it out to you, reminded you of it, not that he taught it to you.

So when a hoop is not rolling along but spinning vertically on its axis, some parts of it are rising and the opposite parts are falling; the upper parts are moving in one direction, and the lower parts in the opposite direction. Now imagine such a hoop suspended in the air and spinning rapidly; still spinning, it is dropped vertically onto the ground. Do you think that when it touches the ground it will continue to spin on its axis without rolling along, as it did before?

SIMPLICIO. No.

SAGREDO. What will it do, then?

SIMPLICIO. It will run rapidly along the ground.

SAGREDO. In what direction?

SIMPLICIO. In the direction that its spinning carries it.

SAGREDO. But its different parts are spinning in different directions; the upper parts are moving in the opposite direction to the lower parts, so we have to say which it will obey. Neither the rising nor the falling parts will give way to the other; and the whole hoop can't move downwards, because it's impeded by the earth, or upwards, because of its weight.

SIMPLICIO. The hoop will roll along the ground in the direction towards which its upper parts are moving.

SAGREDO. Why should it not follow the contrary parts, the ones which are in contact with the ground?

SIMPLICIO. Because they are impeded by the roughness of their contact with the ground, and by the unevenness of the ground itself; whereas the upper parts, being in contact with the air which is soft and yielding, meet little or no impediment, and so the hoop will move in their direction.

SAGREDO. So the fact that the lower parts are, as it were,

attached to the ground, makes them stand still, and only the upper parts press forwards.

SALVIATI. This would mean that if the hoop fell onto ice or some other highly polished surface, it wouldn't move forward so well, but it might well continue spinning on its axis, without gaining any forward motion.

SAGREDO. That could well be the case; but certainly it wouldn't roll as freely as it would if it fell onto a moderately rough surface. But I wonder if signor Simplicio can tell me why, if the hoop is dropped when it is spinning rapidly on its axis, it doesn't move forwards in the air as it does when it reaches the ground?

SIMPLICIO. Because if it has air both above and beneath it, neither the upper nor the lower part has anything to attach itself to, and as it has no reason to move either backwards or forwards it falls straight downwards.

SAGREDO. So the simple fact of spinning, without any other impetus, is enough to propel the hoop quite rapidly as soon as it hits the ground. Now for the next point. When a player winds a cord around the hoop, then attaches it to his arm and pulls it back, what effect does the cord have on the hoop?

SIMPLICIO. It makes the hoop spin round, to unwind itself from the cord.

SAGREDO. So when the hoop hits the ground it is already spinning because of the cord. Does it not therefore have in itself a reason for moving more rapidly on the ground than in the air?

SIMPLICIO. Yes, certainly: when it was in the air its only impulse was what it received from the thrower's arm; it had a spinning motion as well, but as we have said, this has no propulsive effect when it is in the air. But when it hits the ground, the spinning motion is added to the motion derived from the thrower's arm, and so its speed is redoubled. It's quite clear to me now why the hoop's speed decreases when it bounces up in the air, because then it no longer derives any propulsion from spinning; and when it hits the ground again this propulsive effect returns, and so it once more travels more rapidly than in the air. The only thing I still don't understand is how this second phase of moving along the ground can be more rapid than the first, because if this were the case it would constantly accelerate and its motion would be infinite.

SAGREDO. I didn't say that this second phase of motion was necessarily more rapid than the first, only that this can sometimes happen.

SIMPLICIO. This is the point which I don't follow, and I'd very much like to understand it.

SAGREDO. This too is something you know already. So tell me this: if you were simply to drop the hoop from your hand, without it spinning at all, what would it do when it hit the ground?

SIMPLICIO. Nothing; it would just remain where it was.

SAGREDO. Think again: could it not happen that it would acquire motion when it hit the ground?

SIMPLICIO. Only if we dropped it onto a stone with a sloping surface, as children do when they play with counters; then it might acquire a spinning motion by hitting the sloping surface obliquely, and this motion could carry it forward along the ground. Apart from this, I can't think of anything else which would prevent it from just standing still where it landed.

SAGREDO. This shows that it is possible for the hoop to acquire a new spinning motion. Is there any reason, then, why a hoop should not bounce up into the air and land obliquely on a stone sloping in the direction in which it's travelling, and so acquire a new spinning motion in addition to what it originally derived from the cord? And in this case, wouldn't its motion be redoubled, making it move more rapidly than when it first hit the ground?

SIMPLICIO. I understand now how this can easily happen. This has made me think of what would happen if the hoop were made to spin in the opposite direction: then, when it hit the ground the spinning motion would have the opposite effect, and would slow down the motion derived from the thrower.

SAGREDO. It would slow it down, and sometimes it would cancel it out altogether, if the spinning motion was fast enough. This is the explanation for the effect which expert tennis players can produce to their advantage, when they deceive their opponent by cutting the ball (this is the expression they use), hitting it obliquely with the racket so as to give it a spinning motion which is contrary to the motion with which they project it. The result is that when the ball hits the ground, instead of bouncing towards the opposing player as it would if it were not spinning, giving him time to return it, it stops dead or at least bounces much less than it normally would, so that the opponent has no time to return it. The same thing happens with players at bowls, where the aim is to come as close as possible to a defined goal. If they are playing on a stony road full of obstacles which would deflect the balls all over the place instead of letting them move towards the goal, the players avoid all these obstacles by deliberately throwing the ball like a plate instead of rolling it along the ground. But if the ball is held in the usual way, with the player's hand underneath it, the player's fingers will give it a spin which, added to the propulsive effect of the throw, will carry it a long way forward when it lands near the goal. To prevent this happening and to make it stop where it lands, they skilfully hold the ball underneath their hand; then when they throw it, the ball acquires a reverse spin, so that when it lands near the goal it stops or rolls only a little way.

But to return to our original problem, from which all these others arose, I say it is quite possible for someone who is moving rapidly to drop a ball from their hand in such a way that, when it lands, the ball not only follows the thrower's motion but outruns it, moving with more speed than the thrower himself. To illustrate this, imagine a carriage with a sloping board fixed to the outside, with its lower edge towards the horses and its upper edge towards the back wheels. Now, if someone in the carriage drops a ball on the sloping board when the carriage is moving at full speed, the ball will acquire a spinning motion as it rolls down the board, and this added to the motion imparted to it by the carriage will propel the ball along the ground much faster than the carriage itself. If you were then to fix another board sloping the other way, it would be possible to counteract the carriage's motion so that, when the ball rolled down the board, it stood still when it hit the ground, or sometimes even ran back in the opposite direction. But we've spent too long on this digression, and if signor

Simplicio is now satisfied that we have answered the first objection against the motion of the Earth, based on the motion of falling objects, we can pass on to the others.

SALVIATI. Our digressions so far are not really so remote from the subject in hand as to be considered totally unrelated to it; and in any case, our discussions are prompted by the ideas which occur to all three of us, not just to one person's fancy. Besides, we are conversing for our own enjoyment, and we're not obliged to be as rigorous as someone speaking in a professional capacity and giving a methodical exposition of a subject, perhaps even with the intention of publishing it. I wouldn't want our narrative poem* to follow the principle of unity so strictly that it left us no room for self-contained episodes, which we should be able to introduce on the slightest pretext. It should be as if we were here to tell stories, and I should be entitled to tell whatever story was prompted in my mind by listening to yours.

SAGREDO. That suits me very well. And since we've allowed ourselves this freedom, perhaps I could ask you, signor Salviati, another question before we move on. Have you ever had occasion to speculate about the line followed by an object falling from the top of a tower? If you have, I'd very much like to hear your thoughts on the matter.

SALVIATI. Yes, I have given it some thought. I have no doubt that, provided one was certain about the nature of the motion with which a heavy object falls towards the centre of the Earth, this could be combined with the circular motion of the Earth's daily revolution to define exactly the kind of line followed by the centre of gravity of a falling body: it would be a composite of these two motions.

SAGREDO. I think we can believe with absolute certainty that simple movement towards the centre as the result of gravity follows a straight line, just as it would if the Earth were at rest.

SALVIATI. Indeed, not only can we believe this, but experience makes it quite certain.

SAGREDO. How can we be assured of it by experience, if we only ever see the composite motion, of circular and downward motion combined?

SALVIATI. Actually, signor Sagredo, we only see the simple downward motion; the other, circular motion which is common to the Earth, the tower and ourselves, is imperceptible to us, as if it didn't exist. The only motion which is apparent to us is the one that we don't share, which is that of the falling stone; and our senses tell us that this follows a straight line, since it remains parallel to the side of the tower, which is built vertically on the surface of the Earth.

SAGREDO. Yes, you're right. How foolish of me not to have thought of such a simple fact. But since this is so obvious, what else do you think we need to know to understand the nature of this downward motion?

SALVIATI. It's not enough to know that it is in a straight line; we also need to know whether it is uniform or not—that is, whether it maintains a constant speed or whether it accelerates or decelerates.

SAGREDO. Surely it's clear that it continually accelerates?

SALVIATI. That's still not enough; we need to know at what rate it accelerates. This is a problem which I don't believe any philosopher or mathematician has yet been able to solve,* even though philosophers—especially Aristotelians—have written whole vast volumes on the subject of motion.

SIMPLICIO. Philosophers generally concern themselves with universals: they establish definitions and the most common manifestations of them, and leave the minutiae and trivial details—which are more curiosities than anything else—to mathematicians. So Aristotle confined himself to providing an excellent definition of motion in general, and to identifying the principal attributes of local motion: that it can be natural or forced, simple or composite, uniform or accelerating. As regards accelerating motion, he was content simply to provide an explanation for acceleration, leaving it to mechanics or other inferior artisans to work out the rate of acceleration and other such incidental details.

SAGREDO. Yes of course, my dear Simplicio. But you, signor Salviati, have you ever descended from the Peripatetic throne to amuse yourself with this question of the rate of acceleration of falling objects?

SALVIATI. I haven't needed to think about it, because our

mutual friend the Academician has shown me a treatise of his on motion,* in which he demonstrates the answer to this and many other questions. But it would be too much of a digression to interrupt our present discussion, which is itself a digression, in order to explain this; it would be like having a play within a play.

SAGREDO. I am content to excuse you from telling this story now, on condition that this is one of the subjects which we set aside to cover in a separate discussion, as I would very much like to understand it. In the meantime, let's return to the line followed by an object falling from the top of a tower to its base.

SALVIATI. If rectilinear motion towards the centre of the Earth were uniform, the Earth's circular motion towards the east also being uniform, then the composite of these two motions would be a spiral, as described by Archimedes in his book on spirals.* This is the line followed by a point moving uniformly along a straight line which is itself rotating uniformly around one of its extremities as the fixed centre of its revolution. But since the rectilinear motion of a falling object continually accelerates, the line of its composite motion must diverge at an ever-increasing rate from the circumference of the circle which the stone's centre of gravity would have followed if it had remained at the top of the tower. This divergence must initially be very small, in fact tiny or indeed minute, since a falling body starting from a state of rest—that is, passing from a state where it has no downward motion to one where it is moving directly downwards—must pass through all the infinite degrees of slowness between a state of rest and any given speed, as we have already discussed and established at length.*

Given, then, that a falling object accelerates in this way, and given that its motion would terminate at the centre of the Earth, the line followed by its composite motion must be such that while its distance from the top of the tower (or rather, from the circumference of the circle traced by the top of the tower due to the rotation of the Earth) grows at an ever-increasing rate, this distance decreases by degrees ad infinitum the less the falling body has moved away from the point where

The line followed by a naturally falling object, assuming the Earth's rotation on its axis, would probably be the circumference of a circle.

it was at rest. Moreover, the line of this composite motion must terminate at the centre of the Earth.* Making these two assumptions, I have drawn a circle BI around a centre A with a radius AB to represent the terrestrial globe. Then I have extended the radius AB to C to represent the tower, its height being BC; as the tower moves on the surface of the Earth following the circumference BI, its top describes the arc CD. Now I divide the line CA at its middle point E, and I draw a semicircle CIA with E as its centre and EC as its radius. I think it highly probable that this semicircle would be the line followed by a stone falling from the top of the tower C, as the composite of the Earth's circular motion and the stone's own rectilinear motion. To illustrate this, we can mark several equal sections of the circumference CD, calling these CF, FG, GH, and HL. Now let us draw a straight line from each of the points F, G, H, and L to the centre, A. The part of each of these lines enclosed by the circumferences CD and BI will always represent the tower CB, as the Earth's rotation carries it towards DI; and the point where each line is intersected by the semicircular arc CI will be the point which the falling stone has reached at a given moment. It will be seen that these points diverge from the top of the tower at an increasing rate, which is why the stone's vertical motion down the side of the tower appears to be constantly accelerating. It can be seen, too, from the infinite acuteness of the angle where the two circles DC and CI meet, that the distance of the falling object from the circumference CFD—that is, from the top of the tower—is initially very small, which is the same as saying that its motion is initially very slow, and more and more so ad infinitum the closer it is to point C, i.e. to a state of rest. Finally, this also shows that its motion would ultimately end at the centre of the Earth, A.

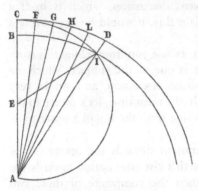

SAGREDO. I understand all this perfectly, and I don't see how the centre of gravity of a falling object could follow any other line than this.

SALVIATI. That's not all, signor Sagredo; I have three more little reflections to offer you, which I think will not displease you. The first is that, if we consider it carefully, the real motion of a falling object is none other than simple circular motion, just as it was when it was at rest at the top of the tower. The second is even more intriguing: it moves neither more nor less rapidly than if it had stayed where it was at the top of the tower, for the arcs CF, FG, GH, etc., which it would have moved through at the top of the tower, are exactly equal to the corresponding arcs below CF, FG, GH, etc., on the circumference CI. And the third marvel follows from this: that the true actual motion of the stone does not accelerate but is always equal and uniform, because it moves through the equal arcs on the circumference CD and the corresponding arcs on the circumference CI all in the same space of time. This leaves us free to look for new causes for acceleration and other motions, because a moving object always moves in the same way, i.e. in a circle, and at the same constant speed, whether it stays at the top of the tower or whether it falls. Now, what do you think of this fancy of mine?*

An object falling from the top of a tower moves on the circumference of a circle.

It moves neither more nor less than if it had remained at the top of the tower.

It moves with equal, not accelerating motion.

SAGREDO. I can't find words to express how marvellous I find it, and I can think of nothing at present to suggest that the facts are other than you have described them. If only all philosophers' demonstrations were half as convincing as this! But to satisfy my curiosity, I would very much like to hear the proof that these arcs are all equal.

SALVIATI. The proof is very easy. Draw this line, IE. Since the radius of the circle CD, namely the line CA, is twice the length of CE which is the radius of the circle CI, the circumference of CD must be twice the circumference of CI, and any arc of the larger circle must be twice the length of a similar arc of the smaller circle. Now the angle CEI, with its vertex at the centre E of the smaller circle and subtending the arc CI, is double the angle CAD, with its vertex at the centre A of the larger circle and subtending the arc CD. Therefore the arc CD is half the arc of the larger circle which would be similar to the arc CI, and hence the arcs CD and CI are equal. The same proof could be applied to every other part.

However, I don't want to state categorically for now that the motion of falling objects occurs in exactly this way; but I will say that if the line followed by a falling object is not exactly as I have described it, it is very close to it.

SAGREDO. I have been reflecting on another extraordinary fact. If all these considerations are valid, then there is no place for motion in a straight line and nature has no use for it. It doesn't serve even that purpose which you allowed for it at the outset,* of restoring to their proper place any parts of an integral body which had become separated from the whole and were therefore wrongly disposed, since this too is brought about by circular motion.

Rectilinear motion seems to be entirely excluded in nature.

SALVIATI. This would be a necessary consequence if we had conclusively proved that the terrestrial globe has a circular motion; but I don't claim to have proved this. All we have done so far, and will continue to do, is examine the strength of the arguments which philosophers have put forward to prove the immobility of the Earth. The first of these, based on the motion of falling objects, is open to the objections we have been discussing. I don't know how much weight these carry with signor Simplicio; so I think it would be good to hear whether he has anything to say in reply to them, before we go on to test the other arguments.

SIMPLICIO. As regards this first argument, I must confess that I've heard a number of subtle points which hadn't occurred to me; and as they are new to me I'm not able to respond to them here and now. But in any case, I don't think this argument based on falling objects is one of the strongest in favour of the immobility of the Earth. I'll be interested to see what we shall make of the question of shots fired from a cannon, especially when they are in the opposite direction to that of the daily motion.

SAGREDO. I wish I could understand the flight of birds with no more difficulty than shots from a cannon and all the other examples we've mentioned above! But how birds can fly freely in all directions and, even more, can stay in the air for hours at a time, is something that baffles me completely. I just can't understand how they don't lose track of the motion of the Earth as they fly back and forth, or how they can keep up with its velocity, which must be many times greater than the speed of their flight.

SALVIATI. Your doubts about this are well founded; in fact I wonder whether Copernicus himself was able to solve them to his entire satisfaction, or whether that was the reason why he said nothing about them. Though it's true that he dealt very briefly with the other objections to his theory, I think because, with his great genius,* his mind was on greater and higher things—rather as lions pay little attention to the importunate barking of small dogs. I suggest, then, that we leave the example of birds until last, and in the meantime try to resolve signor Simplicio's other doubts, following our usual method of showing him that he already has the answers to hand although he is not aware of them. So let us begin with shots fired from a cannon: if two shots were fired from the same cannon, with the same ball and the same amount of powder, one eastwards and the other westwards, on what grounds does he believe that (if it is the Earth which rotates every twenty-four hours) the shot fired to the west would travel much further than the one fired towards the east?

SIMPLICIO. I believe that this would happen because, when a shot is fired towards the east, the cannon follows the ball after it has been fired, being carried along by the Earth as it moves rapidly in the same direction; therefore the ball falls to Earth only a short distance from the cannon. But when a shot is fired towards the west, by the time the ball falls to Earth the cannon has moved some considerable distance towards the east. Therefore the length of the shot—that is, the distance between the ball and the cannon—in the second case will appear to exceed the first by the distance which the Earth, and therefore the cannon, has travelled in the time in which the two balls were in the air.

The reason why a cannon ball fired in a westerly direction should appear to travel further than one fired towards the east.

SALVIATI. I would very much like to devise an experiment corresponding to the motion of these projectiles, as we did with the ship when we were discussing falling objects. I'm just trying to think how it could be done.

An experiment with a moving carriage to illustrate the difference between the shots.

SAGREDO. I think a suitable proof could be set up by taking an open carriage and mounting a crossbow on it at half elevation, so as to maximize the length of the shot. Then, as the horses were running, you would fire one shot in the direction in which they were travelling, and then another in the opposite direction. You would have to make a careful note of the carriage's position at the moment when the bolt landed in each case, and from this you would be able to see how much one shot proved to be longer than the other.

SIMPLICIO. I think this experiment would serve very well, and I have no doubt that the length of the shot—that is, the distance between the arrow and the position of the carriage at the moment when the arrow landed—would be notably less in the case of the shot fired in the direction the carriage was travelling than with the shot fired in the opposite direction. Suppose, for example, that the length of the shot itself was three hundred *braccia*, and that the carriage travelled a hundred *braccia* in the time that the arrow was in the air. With the shot fired in the direction the carriage was travelling, the carriage would have covered one hundred of the three hundred *braccia*, and therefore when the arrow landed its distance from the carriage would be only two hundred *braccia*. But with the shot fired in the opposite direction, the carriage would be moving away from the arrow, and therefore when the arrow had flown its three hundred *braccia* the carriage would have travelled one hundred *braccia* in the other direction, and so the distance between them would be four hundred *braccia*.

SALVIATI. Would there be any way of making the two shots equal?

SIMPLICIO. Only by making the carriage stand still.

SALVIATI. Of course, but I meant with the carriage travelling at full speed.

SIMPLICIO. You would have to increase the tension of the bow when you were shooting in the direction of travel, and reduce it when you were shooting the other way.

SALVIATI. So then, there is another way of doing it. How much would you have to increase the tension of the bow in one case, and reduce it in the other?

SIMPLICIO. In our example, where we supposed that the bow shot the arrow three hundred *braccia*, we would have to increase the tension so as to shoot four hundred *braccia* for the shot in the direction of travel, and reduce it so as to shoot only two hundred *braccia* for the other. This would mean that both shots would be three hundred *braccia* in relation to the carriage, because the hundred *braccia* travelled by the carriage would be subtracted from the four hundred-*braccia* shot and added to the one which was only two hundred, making three hundred for both.

SALVIATI. And what effect does the greater or lesser tension of the bow have on the arrow?

SIMPLICIO. When the tension of the bow is increased it shoots the arrow with a greater velocity, and when its tension is reduced it shoots with less. So the same arrow travels a greater distance in one case than the other, in proportion to the greater or lesser velocity with which it is released from the bow in each case.

SALVIATI. So, for the arrow to travel an equal distance from the moving carriage in both directions, in the first case in your example it would need to be shot with, say, four degrees of velocity, and in the second case with only two. But if it was shot with the bow at the same tension, it would have three degrees every time.

SIMPLICIO. That's correct; and that's why, if the arrows are shot with the bow at the same tension from a moving carriage, the shots can never be equal.

SALVIATI. I forgot to ask at what velocity the carriage is assumed to be travelling in this particular experiment.

SIMPLICIO. The velocity of the carriage must be assumed to be one degree, in comparison with the velocity imparted by the bow, which is three.

SALVIATI. Yes indeed; this makes the figures add up correctly. But surely, when the carriage is moving, everything in the carriage is also moving at the same velocity?

SIMPLICIO. Undoubtedly.

SALVIATI. And this applies to the bow, the arrow, and the bowstring from which the arrow is shot.

SIMPLICIO. Yes.

SALVIATI.   So when the bolt is shot in the direction in which the carriage is travelling, the bow imparts its three degrees of velocity to a bolt which has one degree already, because this is the speed at which the carriage is carrying it in the same direction; and therefore the bolt is released from the bowstring with four degrees of velocity. When, on the other hand, the bolt is shot the other way, the same bow imparts the same three degrees of velocity to a bolt which is moving with one degree in the opposite direction, and therefore it leaves the bowstring with only two degrees of velocity. But you yourself have already said that for the two shots to be equal, the bolt would need to be shot with four degrees of velocity in one case and two degrees in the other; so without any need to change the tension of the bow, it is the carriage itself which makes the necessary adjustment. The experiment will serve to confirm this to those who are unwilling or unable to be convinced by reason.

Solution of the argument based on shots fired eastwards and westwards from a cannon.

If you now apply this reasoning to the shots fired from a cannon, you will find that, whether the Earth is in motion or at rest, shots fired with the same force will always be equal, regardless of the direction in which they are fired. The error which Aristotle, Ptolemy, Tycho, you, and everyone else have made, is based on the fixed and deep-seated impression that the Earth is at rest, which you are incapable of shedding even when you want to speculate about what would happen if the Earth were in motion. In the same way in our earlier discussion, it didn't occur to you that when the stone is at the top of the tower, the question of whether or not it is in motion depends on whether or not the terrestrial globe is in motion. Because you have it fixed in your mind that the Earth is at rest, you always talk about the falling stone as if it were starting from a state of rest; whereas you ought to say: 'If the Earth is at rest, the stone starts from a state of rest and falls vertically; but if the Earth is in motion, the stone is also in motion with the same velocity, and it starts not from a state of rest but from one of motion equal to that of the Earth. In this case, its downward motion combines with the Earth's motion to form an oblique motion.'

SIMPLICIO.   But for heaven's sake, if it moves obliquely how is it that I see it falling vertically in a straight line? This is simply denying the clear evidence of our senses; and if we can't believe our senses, what other basis is there for our philosophy?

SALVIATI. As far as the Earth, the tower, and we ourselves are concerned, since we all share the same daily motion, together with the stone, the daily motion is as if it did not exist: it is insensible, imperceptible, and has no effect whatever. The only motion which is observable to us is the one which we do not share, namely the motion of the stone falling down the side of the tower. You are not the first to experience great reluctance in accepting this ineffectiveness of motion among things which have it in common.

SAGREDO. This reminds me of a fanciful idea which came to me one day as I was sailing to Aleppo, where I was going as our country's ambassador. It might help to explain this ineffectiveness of motion in common, which makes it seem as if it did not exist to those who participate in it; and with signor Simplicio's permission, I would like to tell him about the fancy which occurred to me then.

A striking case cited by Sagredo to show the ineffectiveness of motion among things which have it in common.

SIMPLICIO. Please do; I'm not just willing, but eager to hear this novelty.

SAGREDO. If the nib of a pen which was on the ship all the time it was sailing from Venice to Alexandretta* had been able to leave a visible mark of where it had travelled, what kind of trace, or record, or line would it have left?

SIMPLICIO. It would have left a line stretching from Venice to its destination; not a perfectly straight line—or rather, not a perfect arc of a circle—but fluctuating here and there depending on the varying motion of the vessel. But these deviations from a straight line by a few *braccia* to the right or left, up or down, over a distance of many hundreds of miles, would have been insignificant over the whole length of the line and would have been almost imperceptible. So it would not be seriously misleading to say that the line would have been part of a perfect arc.

SAGREDO. So taking away the motion of the waves, if the ship had travelled the whole way over a perfectly calm sea, the absolutely true, real motion of the pen nib would have been a perfect arc of a circle. What if I had had the pen in my hand the whole time, and had occasionally moved it an inch or two this way or that: what difference would that have made to its principal very long line?

SIMPLICIO. Less than if a line a thousand *braccia* long had occasionally deviated from absolute straightness by the breadth of a flea's eye.

SAGREDO. Suppose, then, that an artist had used this pen to draw on a piece of paper, starting when the ship left port and continuing until it reached Alexandretta. He could have used the pen's motion to draw the whole story of the journey, with lots of figures perfectly outlined and sketched in a thousand different ways, with different countries, buildings, animals, and all kinds of things; and yet the real, true, essential movement of the pen nib would have been nothing more than one very long, simple line. As far as the artist was concerned, his drawing would have been exactly the same if the ship had never moved; and yet the only trace of the pen's very long journey would be those lines drawn on a sheet of paper. The reason for this is that the major motion from Venice to Alexandretta was common to the paper, the pen, and everything else on the ship; whereas the small movements up and down, left and right, which the artist's fingers transmitted to the pen and not to the paper, were peculiar to the pen, and left their marks on the paper because the paper was fixed in relation to them. In the same way, since the Earth is moving, the motion of the falling stone actually extends over several hundred or even several thousand *braccia*, and if it could trace the course which it followed in motionless air or on some other surface, it would leave a very long oblique line. But that part of its motion which is common to the stone, the tower, and us is imperceptible to us, as if it did not exist; the only part of its motion which we can observe is the part in which neither we nor the tower participate, which is the stone's motion as it falls down the height of the tower.

SALVIATI. A very subtle illustration of this point, which many people find very hard to understand. Now, if signor Simplicio has nothing more to say in response, we can move on to the other experiments, which will be a good deal easier to explain in the light of what we have said so far.

SIMPLICIO. No, I've nothing to add. I was absorbed by this example of drawing, and by the thought that all those lines drawn in all directions, this way and that, up and down, back and forth, with a thousand-and-one twists and turns, are in reality nothing more than fragments of a single line all drawn in the same direction. The pen may move a little to the right or left, or move more rapidly one moment and more slowly the next, but these are only minimal deviations from a single straight line. It occurs to me that the same is true when we write a letter, and I was thinking of those expert calligraphers who show off their dexterity by embellishing a letter in a single stroke with innumerable flourishes without lifting their pen from the paper. If they were on a rapidly moving ship, they would produce one of their decorations from the motion of the pen, which is essentially a single line all drawn in the same direction, simply by means of minimal deviations or deflections from a perfectly straight line. I'm most grateful to signor Sagredo for suggesting this thought to me. So let us continue, and I shall be all the more attentive in the hope of hearing more such ideas.

SAGREDO. If you should be curious to hear other such witticisms, which not everyone can appreciate, there are plenty of others, especially in this matter of navigation. What do you think of the fine idea which occurred to me on this same voyage, when I realized that the crow's nest on the ship's mast had travelled further than its foot, without the mast bending or breaking? Clearly the top of the mast is further from the centre of the Earth than its foot, and therefore it travelled through an arc of a larger circle than the foot.

Ironic statement of foolish subtleties taken from an encyclopedia.

SIMPLICIO. So when a man walks, his head travels further than his feet?

SAGREDO. You've understood perfectly, and have worked out the principle for yourself. But we mustn't interrupt signor Salviati.

SALVIATI. I'm glad to see signor Simplicio exercising his mental skills—if indeed this idea is his own, and he hasn't learnt it from a certain handbook of assertions* which contains various others no less acute and amusing. But we should continue our discussion of a shot fired from a cannon set up at right angles to the horizon, in other words vertically into the air, and the fact that the ball falls straight back down onto the cannon again. If, the argument goes, the Earth has carried the cannon several miles towards the east during the long interval in which the ball was separated from it, the ball should fall down to earth some distance to the west of the cannon; but this does not happen, and therefore the cannon has waited for the ball and has not moved. The answer to this objection is

the same as with the stone falling from a tower: the misunderstanding and the fallacy in the argument consist in presupposing the truth of what it set out to prove. The objector is always firmly convinced that the cannon-ball starts from a state of rest when it is fired from the cannon; but it can only start from a state of rest if we presuppose that the terrestrial globe is at rest, which is precisely the conclusion which is being tested. So I reply that those who maintain the mobility of the Earth point out that the cannon and the ball inside it both share in the motion of the Earth; indeed, like the Earth, they derive this motion from nature. Therefore the ball does not start from a state of rest, but shares in this circular motion around the centre; and this motion is neither negated nor diminished when it is fired from the cannon. So the ball follows the Earth's general motion towards the east, and hence remains continually above the cannon both as it goes upwards and as it comes down. You would see the same effect if you tested it by shooting a ball straight up in the air from a catapult on the deck of a ship: it would come back down in the same place, whether the ship was moving or not.

SAGREDO. This answers the objection entirely. But since I see that signor Simplicio appreciates cunning arguments

to trap the unwary, let me put this to him. Supposing for a moment that the Earth is at rest, and the cannon is aimed vertically straight up into the sky, does he have any difficulty in accepting that such a shot is truly vertical, and that the ball will follow the same straight line both when it is shot upwards and when it comes down, always assuming that any external or accidental impediments are eliminated?

SIMPLICIO. Yes, as I understand it, that is exactly what would happen.

SAGREDO. If instead of being aimed vertically, the cannon were set up at an incline in a given direction, what motion would the ball follow then? Would it still go vertically straight up in the air and follow the same line when it comes down, as with the other shot?

SIMPLICIO. No, it wouldn't. It would follow a straight line in the direction in which the barrel of the gun was pointing, except in so far as its own weight made it decline towards the Earth.

SAGREDO. Therefore the direction of the gun's barrel determines the motion of the ball, which does not deviate from this line, or would not if it were not pulled down by its own weight. So then, if the barrel is aimed vertically and the ball is shot straight up in the air, it will come down following the same straight line, because the motion due to the heaviness of the ball pulls it down in the same vertical line. When the ball leaves the cannon it continues to travel in the same direction as that portion of its journey which it made inside the barrel, does it not?

Projectiles continue their motion in a straight line following the direction of the motion which they shared with the projector before they were separated from it.

SIMPLICIO. So it seems to me, yes.

SAGREDO. Now imagine the barrel aimed vertically, and the Earth rotating in its daily motion on its axis and carrying the cannon with it. What will be the motion of the ball inside the barrel when it is fired?

SIMPLICIO. It will move vertically in a straight line, if the barrel is aimed vertically.

SAGREDO. Think carefully, because I don't think it would be vertical at all. It would be vertical if the Earth were at rest, because then the ball would have no other motion apart from that imparted to it by the firing of the cannon. But if the Earth rotates, then the ball in the cannon also has this daily motion. This means that when it is fired, it travels from the breech of the cannon to its muzzle with two motions, the combined effect of which is that the ball's centre of gravity travels in an inclined line. This illustration will make it clearer: AC is the upright barrel of the cannon, and B is the ball inside it. Clearly if the cannon is fired when it is motionless, the ball will emerge from the muzzle A, its centre having travelled the length of the barrel along the vertical line BA, and it will

Given the rotation of the Earth, a ball shot from a gun aimed vertically will not follow a vertical line, but an inclined one.

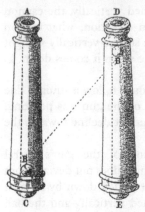

continue in this direction after it has left the cannon, travelling straight upwards. But if the Earth is rotating, and consequently carrying the cannon with it, then when the cannon is fired, it will be carried by the Earth's motion to the new position DE in the time that the ball travels the length of the barrel. So the ball will emerge from the muzzle at D, its centre having travelled along the line BD, which is no longer a vertical line but one inclined towards the east. Since we have established that the ball must continue its motion in the air in the same direction as when it was in the barrel, it will continue to follow the inclination of the line BD, not vertically but inclined towards the east. As the cannon is also moving in the same direction, the ball will be able to follow the motion of the Earth and of the cannon. And so you see, signor Simplicio, how the shot which apparently had to be vertical is in fact not so at all.

SIMPLICIO. I'm not entirely convinced by this argument; what about you, signor Salviati?

SALVIATI. I'm partly convinced, but I have a difficulty which I'm struggling to put into words. It seems to me to follow from what has been said that, if the cannon is vertical and the Earth is in motion, then not only does the ball not come down some distance to the west of the cannon, as Aristotle and Tycho say it would; it should not come down directly above the cannon either, but rather well to the east of it. You have shown that it would have two motions which would combine to impel it in that direction: the motion which it has in common with the Earth, which carries both the cannon and the ball from CA towards ED, and the motion imparted to it by the shot, which impels it along the inclined line BD. Since both of these motions carry it towards the east, the combined motion should be greater than the motion of the Earth.

SAGREDO. No, you're mistaken. The motion which carries

the ball towards the east comes entirely from the Earth, and the shot has no part in it; the motion which impels it upwards comes entirely from the shot, and has nothing to do with the motion of the Earth. The proof of this is that, if you did not fire the cannon, the ball would never come out of the muzzle or rise by a hair's breadth; and if you could stop the Earth and then fire the cannon, the ball would shoot straight upwards without any inclination at all. The ball therefore has two motions, one upward and the other circular, which combined to produce the oblique motion BD. The upward motion derives wholly from the shot; the circular motion derives wholly from the motion of the Earth, and is equal to it. Because this motion is equal to the motion of the Earth, the ball always remains vertically above the muzzle of the gun and eventually falls back down to it. And because it always remains above where the gun is aimed, it will also always appear to be overhead to anyone standing near the gun, and hence we perceive it as rising vertically straight up into the sky.

SIMPLICIO. I have another difficulty. The ball travels the length of the gun at such a speed that it seems impossible for the gun to move in that brief moment from CA to ED. How can it impart such a degree of inclination to the oblique line CD that the ball is able to keep up with the motion of the Earth when it is in the air?

SAGREDO. You make several mistakes here. First of all, I think the inclination of the line CD is much greater than you imagine, because I have no doubt that the speed of the Earth's motion—even at our latitude, never mind at the equator—far exceeds that of the cannon-ball when it travels the length of the barrel. So the distance CE cannot fail to be greater than the length of the gun barrel, and hence the inclination of the oblique line must be more than forty-five degrees. But in any case, it makes no difference whether the Earth's velocity is greater or less than that of the cannon-shot, because if the Earth moves more slowly and therefore the inclination of the diagonal is less, a smaller inclination will be all that is needed for the ball to maintain its position above the cannon. In fact, if you think carefully about it, you will see that when the Earth's motion moves the cannon from CA to ED, it bestows exactly the degree of inclination on the diagonal CD that is needed to compensate for the shot. But you also make a second mistake, by attributing the ability of the ball to keep pace with the Earth's motion to the impetus of the shot. This is the same error which signor Salviati seemed to fall into just now; because the motion by which the ball keeps pace with the Earth is the perennial, perpetual motion which it indelibly and inseparably shares with every other terrestrial object, and which it naturally possesses and will possess in perpetuity.

SALVIATI. We'd better give in, signor Simplicio; it is indeed just as he says. But this discussion has helped me to resolve a puzzle about wildfowling, which is how marksmen with an arquebus can hit and kill birds in flight. I had always imagined that in order to hit a bird they aimed a certain distance in front of it, allowing for how fast the bird was flying and how far away it was. Then, when they fired the gun and the bullet followed their aim, it reached the same point at the same time as the bird in flight, and so hit it. But when I asked one of them if this was indeed what they did, he said not, and that their technique was much easier and more certain than that. He told me that they use exactly the same method as they would to shoot a bird which is sitting still. They take aim at the bird in flight, and follow it with their gun as it moves, keeping it in their sights until they shoot; and this enables them to hit it as they would a bird sitting still. So the movement of the gun, slow as it is, as they turn it to keep it trained on the bird, must be imparted to the bullet and combined with the impetus imparted by the shot. Thus the bullet receives the upward motion from the shot, and the declining motion following the bird's flight from the barrel of the gun, just as we said just now with the cannon-ball. The cannon-ball receives its upward motion from the shot, and its inclination towards the east from the motion of the Earth; the two combined make it follow the motion of the Earth while appearing to an observer to be simply moving straight upwards, and to follow the same line when it comes down. In the same way with the huntsman's bullet, keeping the gun continuously trained on the target means that the shot hits the mark; and to keep the gun trained on the target, if the target is stationary then the gun must be

How marksmen are able to kill birds in flight.

held still; if the target moves, the gun must move to keep it in its sights.

The response to the objection based on cannon-shots fired at a target to the north or south follows the same principle. It is objected that if the Earth was in motion, such shots would always veer towards the west, because in the time that the ball was in the air between leaving the gun and reaching its target, the target would move eastwards, leaving the ball to the west of it. My response is to ask whether a cannon which is trained on a target and then left in place will continue to be focused on the same target, regardless of whether the Earth moves or not. The answer must be that the aim of the cannon does not change at all: if the target is stationary, so is the cannon; if the target moves because it is carried along by the Earth, the cannon moves in the same way. And if it remains trained on the target the shot will always hit the mark, as will be clear from what has been said above.

Solution to the objection based on cannon-shots fired towards the north and south.

SAGREDO. Allow me to interrupt you for a moment, signor Salviati, and to put forward a thought which has occurred to me about these marksmen hitting birds in flight. I don't doubt that their technique is as you describe it, and that the result is that they succeed in hitting the bird. But this doesn't seem to me to be quite the same as the example of shots fired from a cannon, which have to hit their target whether cannon and target are both in motion or whether they are both at rest. The differences seem to me to be as follows. When the cannon is fired, both it and the target are moving at an equal speed, because they are both carried along by the motion of the terrestrial globe. It's true that sometimes the cannon is sited nearer to the pole than the target, so that its movement is somewhat slower because it is turning in a smaller circle; but the short distance between the cannon and the target makes this difference imperceptible. When a marksman is shooting at a bird, on the other hand, the motion of the arquebus as it follows the bird's flight is far slower than the bird itself; so I don't think it is possible that the limited motion which the turning gun-barrel imparts to the bullet inside it can be multiplied when the bullet is in the air—certainly not to the extent of equalling the speed of the bird's flight, and hence remaining directed at the bird. Rather, the bird must fly ahead of the bullet and leave it behind. What's more, there is no suggestion that the air through which the bullet passes is moving with the speed of the bird; whereas in the case of the cannon, the gun, the target, and the intervening air all share equally in the universal daily motion. So I think there must be several reasons why the marksman succeeds in hitting the bird. He must not just follow the bird's flight with his gun-barrel but anticipate it a little, aiming slightly ahead of it. Then, I believe they shoot not just with one bullet but with a good number of small pellets, which spread out in the air and cover quite a wide area. And there is also the great speed with which the pellets are fired from the gun and fly towards the bird.

SALVIATI. See how signor Sagredo's wit flies ahead of mine and leaves me behind; I might well have noticed these differences, but only after long and careful thought. But to return now to our main topic, we still have to consider what happens with cannon-shots fired horizontally towards the east and towards the west. It is argued that if the Earth moved, the former would always pass above their target and the latter would be below it, because the daily motion of the Earth means that the east is always sinking below the tangent parallel to the horizon, and the west is always rising. This is why the stars appear to rise in the east and to set in the west. Therefore a shot aimed along this tangent in an easterly direction would be too high, because the target sinks while the ball is travelling, and one aimed in a westerly direction would be too low, because the target rises as the ball moves along the tangent. The answer to this objection is the same as to the others: that just as the motion of the Earth means that the target to the east is continually falling in relation to a fixed tangent, so the cannon is also continually declining for the same reason. Hence it remains trained on the same target, and the shot hits the mark. But this may be an opportune moment for me to point out that the followers of Copernicus have perhaps been too generous in what they concede to their opponents. They accept as established fact some experiments which their adversaries have never actually carried out, such as the example of objects falling from the mast of a moving ship, and

Response to the argument based on shots fired horizontally towards the east and west.

Solution of the objection based on shots fired towards the east and west.

The followers of Copernicus are too ready to admit the truth of some dubious propositions.

many others. I'm sure that this claim to prove that cannon-shots fired to the east are too high, and those to the west too low, is one such; and since I don't believe they have ever made the experiment, I would like them to tell me what difference they would expect to find between the same shots assuming the Earth to be at rest or assuming it to be in motion. Perhaps signor Simplicio would respond on their behalf.

SIMPLICIO. I can't claim to give such a well-founded reply as someone who understands these things better than me. But I think they would say that the result would be as has already been described: that if the Earth moved, shots fired to the east would always be too high, etc., since it seems probable that the ball's path would move along the tangent.

SALVIATI. And if I said that this is in fact what happens, how would you set about challenging my claim?

SIMPLICIO. We would have to make the experiment to resolve the matter.

SALVIATI. But do you think we could find a gunner so expert that he could undertake to hit the target with every shot at a distance of, say, five hundred *braccia*?

SIMPLICIO. No, I don't; I doubt whether anyone, however expert, could promise to be accurate to within less than a *braccio* over this distance.

SALVIATI. So if shooting is so unreliable, how can we assure ourselves of the point which is in doubt?

SIMPLICIO. We could assure ourselves of it in two ways: first, by firing a large number of shots; and second, because the velocity of the Earth's motion is such that the shot's deviation from the target would, in my view, be very large.

SALVIATI. Very large, meaning much more than a *braccio*, since we grant that this much and even more would be the normal margin of error even if the Earth is at rest.

SIMPLICIO. I have no doubt that the variation would be significantly more than this.

SALVIATI. I suggest we make a very rough calculation, just for our own amusement. If, as I hope, the figures add up, it will also serve to warn us on other occasions not to be too ready to take figures on trust, or to assent straight away to whatever we are invited to imagine. To give every possible advantage to the

Calculation of how far cannon-shots would have to deviate from their target, given the motion of the Earth.

followers of Aristotle and Tycho, let's assume that we are on the equator, and that we fire a horizontal shot from a long-range cannon at a target at a distance of five hundred *braccia*. Let's start by estimating—approximately, as I have said—the time that elapses from when the ball leaves the gun to when it reaches the target. We know that this time is very short, certainly no longer than it would take someone to walk two paces. This is less than a second, assuming that a person walks three miles in an hour: three miles is nine thousand *braccia*, and an hour contains three thousand six hundred seconds; so in a second the walker covers two-and-a-half paces, and therefore a second is longer than the time that the cannon-ball is in the air. Then, the daily revolution of the Earth is twenty-four hours, and therefore the western horizon rises by fifteen degrees in an hour, which is fifteen minutes of arc in one minute of time, or fifteen seconds of arc in one second of time. If we say that the time taken by the shot is one second, then in this time the western horizon, and also the target, rises by fifteen seconds of arc. This is fifteen seconds of a circle whose radius is five hundred *braccia*, this being what we assumed as the distance from the cannon to the target. Now let us look in the table of arcs and chords to find the chord of fifteen seconds for a radius of five hundred *braccia*. Here it is in Copernicus' book,* where we can see that the chord of one minute of arc is less than thirty parts of a radius of a hundred thousand; therefore the chord of one second of arc will be less than half of one such part, i.e. less than one part of two hundred thousand, and hence the chord of fifteen seconds will be less than fifteen parts of two hundred thousand. But fifteen parts of two hundred thousand is still more than fourth-hundredths of five hundred; so the distance by which the target will have risen while the cannon-ball is in the air will be less than four-hundredths, i.e. a twenty-fifth, of a *braccio*, in other words about an inch. Now since such a variation—that is, of shooting an inch lower than one would if the Earth were at rest—in fact occurs in all cannon-shots, how, signor Simplicio, are you going to convince me experimentally that it does not occur? Is it not clear that in order to refute me, you would first have to find a way of shooting at a target with such a degree of accuracy that shots never missed by as much as a hair's breadth? As long as shots can vary by a *braccio*, as in fact they do, I will always be able to say that this variation includes the inch which is due to the motion of the Earth.

SAGREDO. Forgive me, signor Salviati, but I think you're being too generous. I would say to the Aristotelians that even if every shot hit the very centre of the target, that would still not invalidate the motion of the Earth. After all, gunners are always trained to aim their cannon so as to hit the target with the Earth in motion; if the Earth were to stand still, their shots would not hit the mark, but those towards the west would be too high and those towards the east too low. Let signor Simplicio try to convince me otherwise.

A very subtle demonstration that, assuming the motion of the Earth, there is no reason for a cannon-shot to vary any more than if the Earth is at rest.

SALVIATI. That's a subtlety worthy of signor Sagredo. But the point is that since this variation depending on whether the Earth is in motion or at rest must necessarily be extremely small, it is bound to remain submerged among the much larger variations which occur all the time because of any number of accidental factors. So let us concede this to signor Simplicio, simply as a warning of how cautious we must be in accepting as true many experiments which those who cite them have never carried out, although that doesn't stop them boldly bringing them out when they serve to bolster their case. I say let us grant this to signor Simplicio, because the plain truth is that the outcome of these cannon-shots would be exactly the same whether the terrestrial globe is in motion or at rest. The same will be true of all the other experiments which have been or could be cited; they all appear at first sight to have some semblance of truth, in so far as the old idea that the Earth is at rest continues to perpetuate our misunderstandings.

The need to be very cautious in accepting as true experiments which those who cite them have never carried out.

Experiments and arguments against the motion of the Earth appear conclusive in so far as they perpetuate misunderstandings.

SAGREDO. For my part, I'm fully convinced by the arguments so far. It's quite clear to me now that, once our imagination has grasped the fact that this general participation in the daily rotation is common to all terrestrial things, and that this is their natural state—just as in the old world view their natural state was assumed to be a state of rest around the centre—then one has no difficulty in seeing through the false reasoning and misunderstandings which made the objections raised appear convincing. I just have one remaining doubt, as

I mentioned earlier, about the flight of birds. As living creatures, they have the faculty of moving at will in any number of ways and of staying airborne, separated from the Earth, for long spells at a time. They fly about in a completely random way, and I can't understand how in all this confusion of motion they don't lose their sense of the primary common motion, or how having lost it they find it again and compensate for it in their flight. And how do they keep up with the towers and trees which rush so precipitously towards the east? I say precipitously, because at the Earth's greatest extent this is almost a thousand miles per hour, and I don't think a swallow's flight is even fifty miles per hour.

SALVIATI. The birds would certainly struggle to keep up with the motion of the trees if they had to rely on their own wings; and if they no longer shared in the universal rotation, they would be so far left behind, and they would appear to be hurtling so fast towards the west, that it would look as if they were flying much faster than an arrow—if indeed you could see them at all. In fact, I don't think you would be able to see them, any more than we can see a cannon-ball in flight when it is propelled by the force of the gun's charge. But the truth is that the birds' own motion—their flight, that is—has nothing to do with the universal motion, which neither helps nor hinders them. This motion remains constant for the birds because of the air in which they fly, which naturally follows the spinning of the Earth, carrying with it the clouds, the birds and anything else that is suspended in it. So the birds don't need to worry about following the motion of the Earth; they can simply remain oblivious of it.

SAGREDO. I have no problem at all in understanding how the air carries the clouds along with it. They are easily moved, since they are so light and have no disposition to any contrary motion, and in any case they are matter which shares in the conditions and properties of the Earth. But the birds, as living creatures, are capable of motion which may even be contrary to the daily motion, and I find it quite hard to see how the air can restore this motion to them once they have interrupted it. What's more, they are solid bodies which have weight, and we've already seen how stones and other weighty bodies resist the impetus of the air, and even if they succumb to it they never acquire the velocity of the wind which drives them.

SALVIATI. Signor Sagredo, let's not underestimate the strength of air in motion, which has the power when it moves rapidly to move fully laden ships, to uproot forests, and to blow down towers; and yet even these violent actions of the wind can't be said to come anywhere near the velocity of the daily revolution.

SIMPLICIO. Ah, so air in motion can maintain the motion of projectiles, as Aristotle said. I thought it was very strange that he should have erred on this point.

SALVIATI. No doubt it would be able to, if it could maintain motion in itself. But when the wind drops, ships no longer move and trees are no longer uprooted; so, since the air is no longer in motion when the thrower lets go of the stone and stops moving his arm, the fact remains that it must be something other than the air which keeps the projectile in motion.

SIMPLICIO. What do you mean, when the wind drops the ship no longer moves? On the contrary, we can see that without any wind, and even with its sails furled, a vessel can carry on moving for miles at a time.

SALVIATI. But signor Simplicio, it contradicts your argument if, when the air which was driving the ship by filling the sails is no longer in motion, the ship continues on its way without any help from the medium.

SIMPLICIO. It could be said that the water is the medium which drives the ship and maintains its motion.

SALVIATI. It could, but it would be the opposite of the truth. The truth is that the water strongly resists being divided by the body of the vessel, and noisily opposes it, largely negating the velocity which the boat would acquire from the wind if the obstacle of the water were not there. I don't think, signor Simplicio, you can ever have observed how the water foams as it runs past the side of the boat, when the boat is propelled rapidly through still water by oars or the wind. If you had noted this effect you would not have thought of putting forward such a foolish suggestion now. It's becoming clear to me that you are one of that flock who, when you want to find out about such effects and to understand natural phenomena, withdraw to your study and look through your indices and guides to see if Aristotle said anything about them, instead of going out and observing boats, or catapults, or cannons. And once you've satisfied yourselves that you've established the true meaning of the text, you look no further, and it doesn't occur to you that there is any more to be known on the subject.

SAGREDO. What a happy and enviable position to be in! If all humans naturally desire knowledge,* and if being in a position is the same as believing oneself to be in it, then such people enjoy a huge privilege. They are able to persuade themselves that they know everything, and make fools of those who are aware of how little they know and who, recognizing that they know only a tiny part of what there is to be known, wear themselves out with long hours of study and reflection, and labour over experiments and observations.

Great and enviable happiness of those who are convinced that they know everything.

But please let's return to our discussion of birds. You were saying that the air, moving at great speed, could restore to the birds whatever part of the daily rotation they might lose in the course of their erratic flying hither and thither. My reply to this is that air in motion does not seem to be able to impart to a solid heavy body a speed equal to its own; and since the motion of the air is the same as that of the Earth, the air doesn't seem sufficient to restore the motion which the birds might have lost in their flying about.

SALVIATI. Your argument appears very plausible, and the doubt you express is not a trivial one. But if we look beyond appearances, I don't think that in essence it is any more substantial than the others which we have already considered and disposed of.

SAGREDO. It's clear that if it can't be shown to be necessarily conclusive then it's invalid, because only then can no valid counter-argument be produced against it.

SALVIATI. It seems to me that the greater difficulty you have with this example than with the others derives from the fact that birds are living creatures, and that therefore they can use their strength at will to resist the primary motion which is inherent in all terrestrial things. Hence, we can see birds flying upwards as long as they are alive, although this motion is impossible for them as heavy bodies; and indeed dead birds can only fall to the ground. You conclude, therefore, that the principles applying to all kinds of projectiles which we have discussed above, cannot apply to birds. This is quite true, which is why we don't see other projectiles behaving as birds do. If you were to let go of a dead bird and a live one from the top of a tower, the dead bird would behave in the same way as a stone: it would follow first the general daily motion, and second, as a heavy body, a downward motion. But there would be nothing to stop the live bird, while still participating in the daily motion, from using its wings to fly in whatever direction it liked; and this new motion would be apparent to us because it would be specific to the bird and we would have no part in it. And if it chose to fly towards the west, there would be nothing to stop it from using its wings once more to fly back to the top of the tower. This is because, in the last analysis, the effect of the bird's flying towards the west was simply to subtract one degree of velocity from the, say, ten degrees of the daily motion, of which the bird would still have nine degrees as it was flying. Then, if it settled on the ground, it would have the common ten degrees again; and by flying back towards the east it could add one more degree, and these eleven degrees would carry it back to the tower. So in short, if we examine the effects of birds' flight more closely, we find that the only way in which they differ from the motion of other projectiles in any direction is that whereas other projectiles are moved by an external projecting force, birds' flight derives from an internal principle.

Solution to the objection based on birds flying contrary to the motion of the Earth.

And now, to set the seal on the invalidity of all the experiences which have been cited so far, I think this is the time and place to show how they can all be very easily put to the test. Shut yourself in with a friend in the largest room you can find under cover on some large ship. In the room, see that you have some flies, butterflies, and other such flying creatures. Have also a large tank of water, with some fish swimming in it. Suspend from the ceiling a small bucket, so that it lets water fall a drop at a time into another container with a narrow opening placed below it. Then, while the ship is not moving, observe carefully how the flying creatures fly at the same speed towards every part of the room; the fish swim equally in every direction; the

Experiment to show the invalidity of all the examples cited as objections to the motion of the Earth.

drops of water all fall into the container below. If you throw something to your companion, you will not need to throw it with more force in one direction than another, provided the distances are the same. Jump with both feet across the floor, and you will jump the same distance in any direction. Observe all these effects carefully, even though as long as the ship is stationary there is no reason to expect them to be otherwise than as they are. Then have the ship move, at whatever speed you choose; and provided its motion is uniform and not rocking to and fro, you will not notice the slightest change in any of these effects, and none of them would give you any indication of whether the ship was moving or at rest. If you jump across the floor, you will jump the same distance as before, and however fast the ship is travelling you will not jump any further towards the stern than towards the bow, even though the floor will have moved in the opposite direction while your feet were off the ground. If you throw something to your companion, you will not have to throw it any harder to reach him whether he is nearer the bow and you nearer the stern, or vice versa. The drops of water will still fall into the container below as they did before, even though the ship travels several feet as the drop is falling. The fish will swim equally easily towards the forward-facing end of the tank as towards the aft-facing end, and will readily come to food wherever it is placed on the edge. Finally, the flies and butter-flies will continue to fly equally in all directions, and you will not notice any fewer of them in the forward part of the room, as if they were struggling to keep up with the rapid course of the ship, even though they had been out of contact with it by flying around for a long time. And if you were to produce a little smoke by burning a few grains of incense, you would see it rise in the air and form a small cloud, without moving any more in one direction than the other. The reason why none of these effects changes is that the motion of the ship is common to everything in the room, including the air, which is why I said that it should be under cover. If you were on deck in the open air, where the air is not following the course of the ship, you would see more or less notable differences in some of the effects I have mentioned. The smoke, undoubtedly, would be left behind, as would the air itself; the flies and butterflies, too, would encounter resistance from the air, and so would not be able to keep up with the motion of the ship once they had been separated from it for a significant length of time. If they stayed close to the ship, however, they would be able to follow it without difficulty, because the ship, being an irregularly shaped construction, carries some of the air close to it along with it—in the same way as we sometimes see flies and horse-flies following the horses, flying alongside different parts of their bodies, when a stage-coach is travelling at full speed. But with the falling drops of water there would be very little difference, and with jumping and throwing heavy objects it would be quite imperceptible.

SAGREDO. It never occurred to me to put this to the test when I was travelling on a ship, but I'm quite sure that these observations would be exactly as you have said. In fact, I can recall dozens of times when I was in my cabin and I had to ask whether the ship was moving or not; and sometimes if I was daydreaming I thought it was moving in one direction when in fact it was moving in the other. So I'm entirely convinced that there is no substance to the examples that have been cited as evidence against rather than in favour of the rotation of the Earth. There remains the question raised by the observation that a rapid spinning motion has the effect of extruding and dispersing any material adhering to the device which is spinning. It seemed to many, including Ptolemy,* that if the Earth rotated at such a speed, stones and living creatures would be hurled out towards the stars, and there would be no cement strong enough to attach buildings to their foundations without them also being swept away.

SALVIATI. Before I come to the solution of this objection, I can't resist recalling how often I've had occasion to laugh at the reaction of almost everyone when they first hear it suggested that the Earth which they believed to be fixed and immobile is actually in motion. Not only had they never doubted that the Earth was at rest, but they were also firmly convinced that everyone else shared their belief that it had been created immobile and had remained so throughout all the past centuries. With this idea firmly fixed in their mind, they are then

Foolishness of those who believe that the Earth began to move only when Pythagoras began to say that it moved.

astonished when they hear that someone says that it moves, as if that person had originally considered it immobile and had then foolishly imagined that it began to move only when Pythagoras or whoever else it was first suggested that it moved. It doesn't surprise me that the common masses, with their shallow understanding, should make such a stupid mistake— as if those who accept the motion of the Earth imagined that it was motionless from its creation up to the time of Pythagoras, and that it only became mobile when Pythagoras described it as such—but for thinkers like Aristotle and Ptolemy to have fallen into such a childish error seems to me to show extraordinary and quite inexcusable simple-mindedness.

SAGREDO. Do you mean to say, signor Salviati, that you think that when Ptolemy argued for the immobility of the Earth, he thought he had to argue against those who conceded that it had been immobile up to the time of Pythagoras, and claimed that it had only begun to move when Pythagoras attributed motion to it?

When Aristotle and Ptolemy argue against the mobility of the Earth, they appear to argue against those who believed that it had long been at rest, and had only begun to move at the time of Pythagoras.

SALVIATI. No other interpretation is possible, if we consider the argument he uses to refute their claim. He refutes it by saying that buildings would be destroyed, and that stones, animals, and humans themselves would be hurled into the sky. But since this destruction and being thrust into the sky could only happen to buildings and animals which were on the Earth in the first place, and since humans could only settle on the Earth and construct buildings when the Earth was at rest, it's clear that Ptolemy must be arguing against those who conceded that Earth had once been at rest. Only then could animals and stones and builders have established themselves there, and constructed palaces and cities; and then it must suddenly have started to move, causing the ruin and destruction of buildings, animals, etc. But if Ptolemy had set out to argue against those who said that the Earth had been spinning ever since it was created, he would have done so by saying that if the Earth had always been in motion it would never have been possible for animals, humans, or stones to settle there, much less to construct buildings, found cities, etc.

SIMPLICIO. I'm not sure that I see this inconsistency which you attribute to Aristotle and Ptolemy here.

SALVIATI. Ptolemy argues either against those who consider that the Earth has always been in motion, or against those who say it was once at rest and then started to move. If he was arguing against the former, he should have said: 'The Earth has not always been in motion, because if it had there would never have been humans, animals, or buildings on the Earth, as the globe's spinning would have made it impossible for them to remain there.' But what he actually says is: 'The Earth does not move, because the animals, humans, and buildings which are on the Earth would collapse.' This presupposes that the condition of the Earth was once such that animals and humans were able to live there and construct buildings, and consequently that it was once at rest, this being the condition in which animals could live there and houses and buildings could be constructed. Now do you see what I meant?

SIMPLICIO. Yes, though I'm not entirely convinced by it. But this doesn't affect the merits of the case, and an inadvertent slip on Ptolemy's part isn't enough to establish the motion of the Earth if in fact it's at rest. Now let's set joking aside and come to the heart of the objection, which seems to me to be insoluble.

SALVIATI. And I, signor Simplicio, far from wanting to solve it, intend to pull the knot even tighter. I will provide sense evidence to demonstrate that when heavy bodies are spun rapidly around a fixed centre they acquire impetus to move away from that centre, even if their natural propensity would be to move towards it. Suppose we take a small bucket with some water in it, and tie one end of a rope to it. Then, holding firmly on to the other end of the rope, make it go rapidly round in a circle, with the rope and your arm as the circle's radius and your shoulder-joint as its centre. You will find that, whether the circle is horizontal, or vertical, or at any inclination you please, the water will not fall out of the container; rather, the person turning it will feel a constant tension on the rope, pulling it away from their shoulder. If you make a small hole in the bottom of the bucket, you will see water spurting out from it equally whether it is towards the sky, laterally, or towards the ground. If you were to replace the water with small stones and spin it in the same way, you would

Rapid spinning has the effect of extruding and dispersing.

feel the same force pulling on the rope. Or again, you can see boys throwing stones a great distance by whirling round a piece of stick with the stone in a pocket at one end. All these examples demonstrate the truth of the principle that a spinning motion, provided it is fast enough, endows an object with an impetus towards the circumference. This explains why, if the Earth spins on its axis, the motion on its surface, which—especially near the equator—is incomparably faster than those we have cited, ought to extrude everything outwards towards the sky.

SIMPLICIO. The objection seems to me very well established, and I think it will be hard for anyone to disprove it and untie the knot.

SALVIATI. Untying the knot depends on some facts which you know and assent to no less than I do; but since you don't recall them, you can't see how it can be untied. So I shan't teach you these things, because you know them already, but simply remind you of them, so that you can resolve the objection for yourself.

SIMPLICIO. I've noted your method of reasoning several times, and it has reminded me of what Plato says, that *nostrum scire sit quoddam reminisci*—our knowledge is a kind of remembering.* So please explain your thinking, and remove this doubt for me.

Our knowledge is a kind of remembering, according to Plato.

SALVIATI. I will show you my view of what Plato says by what I say and by what I do. I've already done so by what I've done several times in the course of our discussions, and I'll follow the same approach with the example we are dealing with here. This may serve to help you understand my view of how we gain knowledge, provided we have time for another day's discussion and we don't try signor Sagredo's patience with this digression.

SAGREDO. On the contrary, I shall be delighted to hear it. I remember that when I studied logic, I was never entirely convinced by this much-vaunted powerful demonstration of Aristotle's.

SALVIATI. Then let us proceed. First, I would like signor Simplicio to tell me what kind of motion the boy imparts to the stone in the pocket of his sling when he moves to throw it.

SIMPLICIO. The stone in the sling moves with a circular motion; that is, it follows an arc of a circle whose centre is the thrower's shoulder-joint and whose radius is the stick and the thrower's arm.

SALVIATI. When the stone is released from the sling, what is its motion then? Does it continue with its previous circular motion, or does it follow a different course?

SIMPLICIO. It clearly doesn't continue its circular motion, since this wouldn't allow it to move away from the thrower's shoulder, whereas we see that it travels a long way.

SALVIATI. What motion does it follow, then?

SIMPLICIO. Let me think about it for a moment; I haven't tried to imagine it before.

SAGREDO. A word in your ear, signor Salviati: here is *quoddam reminisci*, something remembered, in action, to be sure. You're giving it a lot of thought, signor Simplicio!

SIMPLICIO. It seems to me that the motion which the stone receives when it leaves the sling can only be in a straight line; in fact, the extra impetus which is imparted to it must be in a straight line. I was bothered by seeing it moving in an arc; but since this arc always tends downwards and not in any other direction, I understand that this must come from the weight of the stone, which naturally pulls it downwards. So I say that the impetus imparted to it is definitely in a straight line.

The motion imparted by the projector is solely in a straight line.

SALVIATI. Yes, but what straight line? An infinite number of straight lines in all directions can be produced by the end of the sling and the point at which the stone is separated from the sling.

SIMPLICIO. It moves in the direction of the motion which the stone had when it was moving with the sling.

SALVIATI. But you have already said that the motion of the stone when it was in the sling was circular; and circular motion cannot also be in a straight line, since no part of a circular line is straight.

SIMPLICIO. I don't mean that it is projected in the direction of the whole of the circle, but only of the last point at which its circular motion ceased. I know what I mean but I can't express it properly.

SALVIATI. It's clear to me too that you understand the fact, but you don't have the proper terms to express it. I can teach you these—teach you words, that is, not the truths themselves, which are facts. So to show you that you know the fact but only lack the terms to express it, tell me: when you shoot a bullet with an arquebus, in what direction does it acquire the impetus to travel?

SIMPLICIO. It acquires the impetus to travel along the straight line which continues the direction of the gun's barrel; it deviates from this neither to left or right, nor up or down.

SALVIATI. Which is as much as to say that it forms no angle with the straight line of travel made by the gun's barrel.

SIMPLICIO. Yes, that was what I meant.

SALVIATI. If, then, the line followed by the projectile continues the circular line which it followed when it was in contact with the projector, without forming an angle with it, and if it must pass from circular motion to motion in a straight line, what must this straight line be?

SIMPLICIO. It can only be the line touching the circle at the point of separation. It seems to me that any other line, if it were extended, would intersect the circumference, and therefore would form an angle with it.

SALVIATI. You have argued very well, and shown yourself to have the makings of a geometrician. So here are the words which you can commit to memory to express your understanding of the fact: that a projectile acquires the impetus to move along the tangent to the arc followed by the projector at the point at which the projectile separates from the projector.

SIMPLICIO. I understand this entirely, and it is exactly what I meant.

SALVIATI. On a straight line which touches a circle, what is the closest point to the centre of the circle?

SIMPLICIO. It must clearly be the point of contact, since this is a point on the circumference of the circle, and all the other points on the line are outside it; and every point on the circumference is the same distance from the centre.

SALVIATI. Therefore when an object moves away from the point of contact along a tangential straight line, it moves continually away from the point of contact and from the centre of the circle.

SIMPLICIO. Certainly.

SALVIATI. Now, if you have kept in mind the propositions which you have explained to me, bring them together and tell me what can be deduced from them.

SIMPLICIO. I don't think my memory is so defective that I can't recall this. From what has been said we can deduce that when a projectile which is moved rapidly in a circle separates from the projector, it retains the impetus to continue its motion along the straight line which touches the circle followed by the projector at the point of separation. This motion takes the projectile continually further away from the centre of the circle followed by the projector.

A projectile moves along the tangent to its previous circle of motion at the point of separation.

SALVIATI. You now know the reason why heavy objects adhering to the surface of a rapidly turning wheel are extruded and launched outside the wheel's circumference, at an ever-increasing distance from the centre.

SIMPLICIO. Yes, this is quite clear to me. But this new understanding increases rather than diminishes my incredulity that the Earth can rotate at such a great speed, without extruding stones, animals, etc., towards the sky.

SALVIATI. You will come to know the rest in the same way as you have come to know everything so far; in fact, you know it already, and if you thought about it you would be able to recall it for yourself. But to save time I will help you to recall it. So far, you have worked out for yourself that the circular motion of the projector imparts motion to the projectile, if it separates from the projector, along the straight line which is tangent to the circle of motion at the point of separation. Its motion continues along this line, taking it continually further away from the projector. Then, you have said that the projectile would continue its motion along this straight line if it were not for its own weight, which gives it an additional inclination downwards, thus making the line of motion into a curve. I think you also knew for yourself that this inclination tends always towards the centre of the Earth, because this is the point to which all heavy objects tend. Now for the next step: can you tell me whether the moving object, continuing its rectilinear motion after its separation from the projector, moves at a constant rate away from the centre (or, if you prefer, from the circumference) of the circle whose motion it had previously shared? Putting the question another way, when a moving object separates from a circle at the point of a given tangent and continues to move along that tangent, does it move away from the point of contact at the same rate as it does from the circle's circumference?

SIMPLICIO. No, it does not. When the tangent is close to the point of contact its distance from the circumference is very small, and it forms a very narrow angle with the circumference; but as it moves further away, its distance from the circumference grows at an ever-increasing rate. So, for example, on a circle with a diameter of, say, ten *braccia*, a point on the tangent two hand-breadths away from the point of contact would be three or four times further away from the circumference than a point only one hand-breadth from the point of contact; and the latter point in turn would be almost four times as far from the circumference as a point half a hand-breadth from the point of contact. So within an inch or two of the point of contact, the separation of the tangent from the circumference is barely visible.

SALVIATI. So, then, the initial distance of the projectile from the circumference of its previous circular motion is very small?

SIMPLICIO. Yes, almost imperceptible.

SALVIATI. Now the projector imparts an impetus to the projectile to move along a tangent straight line, and the projectile would continue to move along this line if it were not pulled down by its own weight. So my next question is this: how long after the moment of separation is it before the projectile begins to decline downwards?

A heavy projectile begins to decline immediately on its separation from the projector.

SIMPLICIO. I think it starts to decline immediately, since if there is nothing to hold it up, its weight cannot fail to act on it.

SALVIATI. So then, if a stone thrown out by a rapidly turning wheel had the same natural propensity to move towards the centre of the wheel as it does towards the centre of the Earth, it would readily return to the wheel—or rather, it would never separate from it. This is because, at the initial moment of separation, its distance from the wheel is so tiny, because of the infinite acuteness of the angle of contact, that the slightest inclination to draw it towards the centre of the wheel would be enough to keep it on the circumference.

SIMPLICIO. I don't doubt that, granted such an impossibility, namely that the inclination of these heavy objects draws them towards the centre of the wheel, they would not be extruded or thrown out.

SALVIATI. I don't assume an impossibility, nor do I need to, because I don't deny that the stones would be thrown out by the wheel. I simply put it as a hypothetical case, so that you could tell me what follows. Imagine now that the Earth is the great wheel which is turning so rapidly that it throws out the stones. You've already told me, quite rightly, that the motion of the projectile will be along the straight line which touches the Earth at the point of separation. Now how perceptibly does this tangent move away from the surface of the terrestrial globe?

SIMPLICIO. I doubt whether the distance is as much as an inch in a thousand *braccia*.

SALVIATI. And you said, did you not, that the projectile is drawn by its own weight to decline from the tangent towards the centre of the Earth?

SIMPLICIO. I did indeed, and now I will say what follows from it. I see perfectly that the stone will not separate from the Earth, because its initial separation would be so minimal that its inclination to move towards the centre of the Earth would be a thousand times stronger; and in this case the centre of the Earth is also the centre of the wheel. So it must be conceded that stones, animals, and other heavy objects cannot be extruded by the rotation of the Earth. But now I have a new difficulty with very light objects, whose inclination to fall towards the centre is very weak: since they do not have the property of being drawn down to the surface, I see no reason why they should not be extruded. And as you know, *ad destruendum sufficit unum*: one objection is enough to destroy an argument.

SALVIATI. We shall deal with this objection as well. But tell me first of all what you mean by light objects: do you mean those materials which are truly so light that they move upwards, or do you mean those which are not absolutely light, but which have so little weight that they do fall to the ground, but only very slowly? Because if you mean those which are absolutely light, I will grant that they are extruded even more than you say.

SIMPLICIO. I mean the latter—things such as feathers, wool, cotton wool, and the like. The slightest force is enough to lift them into the air, and yet we see them at rest quietly on the ground.

SALVIATI. If a feather has a natural propensity, however weak, to fall towards the surface of the Earth, this will be enough to prevent it from being thrown out. You know this as well as I do. So tell me: if the feather were extruded by the spinning of the Earth, what would be its line of motion?

SIMPLICIO. The tangent at the point of separation.

SALVIATI. And if it were to return to the surface, what would its line of motion be?

SIMPLICIO. The line from the feather to the centre of the Earth.

SALVIATI. So there are two motions to be considered: one from the projection, which starts from the point of contact and continues along the tangent; the other from the inclination downwards, which starts from the projectile and follows the secant towards the centre. For the projection to continue, the impetus along the tangent must be greater than the inclination along the secant, must it not?

SIMPLICIO. So it seems to me.

SALVIATI. What do you think is necessary in the motion from the projection if it is to prevail over the motion from the inclination, so that the feather is separated and carried away from the Earth?

SIMPLICIO. I don't know.

SALVIATI. Of course you do. The moving object is the same in both cases, namely the feather; so how can a moving object surpass and prevail over itself in its motion?

SIMPLICIO. The only way it can prevail over or be surpassed by itself is by moving either more rapidly or more slowly.

SALVIATI. You see, you did know this for yourself. So then, for the projection of the feather to continue and for its motion along the tangent to prevail over its motion along the secant, what velocities must the two motions have?

SIMPLICIO. The velocity along the tangent must be greater than the velocity along the secant. Oh, what a fool I have been! Isn't this velocity a hundred thousand times greater than that of a falling stone, let alone of a feather? And I was naive enough to let myself be persuaded that stones couldn't be extruded by the spinning of the Earth! So I take back what I said, and I declare that if the Earth were in motion, stones, elephants, towers, and cities would all necessarily fly off into the sky; and since this does not happen, I say that the Earth is not in motion.

SALVIATI. Oh, signor Simplicio, you are so easily carried away that I shall start worrying more about you than about the feather. Calm down a little and listen. If keeping the stone or the feather in contact with the Earth's surface depended on their downward motion being greater than or equal to the motion along the tangent, you would be quite right: the downward motion along the secant would have to be as rapid as, or more rapid than, the motion eastwards along the tangent. But did you not say just now that a distance of a thousand *braccia* from the point of contact along the tangent would barely give rise to a distance of an inch from the circumference? So it's not enough for the motion along the tangent (that of the daily rotation) to be simply greater than the motion along the secant (the downward motion of the feather). It would have to exceed it by so much that the time taken to carry the feather, say, a thousand *braccia* along the tangent, would not suffice for it to fall even an inch along the secant; and I say that this could never happen, however fast you make one motion and however slow you make the other.

SIMPLICIO. Why could the motion along the tangent not be so rapid that there was not time for the feather to reach the surface of the Earth?

SALVIATI. Try to frame the question in terms of proportions, and I will reply. So tell me by how much you think the velocity of the former motion would have to exceed the latter.

SIMPLICIO. I would say that if, for example, the former motion was a million times faster than the latter, both the feather and the stone would be extruded.

SALVIATI. In saying this you're mistaken, not because of any fault in logic or physics or metaphysics, but simply in geometry. If you knew just the basic elements of geometry, you would know that if you drew a straight line from the centre of a circle to a tangent, it wouldn't matter if it intersected it in such a way that the part of the tangent between the secant and the point of contact* was a million, two million, or three million times greater than the part of the secant between the tangent and the circumference. The nearer the secant comes to the point of contact, the more this ratio grows, ad infinitum. Therefore, however rapid the rotation may be, and however slow the downward motion, there is never any reason to fear that the feather, or any even lighter object, might begin to rise from the surface, because the downward inclination will always be greater than the velocity of the projection.

[. . .]

Computation of the time taken for a cannon-ball to fall from the sphere of the Moon to the centre of the Earth.

SALVIATI. Before anything else we must consider how the motion of falling objects is not uniform, but starts from a state of rest and constantly accelerates. This fact is known and recognized by everyone except the modern author just cited,* who says it is constant and makes no reference to acceleration. But this general recognition is useless unless we know the rate at which this increase in velocity occurs, and this is something which has been unknown to all philosophers up to our own time. Our mutual friend the Academician has been the first to discover and demonstrate it, in some of his writings which are not yet published, but which he has shown in confidence to me

The natural motion of heavy objects accelerates by odd numbers, starting from unity.

and a number of his other friends. He shows that the natural motion of heavy objects accelerates by odd numbers, starting from unity: that is, taking any time interval and repeating it any number of times, if the moving object starting from a state of rest travels, say, one *canna** in the first time interval, in the second interval it will travel three *canne*, in the third interval five, in the fourth interval seven, and so on, following the sequence of odd numbers. This is the same as saying that the distance travelled by a moving object starting from a state of rest increases in double proportion to the time interval in which the distance is measured, or in other words that the distances travelled are in the same proportion to each other as the square of their time intervals.

The distances travelled by a falling object are proportional to the square of the time intervals in which they are travelled.

SAGREDO. I am amazed to hear this. And you say that it can be demonstrated mathematically?

SALVIATI. Purely mathematically; and our friend has discovered and demonstrated not just this, but many other fascinating properties of natural motion, and of the motion of projectiles as well. It's been a great source of pleasure and wonder for me to see and study them all, and to witness the rise of a whole new understanding of a subject on which hundreds of volumes have already been written. Not one of the infinite amazing conclusions in our friend's book has been observed or understood by anyone before.

The Academician's whole new science concerning local motion.

SAGREDO. You make me abandon my desire to pursue further the discussions we've already begun, solely so as to hear some of the demonstrations you mention. So either tell me about them now, or at least promise that you will devote a separate session to them for me, and for signor Simplicio as well, if he would like to hear about the properties and effects of this primary force of nature.

SIMPLICIO. I would indeed, although I don't consider it necessary for a natural philosopher to descend to the level of minute details; it suffices to have a general understanding of the definition of motion, the distinction between natural and violent and between constant and accelerated motion, and so forth. If this had not been enough, I don't believe Aristotle would have omitted to teach us whatever we lacked.

SALVIATI. That may be so, but let's not lose any more time on this now. I promise to devote a separate half-day to satisfying you about it; in fact, I remember now that I promised on another occasion to do just that. So to return to what we were saying, we had begun to calculate the time it would take for a falling object to travel from the sphere of the Moon to the centre of the Earth. To proceed not at random but following a method that will yield conclusive results, we shall first try to establish, by means of an experiment that can be repeated several times, how long it takes for a ball of, say, iron to reach the Earth from a height of a hundred *braccia*.

SAGREDO. We shall have to take a ball of a specified weight, and use the same weight when we calculate the time taken for it to fall from the Moon.

SALVIATI. That doesn't matter at all: balls weighing one, ten, a hundred, or a thousand pounds will all travel the same hundred *braccia* in the same time.

SIMPLICIO. That I can't believe, and no more did Aristotle. He writes that the velocity of falling bodies is in proportion to their weight.

SALVIATI. If you want to maintain that this is true, signor *Aristotle's error* Simplicio, then you must also believe that, if you let fall two *in affirming* balls of the same material, one weighing a hundred pounds and *that falling* the other one pound, at the same moment from a height of a *bodies move in* hundred *braccia*, the larger ball will reach the Earth before the *proportion to* smaller one has fallen by a single *braccio*. Now try to imagine, *their weight.* if you can, seeing the larger ball reaching the Earth when the smaller one is still less than one *braccio* from the top of the tower.

SAGREDO. I don't doubt for a moment that this proposition is false, but I'm not entirely convinced that yours is absolutely true. Still, I believe it, since you affirm it so confidently, and I'm sure you wouldn't do so if you were not assured of it either from experience or by a clear demonstration of it.

SALVIATI. I have both, and I will explain them to you when we have our separate discussion on the subject of motion. But for now, to avoid any further digressions, let's assume that we base our calculation on an iron ball weighing a hundred pounds, which in repeated experiments falls from a height of a hundred *braccia* in five seconds. Now since, as I've said, the distance travelled by a falling object increases in double proportion to the time elapsed, i.e. by the square of the time elapsed, and since a minute is twelve times five seconds, if we multiply the hundred *braccia* by the square of 12, or 144, we get 14,400, and this will be the number of *braccia* that the object will fall in a minute. Following the same rule, since an hour is 60 minutes, if we multiply 14,400 (the number of *braccia* travelled in a minute) by the square of 60, or 3600, this will give us 51,840,000, which will be the number of *braccia* travelled in an hour, or 17,280 miles. To find out the distance it would travel in 4 hours we multiply 17,280 by 16 (this being the square of 4), which is 276,480 miles. This is far more than the distance from the sphere of the Moon, which is 196,000 miles assuming, as the modern author does, that the distance of the sphere is 56 times the semi-diameter of the Earth,* and that the semi-diameter of the Earth is 3,500 of our Italian miles, a mile being 3,000 *braccia*. So you can see, signor Simplicio, that the distance from the sphere of the Moon to the centre of the Earth, which your author said would take more than six days to travel, could in fact be travelled in less than 4 hours, if the calculation is based on experimental evidence and not just by counting on one's fingers. To be precise, the time it would take would be 3 hours, 22 minutes, and 4 seconds.

SAGREDO. My dear Salviati, please don't deprive me of the details of this calculation, for it must be truly remarkable.

SALVIATI. Indeed it is. So, having first established by careful experiment that a given object falls from a height of 100 *braccia* in 5 seconds, we ask: if 100 *braccia* are travelled in 5 seconds, how many seconds will it take to travel 588,000,000 *braccia*, this being the length of 56 times the semi-diameter of the Earth? The rule for this calculation is to multiply the third number by the square of the second, which gives 14,700,000,000. Divide this by the first number, i.e. by 100, and the number we are looking for is the square root of the quotient, which is 12,124. 12,124 seconds are 3 hours, 22 minutes, and 4 seconds.

SAGREDO. I can see what you've done, but I have no idea of the reasoning behind it. But perhaps this is not the time to ask for it.

SALVIATI. On the contrary, I'll tell you even though you don't ask me, as it's very easy. Let's mark these three numbers A, B, and C. A and C refer to distances, B refers to

100	5	588000000	
A	B	C	25
1		14700000000	
22		35956	
241		10	
2422		60	12124
24240			202
			3

time; we are looking for the fourth number, which also refers to time. We know that the proportion of distance A to distance C is the same as between the square of time B and the square of the time we are seeking. So following the golden rule above, we multiply C by the square of B and divide the product by A; the quotient will be the square of the number we want, which will therefore be the square root of the quotient. Now you can see how easy it is to understand.

SAGREDO. So are all truths, once they have been discovered; the difficulty lies in discovering them. I understand perfectly, and I'm grateful to you. If you have any other curious things to say about this topic, please tell me; because to speak frankly, and with no offence to signor Simplicio, I always learn some interesting new truth from what you say, whereas from his philosophers I don't know that I've yet learnt anything of any great significance.

SALVIATI. There is still all too much to be said on this matter of local motion, but as we've agreed, we'll save it for a separate session. So now I shall say something about the author cited by signor Simplicio. He thinks he has given a great advantage to his opponent by conceding that the cannon-ball, in falling from the sphere of the Moon, could reach a velocity equal to that which it would have had if it had remained up in the lunar sphere and had participated in the sphere's daily motion. My reply is that in falling from the sphere to the centre, the ball will acquire a velocity more than double that of the sphere's daily motion; and I shall demon-

If a falling object continued to move at the velocity it had acquired for an equal interval of time with constant motion, it would travel twice the distance that it had already travelled with accelerated motion. strate this on the basis of factual suppositions, not arbitrary ones. As a heavy object falls and continually gains speed at the rate we have established, at any given point in its fall it will have gained a degree of velocity such that, if it were to continue to move constantly at that speed without accelerating any further, it would travel twice the distance it had travelled up to that point in the same time as it had already been falling. So, for example, if the cannon-ball has taken 3 hours, 22 minutes, and 4 seconds to fall from the sphere of the Moon to the centre, then when it reaches the centre it will have gained such a degree of velocity that if it were to continue moving constantly at that speed, without accelerating any further, in the next 3 hours, 22 minutes, and 4 seconds it would travel twice that distance, which is to say the full diameter of the lunar sphere. Now since the distance from the sphere of the Moon to the centre is 196,000 miles, and the ball travels that distance in 3 hours, 22 minutes, and 4 seconds, then—bearing in mind what has already been said—if it continued to move at the speed it had acquired when it reached the centre, in the next 3 hours, 22 minutes, and 4 seconds it would travel twice this distance, i.e. 392,000 miles. But the circumference of the sphere of the Moon is 1,232,000 miles; so if the ball had remained up in the lunar sphere and had participated in its daily motion, in 3 hours, 22 minutes, and 4 seconds it would have travelled 172,880 miles, which is well under half of 392,000. This shows that the motion of the surface of the sphere does not, as the modern author claims, have a velocity that the falling ball could not possibly share, etc.

SAGREDO. This argument would make perfect sense and would convince me, if I were assured of the assertion that a falling object would travel twice the distance in the same time if it were to continue in constant motion at the maximum velocity it had already acquired in its descent. You have already assumed the truth of this proposition on another occasion, but you haven't demonstrated it.

SALVIATI. This is one of the propositions that our friend has demonstrated, as you will see in due course. In the meantime I want to put some conjectures to you, not to teach you anything new, but to dissuade you from a contrary opinion and to show you that what I say is indeed possible. I'm sure you have observed how a lead weight suspended from the ceiling by a fine, long thread, if it is moved from the perpendicular and then let go, will fall and then spontaneously move almost as far from the perpendicular on the other side.

SAGREDO. Yes, I've observed it very well, and I've seen how—especially if the weight is quite heavy—there is so little disparity between the distance it descends and the distance it rises again, that I've sometimes thought that the rising arc was equal to the falling one, and so have wondered whether its oscillations could continue in perpetuity. I could believe that they would if it were not for the impediment of the air,

The motion of a weighted pendulum would be perpetual if all impediments were removed.

which resists being parted by the motion of the pendulum. But it's a very small resistance, as can be seen from the large number of oscillations the pendulum makes before it comes to a stop.

SALVIATI. The motion wouldn't be perpetual, signor Sagredo, even if the impediment of the air were completely removed, because there is another which is much less apparent.

SAGREDO. What is it? I can't think of anything else.

SALVIATI. You will appreciate it when you hear it, but I'll tell it to you later; in the meantime let us continue. I have proposed this observation of a pendulum to you to help you understand that the impetus acquired in the descending arc, where the motion is natural, has in itself the power to impel the weight upwards, with violent motion, by an equivalent distance in the corresponding rising arc; and this power is intrinsic to it, when all external impediments are removed. I think we can also take it as beyond doubt that, just as the velocity increases in the descending arc until the lowest point is reached at the perpendicular, so from the lowest point the velocity decreases in the rising arc until it reaches its highest point; and it decreases in the same proportions as it first increased, so that the degrees of velocity are equal at points which are equidistant from the lowest point.

Hence, arguing by analogy, it seems to me reasonable to believe that if the terrestrial globe were pierced by a hole pass-

If there were a hole passing right through the terrestrial globe, an object falling down this hole would pass the centre and would then rise a distance equal to its descent on the other side.

ing through its centre, a cannon-ball falling down this shaft would gain such an impetus of velocity moving towards the centre, that it would pass the centre and would be impelled upwards by the same distance as it had fallen. Its velocity would decrease after it had passed the centre at the same rate as it increased while it was falling, and I believe the time taken by this rising motion would be equal to the time of its fall. So as the object's velocity gradually decreases from its maximum point at the centre to the point where it stops altogether, it moves through the same distance in the same time as it had travelled as its velocity increased from complete motionless-ness to its maximum. Hence it seems reasonable to suppose that if the object moved constantly at its maximum speed it

would travel both distances in the same time. If we mentally
divide these speeds into ascending and descending
degrees, as in these numbers here, with the ascending
numbers from 1 to 10 corresponding to the time the
object is falling and the descending numbers from 10 to
1 corresponding to its rise, we can see that when they
are all added together they come to the same total as
one of the two sets made up entirely of the highest
number, namely 10. Therefore the total distance
travelled through the whole range of ascending and
descending velocities—which is the whole diameter of
the Earth—must be equal to the distance travelled at
the maximum speed for half of the total of ascending
and descending numbers. I know that I've explained
this very badly, but I hope I've made myself
understood.

SAGREDO. I think I've understood it very well, and I
think I can show that I've understood in a few words.
You mean that if motion starts from a state of rest and
increases in speed by equal amounts, corresponding to
consecutive numbers starting from 1—or rather from
zero, since this represents a state of rest—up to any
number you choose, so that the lowest degree is zero
and the highest is, say, 5, then all these degrees of
velocity with which the object moves add up to 15. But
if the object were in motion with the same number
of degrees, each degree being the maximum, i.e. 5, the
total of all these velocities would be twice as much,
namely 30. So if the object is in motion for the same
length of time but at a constant speed of the highest
degree, 5, then it will travel twice the distance as it did
with accelerated motion starting from zero.

SALVIATI. Your quick and perceptive understanding
has enabled you to explain it all much more clearly than
me, and you prompt me to add something further. In
accelerated motion the increase is continuous, and so
velocity which is constantly increasing cannot be broken
down into a fixed number of degrees; it is changing every
moment, and so its degrees are infinite. So a better illustration

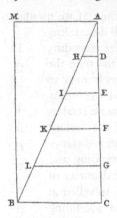

of our meaning would be to draw a triangle, such as this triangle ABC, dividing its side AC into as many equal parts as we choose, AD, DE, EF, FG, and drawing straight lines from D, E, F, and G parallel to the base BC. Let us imagine the sections we have marked on the line AC as equal intervals of time, and the parallel lines drawn from D, E, F, and G as representing degrees of velocity, increasing by equal amounts in equal intervals of time. A represents a state of rest, starting from which the moving object gains the degree of velocity DH in the time AD. In the following time interval its speed will have increased from the degree DH to EI, and so on in each successive time interval, corresponding to the increasing length of the lines FK, GL, etc. But acceleration occurs continuously moment by moment, not in steps from one time interval to the next. So, if the extremity A represents the minimum degree of velocity which is a state of rest, and also the first instant of the time interval AD, it is clear that before the moving object reaches the degree of velocity DH, which it does in the time interval AD, it has passed through an infinite number of smaller degrees, acquiring them in the infinite number of instants in the time interval DA, corresponding to the infinite number of points on the line DA. So if we want to represent the infinite degrees of velocity which come before the degree DH, we must imagine an infinite number of shorter and shorter lines, parallel to DH, drawn from the infinite number of points on the line DA. This infinite number of lines will eventually represent the entire area of the triangle AHD. So we shall see that the uniformly accelerating motion, starting from a state of rest, of a moving object travelling any distance, has absorbed and passed through infinite increasing degrees of velocity, corresponding to the infinite number of lines that we can imagine drawn, starting from point A, parallel to the lines HD, IE, LG, and BC, for as long as the motion continues.

Now let us complete the parallelogram AMBC, and extend to its side BM not just the parallel lines we have marked in the triangle, but the infinite number of lines we have imagined drawn from every point on the side AC. We saw that BC was the longest of the infinite number of lines in the triangle, and that it represented the maximum degree of velocity gained by the object in accelerated motion; and also that the whole area of the triangle represented the sum total of all the velocity with which it travelled a given distance in the time interval AC. So now the parallelogram represents the sum total of an equal number of degrees of velocity, but each one now equal to the maximum value BC; and this total velocity can be seen to be twice the total of the increasing velocities contained in the triangle, since the parallelogram is twice the size of the triangle. So if the falling object, passing through the degrees of accelerated motion represented by the triangle ABC, has travelled a certain distance in a given interval of time, it is reasonable and probable to conclude that if it passes through the uniform velocity represented by the parallelogram, it will travel twice the distance in constant motion as it travelled in accelerated motion.

SAGREDO. I am entirely satisfied. If this is what you call a probable argument, what will your necessary demonstrations be like? If only there were just one such conclusive proof in the whole of common philosophy!

SIMPLICIO. We should not look for perfect mathematical certainty in natural science.

Mathematical certainty is not to be sought in the natural sciences.

SAGREDO. Surely motion is a topic in natural science, but I don't recall Aristotle demonstrating even the smallest detail about it. But let's not embark on any further digressions; and signor Salviati, please keep your promise to explain the reason why the motion of a pendulum ceases, apart from the resistance of the air to being parted.

SALVIATI. Tell me: of two pendulums of different lengths, doesn't the one whose cord is longer oscillate more infrequently?

A longer pendulum oscillates more infrequently than one which is shorter.

SAGREDO. Yes, assuming they move an equal distance from the perpendicular.

SALVIATI. The distance from the perpendicular doesn't make any difference, because the same pendulum will always complete its oscillations in the same time interval, however long or short they are, in other words however much or little it is moved away from the perpendicular. The time intervals may not be exactly the same, but experience will show that the differences are imperceptible, and even if they were substantial this would help rather than hinder our argument. Let us draw a perpendicular AB, with a cord AC fixed at point A, to which we attach a weight, C, and another higher up on the same cord, E. If we move the cord AC away from the perpendicular and let it go, the weights C and E will move through the arcs CBD

The oscillations of a pendulum always have the same frequency, regardless of whether their amplitude is large or small. The cause which impedes the motion of a pendulum and brings it to a stop.

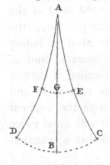

and EGF. Weight E, as it hangs from a shorter length of cord and also, as you say, has moved a shorter distance from the perpendicular, will seek to return more quickly and to oscillate with a higher frequency than weight C. Hence it will prevent weight C from moving as far towards its terminal point D as it would if it were unimpeded; and so, by continually exerting this impediment on each oscillation, it will eventually bring it to a stop. Now the cord itself, even without the weights, is made up of many pendulums with a weight of their own, because each part of the cord is such a pendulum, attached closer and closer to point A and therefore with a tendency to oscillate with greater and greater frequency; and consequently it has the effect of exerting a continuous impediment on weight C. This is clear from the fact that if we observe the cord AC, we shall see that it extends in an arc and not in a straight line; and if we were to replace the cord with a chain, this effect would be even more apparent, especially if weight C were moved a long way from the perpendicular AB. For since a chain is made up of many separate links, each of them quite heavy, the arcs AEC and AFD will be seen to have a significant curve. This is because the parts of the chain that are closer to point A seek to oscillate with greater frequency, and so will not let the lower parts move as far as they otherwise would; and by continuously detracting

The cord or chain to which the pendulum is attached does not extend in a straight line when the pendulum oscillates, but bends in an arc.

from the oscillations of weight C they will eventually bring it to a stop, and would do so even if the air impediment could be removed.

SAGREDO. Here are the books we sent for just now. Signor Simplicio, take them and see if you can find the passage we were unsure about.

SIMPLICIO. Here it is, where the author begins to argue against the daily motion of the Earth, having first refuted its annual motion: 'The annual motion of the Earth compels the Copernicans to assert its daily rotation, because otherwise the same hemisphere of the Earth would always be turned towards the Sun, and the other side would always be in shadow', meaning that one half of the Earth would never see the Sun.

SALVIATI. It seems to me from this first comment that your author has failed to understand Copernicus' position correctly. If he had noted how Copernicus makes the terrestrial globe's axis constantly parallel to itself, he would not have said that one half of the Earth would never see the Sun, but rather that a year would be only a single natural day, meaning that the whole Earth would have six months' daylight and six months' night, as in fact happens for the inhabitants of the polar regions. But let's excuse him this oversight, and pass on to what he says next.

SIMPLICIO. He continues: 'That this revolution of the Earth is impossible we demonstrate as follows.' Then he goes on to describe the illustration that follows, which shows many heavy objects falling and light ones rising, birds flying in the air, etc.

SAGREDO. Please show us. Oh, what delightful figures—the birds, the balls, and what are these lovely things here?

SIMPLICIO. They are balls coming from the sphere of the Moon.

SAGREDO. And what's this?

SIMPLICIO. A snail, of the kind they call *buovoli* here in Venice, which also comes from the sphere of the Moon.

SAGREDO. Yes, of course—and this is the Moon that has such a great influence on these shellfish, or 'armed fish' as we Venetians call them.

SIMPLICIO. Here is the calculation I was telling you about, of the distance that would be travelled in a day, an hour, a minute, and a second, by a point on the Earth's surface at the equator and also at latitude 48 degrees. Then comes the passage which I was afraid I had quoted wrongly, so let's read it: 'With these suppositions, if the Earth revolves it must follow that everything that from the air to the same . . . ', etc. 'If we assume these balls to be equal in weight and size, and if they are placed inside the surface of the lunar sphere and are allowed to fall freely, then if their downward motion is equal in velocity to their circular motion—although this is not the case if ball A', etc.—'they will fall for at least six days (to concede this much to our adversaries), and in this time they will have circled around the Earth six times', etc.

SALVIATI. You've reported this author's argument all too faithfully. This just shows, signor Simplicio, how careful one must be in trying to persuade others to believe what one doesn't believe oneself—because this author can't possibly have failed to realize that he was envisaging a circle with a diameter more than twelve times larger than its circumference, and all mathematicians know that the diameter is less than a third of the circumference. This error means giving a value of more than 36 to a measurement that is actually less than one.

SAGREDO. Perhaps these mathematical proportions, while true in the abstract, don't correspond exactly to all real physical circles—although I think when coopers want to find the diameter of the head they have to make for a barrel, which is real and physical enough, they do use the abstract mathematical rule. So perhaps signor Simplicio would like to defend this author, and tell us whether he thinks that physics can be so very different from mathematics.

SIMPLICIO. I don't think this would be enough to explain such a large discrepancy as this, so all I can say in this case is, *Quandoque bonus* . . . :* sometimes good Homer nods. But assuming that signor Salviati's calculation is more accurate and that the ball's descent would take not more than three hours, it still seems to me that, if it falls from such a great distance as the sphere of the Moon, it would be extraordinary if its natural impulse kept it constantly above the point on the

Earth's surface that was below it when it started to fall. Would it not rather be left a great distance behind?

SALVIATI. Whether it would be extraordinary, or simply natural and normal, would depend on its state before it started to fall. Your author supposes that the ball, while it was on the surface of the sphere of the Moon, would participate in the circular motion every twenty-four hours which it shares with the Earth and everything else within the lunar sphere; in which case the same force that impelled its circular motion before it started to fall would continue to affect it while it was falling. Then, far from failing to keep up with the Earth's motion and being left behind, it ought rather to run ahead of it, because as it approached the Earth it would be revolving in continually decreasing circles; so if the ball maintained the same velocity that it had when it was on the surface of the sphere, it would run ahead of the rotation of the Earth. If on the other hand the ball was not affected by the circular motion when it was in the sphere, then there is no reason why it should continue to fall directly above the point on the Earth that was beneath it when it started. Neither Copernicus nor any of his followers would claim that it should.

SIMPLICIO. But the author presses his point, as you can see: he asks what principle produces this circular motion of heavy and light bodies—whether it is an internal principle or an external one.

SALVIATI. Keeping to the question we're discussing, my reply is that the principle that produced the ball's circular motion when it was in the lunar sphere is the same principle as kept it circulating as it was falling. Whether this was an internal or an external principle I leave for the author to decide as he prefers.

SIMPLICIO. He will prove that it cannot be either an internal or an external principle.

SALVIATI. To which I shall reply that when the ball was in the lunar sphere it was not moving. This will exempt me from having to explain how it remains constantly above the same point as it falls, because this doesn't happen.

SIMPLICIO. Very well; but if heavy and light bodies can have no internal or external principle causing them to move in a circle, then neither does the Earth move in a circle, which proves the author's point.

SALVIATI. I didn't say that the Earth has no external or internal principle causing it to move in a circle; I said that I don't know which of the two it is. The fact that I don't know doesn't mean that it doesn't exist. If this author knows what kind of principle causes the circular motion of the other heavenly bodies, whose motion is not in doubt, then I shall say that the principle which moves the Earth is similar to that which causes Mars, Jupiter, and in his view also the sphere of the fixed stars to move. If he can tell me the moving force of one of these bodies in motion, I shall undertake to tell him the force that moves the Earth. In fact I'll do the same if he can identify for me the force that causes individual parts of the Earth to move downwards.

SIMPLICIO. Everyone knows that the cause of this is gravity.

SALVIATI. You're mistaken, signor Simplicio; you mean that everyone knows that it's called gravity. But I'm not asking you what it's called, but what its true essence is; and of that essence you have no more idea than of the essence which causes the stars to turn. You know only the name that we have given to it, which makes it familiar to us because we see the name a hundred times a day. But we do not really understand the principle or force that makes a stone fall downwards, any more than we do the force that impels it upwards when it separates from its projector, or the force that makes the Moon turn. All we know, as I have said, is the specific name we have given to the first force, which is gravity, and to the second, which we have called, more generically, applied force. As for the force that causes the stars and Moon to turn, we call it intelligence, either assisting or informing;* and for any number of other motions we simply attribute their cause to nature.

We have no more knowledge of the force that moves heavy objects downwards than of that which moves the stars in a circle. Of these causes we know only the names that we have given them.

[...]

SALVIATI. Now let signor Simplicio begin to put forward the difficulties which make him unable to believe that the Earth moves around a fixed centre in the same way as the other planets.

SIMPLICIO. The first and greatest difficulty is the contradiction and incompatibility of being both at the centre and remote from it. If the terrestrial globe does indeed move in the course of a year around the circumference of a circle, namely the zodiac, then it cannot possibly at the same time be at the centre of the zodiac. But the Earth's location at the centre has been proved in many ways by Aristotle, Ptolemy, and others.

SALVIATI. Your logic is impeccable: there's no doubt that if we are to prove that the Earth moves around the circumference of a circle, we must first prove that it is not at the centre of the same circle. So we must now establish whether the Earth is at this centre, as you say it is, or whether it revolves around it, as I say it does. First, however, we need to be clear whether you and I have the same concept of this centre or not. So please state what and where you understand this centre to be.

SIMPLICIO. I understand it to be the centre of the universe, of the world, of the sphere of the stars, of the heavens.

SALVIATI. I could reasonably question whether such a centre exists in nature, since neither you nor anyone else has ever proved whether the world is finite and has a shape, or is infinite and boundless.* But conceding for now that it is finite and has a bounded spherical shape, and therefore has a centre, let us see how credible it is to say that the Earth, rather than any other body, is located at this centre.

No one has ever proved whether the world is finite or infinite.

SIMPLICIO. Aristotle demonstrates in a hundred ways that the world is finite, bounded, and spherical.*

SALVIATI. These hundred ways all come down to just one, which itself is invalid: if I deny his presupposition that the universe is in motion, all his demonstrations fall to the ground,

Aristotle's demonstrations proving that the universe is finite fall to the ground if it is denied that it is in motion.

because he only proves that the universe is finite and bounded on the assumption that it is in motion. But so as not to multiply our points of disagreement, let's concede for now that the world is finite and spherical, and that it has a centre. Since its spherical shape and its centre are posited on its being in motion, it seems reasonable to begin our investigation into the location of this centre with the circular motion of the celestial bodies. Aristotle himself, indeed, reasoned and concluded in

Aristotle defines the centre of the universe as the point around which all the celestial spheres rotate.

this way, saying that the centre of the universe was the centre around which all the celestial spheres rotate, which he believed to be also the location of the terrestrial globe. So tell me, signor Simplicio: if Aristotle were obliged by the clear evidence of experience to modify his account of the disposition and order of the universe, and to admit that one of these two

The question of which of two propositions contradicting his teaching Aristotle would accept, if obliged to choose one of them.

propositions had been mistaken, which do you think he would choose? Would it be locating the Earth at the centre of the universe, or saying that the celestial spheres revolve around this centre?

SIMPLICIO. If this were to arise, I think that the Peripatetics ...

SALVIATI. I'm not asking about the Peripatetics; I'm asking about Aristotle himself. I know very well how the Peripatetics would respond. Humble and reverent servants of Aristotle as they are, they would deny all the experiments and all the observations in the world; they would refuse even to look at them, for fear of having to accept them.* They would maintain that the world is as Aristotle described it, not as nature shows it to be; for if you took away Aristotle's authority, what other basis would they have for arguing? So please tell me what you think Aristotle himself would do.

SIMPLICIO. I must admit that I can't decide which of these two he would regard as the lesser obstacle.

SALVIATI. Please don't use this word 'obstacle' to describe what may prove to be true; the obstacle was placing the Earth at the centre of the celestial revolutions. But as you don't know which way Aristotle would decide, and as I consider him to be a man of great intelligence, let's proceed by considering which of the two choices seems the more reasonable, and assume that this is the one he would have chosen.

Going back, then, to the beginning of our discussion, let's assume for Aristotle's sake that the world is spherical and has a circular motion—although the only physical evidence that we have of its size are the fixed stars. It must therefore have a centre, both for its shape and for its motion. Since, moreover, we can be sure that the sphere of the fixed stars contains many other orbs with their own stars, one inside the other, which also move with a circular motion, we must ask whether it is more reasonable to believe that all these inner spheres move around this same centre, or around another centre which is remote from it. So, signor Simplicio, let us have your opinion on this point.

SIMPLICIO. Provided we need go no further than this single proposition, and we could be sure of not encountering some other difficulty, I would say that it is much more reasonable to suppose that the container and what it contains all move around a common centre, than around different centres.

It is more appropriate that the container and what is contained should move around the same centre than around different centres.

SALVIATI. If, then, it is true that the centre of the world is the same as the centre around which the celestial bodies, i.e. the planets, move, it is quite certain that it is not the Earth but the Sun which is located at the centre of the world. So on this first basic general point, we can say that the centre is occupied by the Sun, and the Earth is as remote from the centre as it is from the Sun.

If the centre of the world is also the centre around which the planets move, the Sun and not the Earth is located there.

SIMPLICIO. On what do you base your argument that the Sun and not the Earth is at the centre of the revolution of the planets?

SALVIATI. On observations which are absolutely clear, and therefore conclusive. The most palpable of those which show that the Sun and not the Earth is at the centre, is that the planets all appear at varying distances from the Earth. These variations are so great that Venus, for example, is six times further away from us at its furthest point than at its closest, and Mars rises almost eight times larger at one point than at another. So you see that Aristotle was a little deceived in thinking that they were always equally distant from us.

Observations from which it can be deduced that the Sun and not the Earth is at the centre of the revolution of the heavens.

SIMPLICIO. What then are the indications that they revolve around the Sun?

SALVIATI. In the case of the three outer planets, Mars, Jupiter, and Saturn, this can be deduced from the fact that they are always close to the Earth when they are in opposition to the Sun, and remote from us when they approach conjunction with the Sun. The extent of this variation in distance is such that Mars appears fully sixty times larger when it is close to us than when it is furthest away. In the case of Venus and Mercury, we can be certain that they revolve around the Sun because they never move very far away from it, and because they appear now above and now below the Sun; the changing shape of Venus proves this. As for the Moon, it is clear that it cannot be separated from the Earth in any way, for reasons which we shall see in more detail as we proceed.

The changing shape of Venus indicates that its motion is around the Sun. The Moon cannot be separated from the Earth.

SAGREDO. I look forward to hearing even more marvellous things arising from this annual motion of the Earth than we saw from its daily rotation.

SALVIATI. You won't be disappointed. As far as the daily rotation was concerned, the only effect it had or could have on the celestial bodies was to make them all appear to move very fast in the opposite direction. But this annual motion, when it is combined with the individual motions of all the planets, produces a host of extraordinary effects which so far have perplexed all the world's greatest intellects. But to return to our first general principles, I repeat that the centre of the celestial revolutions of the five planets—Saturn, Jupiter, Mars, Venus, and Mercury—is the Sun; and it will be the centre of the Earth's motion as well, if we succeed in locating the Earth in the heavens. As for the Moon, it has a circular motion around the Earth, from which, as I've said, it is impossible for it to be separated; but this doesn't prevent it from also revolving around the Sun, sharing the annual motion of the Earth.

The annual motion of the Earth, combined with the motions of the other planets, produces extraordinary appearances.

SIMPLICIO. I'm not yet entirely sure about this structure. Perhaps it will be easier to understand and discuss if we make a drawing of it.

SALVIATI. So we shall; in fact, so that you are both more convinced and more amazed, I would like you to draw it yourself. You will see that although you don't think you understand it, you actually understand it very well; and you will be able to

Drawing of the system of the universe, based on appearances.

describe it point by point simply by answering my questions. So take a sheet of paper and a pair of compasses, and let this blank page be the immense expanse of the universe, in which you will distribute and arrange the various parts as reason dictates to you.

First, since you are firmly convinced that the Earth is located in this universe without needing me to tell you, mark a point wherever you like to show where you think it is located, and assign a letter to it.

SIMPLICIO. Let this be the location of the terrestrial globe, marked A.

SALVIATI. Good. Secondly, you know very well that the Earth is not located within the body of the Sun or even close to it, but is some distance away from it. So place the Sun wherever you like, as distant from the Earth as you think fit, and mark it with a letter as well.

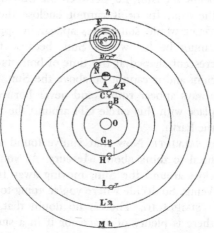

SIMPLICIO. There it is: let this be the location of the solar body, marked O.

SALVIATI. Now that we've established these two, we need to consider how to place Venus in such a way that its position and motion conform to what the evidence of our senses shows us. So call to mind what you know about what happens with this star, either from our earlier discussions or from your own observations, and place it where you think it should be.

SIMPLICIO. I will assume that the appearances which you have described, and which I have also read in my handbook of proofs, are correct: namely, that this star's elongation from the Sun never exceeds a fixed interval of about forty degrees, so that not only does it never reach a point of opposition to the

Sun, but not even a quarter or a sixth. Furthermore, it appears almost forty times larger at some times than at others: it is at its largest when it is retrograde and moving towards its evening conjunction with the Sun, and at its smallest when it is moving forwards towards its morning conjunction. Finally, when it appears at its largest it has a horned shape, and at its smallest it appears perfectly round. Assuming, as I say, that all these appearances are correct, it seems to me inescapable to say that

this star revolves in an orbit around the Sun. Its orbit cannot enclose and contain the Earth within it, nor can it be either below the Sun, i.e. between the Sun and the Earth, or above the Sun. Its orbit cannot enclose the Earth, because then Venus would sometimes appear in opposition to the Sun; it cannot be below the Sun, because then it would appear crescent-shaped at the time of both its conjunctions with the Sun; and it cannot be above the Sun, because then it would always appear round and never horned. So I shall mark its place with this circle CH around the Sun, but not enclosing the Earth.

SALVIATI. Now that you've found a place for Venus, you need to think about Mercury. As you know, it always stays close around the Sun, moving away from it much less than Venus. So decide where you're going to place it.

SIMPLICIO. There's no doubt that, as it imitates Venus,

there is plenty of space for it in a smaller circle, inside the orbit of Venus, and also moving around the Sun. A conclusive reason for this, and especially for its closeness to the Sun, is its brightness, which exceeds that of Venus and the other planets. So on this basis we can mark its circle here, with the letters BG.

SALVIATI. Now, where shall we place Mars?

SIMPLICIO. Mars comes into opposition to the Sun, so its orbit must enclose the Earth. But I see that it must enclose the Sun as well, because if it passed below the Sun and not

above it, it would appear horned when it is in conjunction with the Sun, in the same way as Venus and the Moon, whereas in fact it always appears round. So its orbit must enclose the Sun as well as the Earth. Since I remember you saying that when Mars is in opposition to the Sun it is 60 times bigger than when it approaches its conjunction, I think that these appearances will correspond very well with an orbit around the Sun which also encloses the Earth, which I will draw here and mark with the letters DI. D is the point where Mars is closest to the Earth and is in opposition to the Sun; when it is at point I it is in conjunction with the Sun and at its greatest distance from the Earth. And since we can observe the same appearances in Jupiter and Saturn, but with Jupiter showing much less variation than Mars, and Saturn even less than Jupiter, I think we can also satisfactorily place these two planets in two circles around the Sun. I will mark this first one, for Jupiter, EL, and another outside it for Saturn, FM.

The orbits of Jupiter and Saturn also enclose both the Earth and the Sun.

SALVIATI. You've distinguished yourself so far. Now since, as you see, the distance by which the three outer planets approach and recede measures twice the distance between the Earth and the Sun, this produces a greater variation in the distance of Mars than of Jupiter, since Mars's orbit DI is smaller than Jupiter's orbit EL. Similarly the variation in the distance of Saturn is less than that of Jupiter, because Jupiter's orbit EL is smaller than Saturn's orbit FM. This all corresponds exactly with the appearances.

The approach and recession of the three outer planets is double the distance of the Sun.

The apparent difference in size is less in Saturn than in Jupiter, and less in Jupiter than in Mars; the reason for this.

Now it remains for you to decide the place you will assign to the Moon.

SIMPLICIO. Following the same procedure, which seems to me absolutely conclusive, we can see that the Moon comes into conjunction with and opposition to the Sun, and hence we must say that its orbit encloses the Earth. But there is no evidence that it encloses the Sun, because if it did it would not appear as a crescent when it approaches conjunction with the Sun, but would always be round and full of light. Moreover, it would not be able to eclipse the Sun from us, as it often does, by coming between us and the Sun. Therefore we must assign it an orbit around the Earth, which I shall mark here NP, where P is the point at which it appears to us from the Earth, A, to be in conjunction with the Sun, so that sometimes it eclipses the Sun. When it is at N we see it in opposition to the Sun; in this position it can sometimes fall under the Earth's shadow and be obscured.

The Moon's orbit encloses the Earth, but not the Sun.

SALVIATI. And now, signor Simplicio, what are we to do with the fixed stars? Should we show them scattered across the vast abyss of the universe, at varying distances from any defined point, or should we place them on the surface of a sphere, so that each of them is equally distant from its centre?

SIMPLICIO. I would rather take a middle path and assign them to a sphere around a specified centre, made up of two spherical surfaces, one of them very high and concave and the other lower and convex. I would place the innumerable host of fixed stars between these two surfaces, at varying heights. This could be defined as the sphere of the universe, which contains within it the orbits of all the planets we have already drawn.

Probable position of the fixed stars. How we should conceive the sphere of the universe.

SALVIATI. So now, signor Simplicio, we have arranged the celestial bodies exactly as Copernicus placed them, and you have done this with your own hand. What's more, you have assigned to each its proper motion, except for the Sun, the Earth and the fixed stars. You have given Mercury and Venus a circular motion around the Sun which does not enclose the Earth. You have made the three outer planets, Mars, Jupiter, and Saturn, also revolve around the Sun, enclosing the Earth within their orbits. And you have shown the Moon with no motion other than its orbit around the Earth, without enclosing the Sun. In all of these motions, too, you are in agreement with Copernicus.

It remains now for us to decide how to distribute three things among the Sun, the Earth, and the sphere of the fixed stars. These are a state of rest, which appears to belong to the Earth; the annual motion under the zodiac, which appears to belong to the Sun; and the daily motion, which appears to belong to the fixed stars, with the rest of the universe apart from the Earth participating in it. Now since it is the case that all the planetary spheres—Mercury, Venus, Mars, Jupiter, and Saturn—revolve around the Sun as their centre, it seems reasonable to assume that it is the Sun, rather than the Earth, that is at rest; for in a moving sphere it is reasonable to suppose that the centre is at rest rather than anywhere else. The Earth is placed between two moving bodies, namely Venus and Mars, one of which completes its revolution in nine months and the other in two years; so we can appropriately attribute the annual

A state of rest, annual motion, and daily motion to be distributed among the Sun, the Earth, and the firmament of fixed stars.

In a moving sphere, it seems more reasonable for the centre to be at rest than any other part.

motion to the Earth, and leave the state of rest to the Sun. This being the case, it necessarily follows that the daily motion, too, must belong to the Earth. Otherwise, the Sun being at rest, if the Earth did not turn on its axis but had only its annual motion around the Sun, our year would consist of just one day and one night of six months each, as has been said on another occasion.* See, too, how neatly this removes the need for the precipitous motion of the universe every 24 hours, and how the fixed stars, which are themselves so many Suns, enjoy perpetual rest in the same way as our own Sun. Finally, see how easily this first sketch is able to account for so many great appearances in the celestial bodies.

If the annual motion is attributed to the Earth, the daily motion should be assigned to it as well.

SAGREDO. I see this very well. But just as you take this simplicity as grounds for the strong probability of this system, others may draw the opposite conclusion. Since this model was put forward in ancient times by the Pythagoreans, and since it fits the appearances so well, they might reasonably wonder why it has attracted so few supporters in the course of thousands of years, and why Aristotle himself rejected it. And more recently Copernicus himself has met with the same fate.

SALVIATI. Signor Sagredo, you wouldn't be so surprised that this view has found so few followers if you had encountered, as I have very many times, the kind of stupidities that suffice to make the common people stubborn and resistant to even listening, much less assenting, to these novel ideas. I don't think we should attach much weight to the views of those who say that since they can't have breakfast one morning in Constantinople and dinner the same evening in Japan, this is conclusive proof to confirm their fixed conviction that the Earth is at rest. Or they say that the Earth is so heavy that it can't possibly climb up above the Sun and then come hurtling down again. There is an infinite number of such people, and there's no point in paying any attention to their foolish ideas or trying to win their assent. When we are dealing with subtle and delicate arguments, we don't need the company of those who can't see beyond generic definitions and are unable to make any kind of distinctions. Besides, what impression can you expect to make with all the proofs in the world on minds which are so obtuse that they can't recognize their own sheer folly?

Utterly childish arguments which suffice to confirm the ignorant in their belief that the Earth is at rest.

My amazement is quite different from yours, signor Sagredo: you wonder that the view of the Pythagoreans has so few followers, but I am astonished how anyone hitherto has embraced and followed it. My admiration for their originality of mind knows no bounds: having accepted it and deemed it to be true, the force of their intellect has so overridden their senses that they have been able to follow what reason dictated to them rather than the manifest evidence of their senses to the contrary. You have already examined the arguments against the daily rotation of the Earth and we have seen how plausible they are, and the fact that they were accepted as conclusive by Ptolemy, Aristotle, and all their followers is a strong testimony to their effectiveness. But the experiences which openly contradict the Earth's annual movement seem even more plainly to argue against it; so I say again that I can't find words to express my admiration for the way in which Aristarchus and Copernicus were able to let their reason conquer their senses and command their assent.

The view of Copernicus shown to be improbable.

In Aristarchus and Copernicus, reason and argument prevail over the manifest evidence of the senses.

FOURTH DAY

SAGREDO. I don't know whether you really are later than usual in returning to continue our discussion, or whether it is just my eagerness to hear signor Salviati's views on such a fascinating subject that has made it seem so. I've been waiting at the window for a good hour, expecting any minute to see the gondola appear which I sent to fetch you.

SALVIATI. I think it's your imagination, rather than our lateness, that has made the time seem longer; and so as not to prolong it any further, let us come to the point without more ado. I shall show how the motions we have already attributed to the Earth for every reason except as an explanation of the ebb and flow of the tides, fit perfectly with this phenomenon as well, and conversely how the ebb and flow of the tides serves to confirm the mobility of the Earth. Nature seems to have allowed this, either because the reality is in fact as it appears, or because she wants to play a trick on us and to mock our foolish fancies. So far we have based our arguments for the mobility of the Earth on appearances in the heavens, since nothing that occurs on Earth seemed to be conclusive one way or the other. In our discussions we have shown at length how all the terrestrial effects which are commonly cited as evidence for the stability of the Earth and the mobility of the Sun and the firmament, would appear exactly the same to us if the Earth were mobile and the Sun and the firmament were at rest. Only the element of water, vast as it is, being fluid and not attached or joined in any way to the terrestrial globe as all its other solid parts are, has a degree of freedom and is almost a law unto itself; so it alone in the sublunary world is able to give us some trace and indication of the mobility or otherwise of the Earth. Having studied the effects and variations in the motion of the tides at great length, drawing partly on what I have seen and partly on what I have heard from others, and having read and heard the vain explanations which many people have produced to explain these variations, I have been drawn to reach two firm conclusions (subject to the necessary

Nature plays a trick on us by making the ebb and flow of the tides endorse the mobility of the Earth. The ebb and flow of the tides and the mobility of the Earth confirm each other.

All earthly effects, except for the motion of the tides, could equally well confirm that the Earth is mobile or that it is at rest.

assumptions). First, if the Earth is immobile, the ebb and flow of the tides cannot occur naturally; and second, if we grant that the motions of the Earth are as we have suggested, then the sea must necessarily be subject to the ebb and flow of the tides exactly as observation shows them to be.

SAGREDO. This is a tremendously important proposition, both in itself and for the consequences which follow from it, so I shall listen with the greatest attention to your exposition and proof of it.

SALVIATI. In questions of natural science, such as this matter which we are discussing, it is our knowledge of the effects which leads us to investigate and find their causes. Without this knowledge we are travelling blind—indeed worse than blind, because we don't even know what it is we are looking for, and a blind man at least knows where he wants to go. So we need first of all to have a clear understanding of the effects whose causes we are trying to establish. And in this case you, signor Sagredo, are much more fully and reliably informed than me, because, besides having been born and lived for a long time in Venice, where the tides are notable for the extent to which they rise and fall, you have also sailed to Syria and, as someone with an alert and curious mind, you must have made many observations in the course of your travels. I, on the other hand, have only been able to observe for a short time what happens here at the extremity of the Adriatic, and also down on the shores of our own Tyrrhenian sea;* so I have to rely for much of my information on the reports of other people, which often differ among themselves and so are very unreliable and bring more confusion than confirmation to our investigations. Still, from the features which we do know for certain, which are also the most important, I think I have been able to discover their true primary causes, although I am not so bold as to claim that I can give correct and adequate explanations for any effects which are new to me and which I have not therefore been able to think about carefully. What I am about to say I put forward simply as a key opening the way to a path which no one has trodden before, in the firm hope that more speculative minds than mine will be able to explore more widely and to penetrate more deeply than I have been able to with this initial discovery. And if it transpires that in other distant seas there are variations which do not occur in our Mediterranean, that will not necessarily invalidate the reason and cause which I shall give, provided it is proved to be a correct and complete explanation for what happens in our sea—for there must in the end be just one true primary cause for effects which are of the same kind. So I shall give an account of the effects which I know to be true, and will explain the reason for them which I believe to be correct; and if you, gentlemen, will cite other effects known to you in addition to mine, we shall test the cause which I put forward to see whether it meets their cases as well.

There are, then, three intervals which can be observed in the ebb and flow of the tides.* The first and most important is the biggest and best-known, the daily motion with which the waters rise and fall in an interval of a few hours; in the Mediterranean these are of roughly six hours each, that is, the water rises for six hours and falls for six hours. The second interval is monthly, and appears to derive from the motion of the Moon: not that the Moon introduces any new motion, but it affects the range of the motion already described, which is significantly different depending on whether there is a full, new or half Moon. The third interval is annual, and appears to derive from the Sun; this also affects just the daily motion, with differences between the range of the tides at the solstices and those at the equinoxes.

The tides ebb and flow with three intervals, daily, monthly, and annual.

Let us first consider the daily interval, as this is the main motion, on which the Moon and the Sun apparently exert their secondary influences with their monthly and annual variations. These hour-by-hour changes can be seen to be of three kinds: in some places the water rises and falls without any lateral motion; in others it moves first towards the east and then back towards the west, without rising or falling; while in other places there is a motion both in height and laterally, as is the case here in Venice, where the water rises as it comes in and falls as it goes out. This happens at the extremity of a gulf which extends from west to east and ends on a beach which allows the water to spread out as it comes in; in a place where its course was intercepted by mountains or very steep banks, it

Variations which occur in the daily interval of the tides.

would rise and fall without any forward motion. Where it is not at an extremity of the sea, the water flows back and forth without any change in level, as happens very strikingly in the Strait of Messina, between Scylla and Charybdis,* where the currents are very rapid because of the narrowness of the channel. In the open sea, and around islands surrounded by open sea such as the Balearics, Corsica, Sardinia, Elba, the coast of Sicily facing Africa, Malta, Crete, etc., changes in level are very small but there are substantial currents, especially where there is a narrow stretch of sea between islands or between an island and the mainland.

Now it seems to me that these effects on their own, confirmed and indisputable as they are, are enough to convince anyone who seeks to stay within the terms of natural science that the mobility of the Earth is, at least, highly probable; for the alternative—that the Mediterranean basin remains motionless and the water it contains behaves in the way it does—is beyond my imagination and, I would guess, that of anyone who delves beneath the surface in considering this matter.

SIMPLICIO. Signor Salviati, these phenomena are not new; they have been studied by countless scholars for centuries, and many have exercised their minds to provide one or another explanation for them. A great Peripatetic philosopher not many miles from here has put forward a cause which he has brought newly to light from a text of Aristotle, whose significance other interpreters had overlooked. He finds that, according to this text, the true cause of the motion of the tides is simply the varying depth of the sea: deeper water, being greater in volume and therefore heavier, displaces the shallow water which then tries to flow back, and it is this constant struggle which causes the ebb and flow of the tides. Then there are many who relate the tides to the Moon, saying that the Moon has a particular domination over water. There is a prelate* who has recently published a treatise arguing that as the Moon moves across the sky, it attracts a body of water which continually follows it, so that the sea is always higher where it lies under the Moon. As for the fact that the water still rises even when the Moon is below the horizon, he says

The cause of the ebb and flow of the tides according to a certain modern philosopher.

The cause of the ebb and flow of the tides attributed to the Moon by a certain prelate.

that the only way to account for this is that the Moon, besides having this natural capacity itself, is able also to transmit the same power to the opposite point in the zodiac. Others, as I expect you know, say that the Moon with its temperate heat has the power to rarefy water so that it rises when it is rarefied.* Then there are also those who . . .

SAGREDO. Please, signor Simplicio, spare us any more: I don't think it is worth wasting time reproducing these theories, let alone wasting words on refuting them. You would be doing an injustice to your intelligence if you were to go along with these or any other such nonsensical ideas, when you have just rid yourself of so many others.

SALVIATI. I am not quite so impatient as you, signor Sagredo, so I don't mind taking a few words to answer signor Simplicio, in case he should find anything probable in the theories he has recounted. So, signor Simplicio, water whose outer surface is higher displaces water which is below it, but water which is deeper has no such effect; and once the higher water has displaced the lower it quickly becomes calm and finds its level again. This Peripatetic philosopher of yours must believe that all the lakes in the world which have no tide, and all the seas where the motion of the tides is imperceptible, have a completely level bed—and I was so naive as to think that, in the absence of evidence from soundings, islands projecting above the surface of the water were a clear indication of the unevenness of the seabed. As for the prelate, you could point out to him that the Moon moves over the whole surface of the Mediterranean every day, but the water only rises at its eastern extremity and here in Venice where we are.* To those who say that the Moon's temperate heat is enough to expand the water, tell them to put a saucepan of water on the fire and put their right hand into it until the heat makes the water rise just an inch, and then take their hand out and write about the expansion of the sea; or at least ask them to explain why the Moon rarefies the water in some places and not others, as for example here in Venice and not in Ancona, Naples, or Genoa. It must be said that there are two kinds of poetic inspiration: there are those who have the gift and skill to invent fables, and there are those who are inclined and ready to believe them.

Girolamo Borro and other Peripatetics explain the tides by reference to the temperate heat of the Moon.

Response to the vain theories adduced as causes of the ebb and flow of the tides.

Islands are an indication of the unevenness of the seabed.

Two kinds of poetic inspiration.

SIMPLICIO. I don't think anyone believes fables once they understand that that is what they are. As for the many opinions about the causes of the tides, I know that any given effect can have only one true primary cause, and therefore I understand very well that only one explanation can be true and all the others are false, like fables. Indeed, it may well be that none of the explanations produced so far is the true one. Actually, I believe that that is the case, for it would be very strange if the truth were so obscure that it completely failed to stand out against the darkness of so many errors. But I will permit myself to say, since we have licence to speak freely among ourselves, that introducing the motion of the Earth as an explanation of the ebb and flow of the tides seems to me every bit as much a fable as any of the others I have heard. And until I am given reasons which are more in keeping with natural phenomena, I will have no hesitation in believing that the tides are a supernatural effect, and as such are a miracle inscrutable to human understanding—as indeed are many others which derive directly from the omnipotent hand of God.

The truth is not so obscure that it fails to stand out against the darkness of errors.

SALVIATI. That's a very prudent line of argument, and one which follows Aristotle's teaching, for as you know, at the beginning of his Mechanics he defines as miraculous those things whose causes are hidden. But that the true cause of the tides is one such, I think the strongest indication is that not one of the explanations which has hitherto been put forward as the true cause can be replicated, by whatever artificial means we may try, to produce a similar effect. Nothing we can do with moonlight or sunlight, or temperate heat, or differences of depth, can artificially make water in a motionless container flow back and forth, or rise and fall, in one place and not in another. But if by simply moving the container, without any artificial aid, I can replicate for you exactly all these changes that we observe in the water of the sea, why should you reject this explanation and fall back on the miraculous?

Aristotle defines as miraculous those things whose causes are unknown.

SIMPLICIO. I will fall back on the miraculous unless you are able to persuade me with other natural causes than moving the basins which contain the water of the sea, because I know that these basins do not move, because the whole of the terrestrial sphere is by its nature immobile.

SALVIATI. But you believe, don't you, that the terrestrial sphere could be made mobile supernaturally, that is, by the absolute power of God?

SIMPLICIO. Of course; who could doubt it?

SALVIATI. Well then, signor Simplicio, since to explain the ebb and flow of the tides we need to introduce a miracle, let's say that the miracle consists in making the Earth move, and that the motion of the sea follows naturally from that. This will be much simpler—I might say, more natural—as miracles go, just as it is easier to set a sphere moving in a circle, as we see many other spheres move, than to make an immense body of water move to and fro more quickly in some places than others, to rise and fall more in some places than others and in some places not at all, and all these different motions in the same container: this would be a whole series of different miracles, whereas making the Earth move is just one. What's more, the miracle of moving the water would require another miracle as a consequence, namely making the Earth stand firm against the impetus of the water, which would be powerful enough to move the Earth first one way and then the other if it were not miraculously held in place.

SAGREDO. Please, signor Simplicio, let's suspend our judgement for a while before we condemn as vain this new opinion which signor Salviati is trying to explain to us, and let's not be too ready to lump it together with the ridiculous older theories. As for miracles, let's not have recourse to them until we have heard what he has to say within the limits of nature—although to my way of thinking, all the works of God and nature are to be seen as miracles.

SALVIATI. I think so too; and saying that the motion of the Earth is the natural cause of the ebb and flow of the tides doesn't make it any less of a miracle. So to return to our discussion, I say again that no one has yet explained how it is that the waters in our Mediterranean basin can move as they do, while the basin itself which contains the water remains at rest. The reasons for this difficulty, which so defy explanation, lie in the daily observable facts which I shall now describe; so listen carefully.

We are here in Venice; the tide is low, the sea is calm, and the air is still. The water begins to rise, and in the space of five or six hours rises by three feet or more. This is not because the original water has become rarefied; it is water which has come from elsewhere, of exactly the same kind as the water that was here before; it has the same salinity, the same density, the same weight. Ships, signor Simplicio, float in it exactly as before, not lying a hair's breadth lower in the water; a barrel of this new water doesn't weigh an ounce more than an equal quantity of the other; it is the same temperature, not a degree different. In short, it is water which we can see has just come in to the lagoon through the gaps and channels of the Lido. Now please tell me where this water has come from, and how it got here. Do you think there are abysses or channels in the bottom of the sea, through which the Earth draws in water and expels it again, like the breath of an immeasurably enormous whale?* In that case why does the water not rise by the same amount in Ancona, Ragusa, or Corfu,* where the variation in the tides is so small as to be barely perceptible? How can anyone possibly pour more water into a motionless container in such a way that the level rises in one defined part of it and not in another? Perhaps you will suggest that this extra water comes from the wider Ocean, flowing in through the Strait of Gibraltar; but this creates even bigger difficulties, without solving the ones we've mentioned already. To start with, what must be the speed with which this water flows, if it comes in through the Strait of Gibraltar and reaches the furthest shores of the Mediterranean in just six hours? This is a distance of two or three thousand miles—and it would have to flow the same distance in the same time when it runs out. What would become of the ships scattered across the sea, not to mention those which were in the Strait itself, with a continuous torrent of an immense volume of water, enough to spread across an area hundreds of miles wide and thousands of miles long in just six hours, all through a channel not more than eight miles wide? What tiger or falcon ever ran or flew at such a speed— enough to travel four hundred miles or more in an hour? No one denies that there are currents running the length of the Mediterranean, but they are slow enough for a vessel equipped

Showing that it is impossible for the ebb and flow of the tides to occur naturally if the Earth is at rest.

with oars to overcome them, albeit with some loss of headway. And there is a further difficulty if the water comes through the Strait of Gibraltar: how, in this case, can it produce such a rise in the sea in places so far away from the Strait, without first causing a similar or greater rise in places which are closer to it? In short, I don't think that the most stubborn resistance or the most subtle ingenuity, staying within the terms of natural science, can ever find a way round these difficulties while still maintaining that the Earth is at rest.

SAGREDO. I'm entirely convinced by this, and I'm eager to hear how these marvels can follow unimpeded from the motions we have already assigned to the Earth.

SALVIATI. For these effects to follow as a consequence of the natural motions of the Earth, they must not only occur unimpeded and unopposed; they must follow easily, indeed necessarily, such that they could not possibly be otherwise. For this is what characterizes true natural effects. So, having established that it is impossible to explain the motions of the sea while at the same time maintaining the immobility of the basin which contains it, let us go on to see whether we can explain the observed effects by positing the mobility of the container.

True natural effects follow easily from their causes.

There are two kinds of motion which can be imparted to a container so that the water it contains acquires the property of flowing towards one end of the container or the other, and of rising and falling in a given place. The first is when one or other end of the container is lowered, making the water flow towards the lower end, rising in one place and falling in another. But any such rising or falling is simply motion away from or towards the centre of the Earth; and such motion cannot be attributed to the concavities in the Earth's own surface, which contain the waters of the sea. No motion we might attribute to the terrestrial globe can cause any part of the Earth's surface to move towards or away from its centre.

A container can move in two ways which make the water contained in it rise and fall.

Concavities in the Earth's surface cannot move towards or away from its centre.

The second kind of motion is when the container, without tilting at all, moves forward, not uniformly but at varying speeds, accelerating at one moment and slowing down at another. The water in the container is not fixed to it like any

Uneven forward motion can make the water in a container move about.

of its solid parts; rather, being liquid, it is as it were free and separated from the container, and not constrained to participate in every movement which the container makes. So the result of this uneven motion is that when the container slows down, the water retains some of the impetus which it has already acquired, and flows towards the front of the container, so that its level necessarily rises there; conversely, when the container gains speed, the water retains some of the earlier slowness and falls behind, flowing towards the back of the container and rising there, before it adapts to the new impetus. We can see these effects very clearly with our own eyes in the example of one of the barges which constantly come from Fusina filled with fresh water to be used in the city. Picture one of these barges moving steadily across the lagoon, with the water it contains perfectly calm. Then its progress is significantly impeded, either because it runs aground or because of some other obstacle. The water which it contains will not lose its forward impetus to the same degree as the barge, but will retain some of it, flowing forwards towards the bow, where it will rise perceptibly, and falling at the stern. If on the other hand the same barge, in the midst of its steady progress, acquires a new burst of speed, the water it contains will not adapt straight away but will retain some of its slowness and will be left behind, flowing towards the stern, where it will consequently rise, and falling at the bow.

There are three things we should note about this effect, which is plain to see and can be tested at any time. The first is that, for water to rise at one end of the container, there is no need for any additional water, nor does the water have to flow all the way from the other end. The second point is that the water in the middle of the barge does not rise or fall appreciably—unless, that is, the barge was moving very rapidly and the impact or other obstacle which impeded it was very powerful and sudden; in this case not only might all the water rush forwards, but much of it might splash out of the barge. The same would happen if the barge was moving slowly, and suddenly received a violent impulse propelling it forwards. But if the barge was moving steadily and was moderately impeded or propelled forwards, the rise or fall of the water in the middle would, as I have said, be imperceptible. In the other parts of the barge it would rise less the nearer it was to the middle, and more the further it was away. The third point to note is that, while the water in the middle rises and falls very little in comparison with the water at either end, conversely it flows much further forwards and backwards in comparison with the water at the ends.

Now, gentlemen, the effect of the barge on the water it contains and of the water on the barge which contains it, is exactly the same as the effect of the Mediterranean basin on the waters it contains, and of the waters on the Mediterranean basin which contains them. So we shall now proceed to show how it is that the Mediterranean basin and every other bay or inlet, in short every part of the Earth, moves with a significantly uneven motion, even though we assign nothing but regular and uniform motion to the globe itself.

The parts of the terrestrial globe accelerate and slow down in their motion.

SIMPLICIO. I'm not a mathematician or an astronomer, but this strikes me at first sight as highly paradoxical. If it's true that the motion of the whole Earth is regular and yet the motion of the parts, even if they remain joined to the whole, can be irregular, this paradox would seem to defy the axiom which states *eandem esse rationem totius et partium*: the proportion of the whole and the parts is the same.

SALVIATI. I'll prove my paradox and leave it to you, signor Simplicio, either to defend the axiom against it or to make the two agree. My proof will be brief and very simple, and will draw on matters which we have dealt with at length in our discussions so far; and I won't say a word in favour of the ebb and flow of the tides.

As we have said, two motions are to be attributed to the terrestrial globe. The first is the annual motion of the Earth's centre on the circumference of its orbit around the ecliptic, following the order of the signs of the zodiac, that is from west to east. The second is the same globe's rotation on its own axis in twenty-four hours, also from west to east, although its axis is somewhat inclined and is not parallel to that of its annual revolution. Now the composite of these two motions, each uniform in itself, results in a motion which varies in different parts of the Earth. This will be easier to understand if I

Showing how the parts of the terrestrial globe move more and less rapidly.

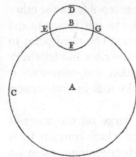

The parts of a circle moving uniformly around its centre move with contrary motions at different times.

explain this by means of a drawing. First, I will draw the circumference of the Earth's orbit, BC, around its centre, A. Then, taking any point B on this circumference, let us draw this smaller circle, DEFG, with B as its centre, to represent the terrestrial globe. We know that this globe, with its centre B, moves around the circumference of its orbit from west to east, i.e. from B in the direction of C. We also know that the terrestrial globe rotates around its centre B, also from west to east, i.e. in the sequence D, E, F, G, in the space of twenty-four hours. But here we must be careful to note that as the globe rotates on its axis, any given part of it will move at different times with contrary motions. This will be clear if we consider that when the parts of its circumference around point D are moving towards the left, in other words towards E, the opposite parts which are around point F are moving towards the right, in other words towards G. So when the parts around D reach point F their motion will be in the contrary direction to what it was when they were at D; and when the parts at point E are as it were descending towards F, those at point G are ascending towards D.

The mixture of the annual and daily motions causes unequal motion in different parts of the terrestrial globe.

Given these contrary motions of the different parts of the surface of the globe as it rotates on its axis, it follows that when this daily motion is combined with the other annual motion, the result must be an absolute motion for each part of the Earth's surface which is sometimes significantly accelerated and sometimes equally slowed down. This will be clear if we look at the part around point D. Its absolute motion will be very rapid, being the product of two motions both in the same direction, namely towards the left: the first is part of the annual motion which it shares with every other part of the globe, while the second is specific to this point D, as the daily rotation also carries it towards the left. So in this case the daily motion augments and accelerates the annual motion. This is the opposite of what happens at the opposite point F:

here, as the common annual motion carries it together with the rest of the globe towards the left, the daily rotation is carrying it towards the right. So here the daily motion works against the annual motion, making the absolute motion produced by the combination of the two much slower. Finally, around points E and G the absolute motion works out as equal to the annual motion, since the daily motion, being upwards and downwards rather than towards the left or right, neither augments nor decreases the annual motion. We must conclude, therefore, that while the motion of the globe as a whole and of each of its parts would be equal and uniform if they had only a single motion—cither just the annual motion or only the daily one— it follows that when the two motions are combined they produce unequal motions for the different parts of the globe, sometimes accelerated, sometimes held back, because the daily rotation is either added to or subtracted from the annual revolution. So if it is true (as experience clearly shows that it is) that acceleration and slowing down of the motion of a container makes water flow back and forth along its length, and rise and fall at each end, surely it must be conceded that this effect can, indeed must, affect the waters of the sea, when the basins in which they are contained are subject to such variations. And this must be especially true of seas whose length extends from west to east, as this is the direction along which the containers do in fact move.

This, then, must be the primary and most powerful cause of the ebb and flow of the tides, without which they would not exist at all. But there is also a wide variety of individual phenomena which can be observed at different times and places, which must derive from other concomitant causes even if these are all linked to the primary cause; so we need to identify and examine the various factors which could be the reason for these effects. The first such phenomenon is that when a significant acceleration or slowing down of the container causes water to flow towards one or other of its extremities, rising at one end and falling at the other, the water does not remain in this state when the primary cause ceases. Rather, its own weight and natural inclination to find its own level make it flow rapidly back again; and being both heavy and

Primary and most powerful cause of the ebb and flow of the tides.

Various phenomena which occur in the ebb and flow of the tides. First phenomenon: water which rises at one extremity returns to equilibrium of its own accord.

a liquid, it does not simply flow back to a state of equilibrium, but its own impetus carries it further, so that it rises where it had originally fallen. Even then it does not stop moving but runs back again, and flows back and forth several times, showing its reluctance to pass instantly from the velocity it had acquired to a state of rest; rather, its motion reduces gradually, a little at a time. It is exactly the same phenomenon as we see with a weight suspended from a cord, which when it is moved from its state of rest—that is, from the perpendicular— returns to the perpendicular of its own accord, but only after going beyond it many times as it swings to and fro. The second

In a shorter container the oscillations are more frequent.

phenomenon to note is that these oscillations in motion occur with greater or lesser frequency—that is, with shorter or longer time intervals—according to the varying length of the basin containing the water. In a shorter space the oscillations are more frequent, and in a longer space they are more infrequent; again this is exactly the same as can be observed in pendulums, where the oscillations of those with a shorter cord are more frequent than in those where the cord is longer. This

Greater depth makes the oscillation of the water more frequent.

leads to the third notable fact, which is that it is not only the varying length of the container that affects the frequency of the water's oscillation. The greater or lesser depth of the water has the same effect: if water is contained in receptacles of equal length but different depths, the deeper water will settle with shorter oscillations, while the oscillations will be less frequent in the water which is shallower. The fourth point which we

Water rises and falls at the extremities of the container, and flows in the middle parts.

should note and observe carefully is that water produces two effects as it finds its level. One is rising and falling alternately at each extremity of the container; the other is flowing back and forth horizontally. These two diverse motions occur differently in different parts of the water. The parts at either end are those which rise and fall the most; those in the middle do not rise or fall at all; of the other parts, those which are nearer to the extremities rise and fall proportionately more than those which are further away. In contrast, the other motion back and forth is more marked in the parts in the middle, and does not occur at all in those at the extremities (unless they rise so much that they spill over the edge and overflow out of their original container; but where the banks hold them back they simply rise and fall). This does not, how-ever, prevent the water in the middle from flowing back and forth, and this motion also occurs proportionately in the other parts, where it flows to a greater or lesser extent in proportion to its distance from the middle.

We should note the fifth particular phenomenon all the more carefully because it is impossible for us to put its effect into practice experimentally. It is this: when containers which we construct artificially, such as the barges we mentioned above, are set in motion at a greater or lesser speed, the acceleration and deceleration are shared in the same way by the container as a whole and by each of its parts. So, for example, if the barge's motion is held back, the forward part is not slowed down more than the part which follows, but every part shares equally in the same deceleration. The same happens with acceleration; if the barge receives a new stimulus to greater speed, the bow and the stern both accelerate in the same way. But in immense basins such as enormously long seabeds, the extraordinary fact is that their extremities do not increase or decrease their motion uniformly, equally or at the same moment in time, even though these basins are simply hollows formed in the solid surface of the terrestrial globe. What happens is that when the motion at one extremity is very much slowed down, because of the combined effect of the daily and annual motions, at the other extremity the two motions are still reinforcing each other to produce very rapid motion. This will be easier to understand if we go back to the drawing which we made just now.* Let us suppose that a stretch of sea is, say, a quadrant in length, such as the arc BC. We saw above that the parts near point B are in very rapid motion, because of the combined effect of the daily and annual motions both moving in the same direction. But at the same time point C is moving very slowly, because the forward motion deriving from the daily motion has been cancelled out. So we can see

Phenomena concerning the motions of the Earth which cannot be illustrated in practice.

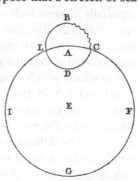

that in a gulf of the sea whose length extends over the arc BC, its extremities are moving at the same time at very unequal rates. The differences would be greatest of all in a stretch of sea extending over a semicircle, such as BCD: here one extremity, at B, would be in very rapid motion; the other extremity, D, in very slow motion; and the middle parts around C would be in moderate motion. Shorter stretches of sea would be less subject to this strange phenomenon of having their different parts affected by motion at different rates at certain times of the day. Now since the earlier example has already shown from experience that acceleration and deceleration cause the water in a container to flow back and forth, even when these are shared equally by every part of the container, what are we to suppose will happen in a container so remarkably placed that the acceleration and deceleration are very unevenly spread among its various parts? Surely we can only say that there must be other, even greater and more extraordinary causes of turbulence in water, which remain yet to be discovered. Many people may consider it impossible to construct machines and artificial containers which could reproduce the effects of such a phenomenon experimentally, but it is not completely impossible; in fact I have devised a mechanical model* which illustrates the detailed effect of these extraordinary combinations of motion. But for our present purpose, what you have so far been able to grasp in your imagination will suffice.

SAGREDO. Speaking for myself, I can well understand how this extraordinary effect must arise in the basins of the sea, especially those which extend a long way from west to east, since this is the direction of the Earth's motions. I can see, too, that as it is almost impossible for us to work out, and is unparalleled among the motions that we can produce, it gives rise to effects which we cannot re-create artificially.

SALVIATI. Now that we have dealt with these points, it is time for us to examine the particular diverse phenomena which can be observed in the ebb and flow of the tides. First, it should not be difficult for us to understand why it is that lakes, ponds, and even small seas have no significant tides. There are two clear reasons for this. The first is that the small size of the

Explanations for the particular phenomena observed in the ebb and flow of the tides. Secondary causes for the lack of tides in small seas and lakes.

basin means that, in acquiring different degrees of velocity at different times of the day, it acquires them with only small differences between all its parts: the leading parts and those which follow, i.e. the eastern and western parts, accelerate and decelerate with very little difference between them. Moreover, these changes come about gradually; the motion of the container is not abruptly impeded by a sudden obstruction or suddenly subjected to a rapid acceleration, so that it and all its parts gradually and evenly acquire the same degrees of velocity. This uniformity means that the water in the basin also receives the same velocity with hardly any turbulence or resistance, and so its rising and falling, and its flowing towards one or other extremity, is barely perceptible. We can see this effect clearly in a small artificial container, where the water will gradually acquire the same degrees of velocity if the container is made to accelerate or decelerate slowly and uniformly. But in basins of the sea which extend a large distance from east to west, the acceleration and deceleration are much more marked and uneven, as one of its extremities will have a very slow motion while the other is moving very fast.

The second reason is the reciprocal oscillation of the water as it settles after the impetus it receives from the motion of its container. As we have seen, this oscillation has a high frequency in a small basin. Now the motion of the Earth imparts motion to the waters only at twelve-hour intervals, since the maximum acceleration and deceleration in the basin's motion take place only once a day. But the second cause of motion, which derives from the weight of the water as it seeks to return to equilibrium, has an interval of only one, two, three hours or more, depending on how small the basin is. The mingling of this motion with the first, which in a small basin is very small in any case, has the effect of making the first motion almost imperceptible. For before the motion deriving from the first cause, which has a frequency of twelve hours, has fully taken effect, it is superseded and counteracted by the secondary movement deriving from the weight of the water, which has a frequency of one, two, three, or four hours, etc., depending on the length and depth of the basin. The secondary motion, running counter to the first, disrupts it and cancels it out before it can reach its maximum or even the midpoint of its extent; and these conflicting motions have the effect of negating the ebb and flow of the tides, or at least of making them much less apparent. This is quite apart from the continuous change produced by the air, which ruffles the surface of the water so that we would be unable to perceive such minor variations in level, of half an inch or less, as might exist in basins or inlets of water which are only a degree or two in length.

The reason why the tides generally ebb and flow with an interval of six hours.

I come, secondly, to the question of why the period of the ebb and flow of the tides commonly appears to be six hours, when the principal cause of the motion of the waters acts at intervals of twelve hours, i.e. once because of the maximum velocity of its motion and once because of its maximum slowness.* The answer is that this could not possibly happen solely as a result of the primary cause, but account must be taken of the secondary causes as well, i.e. those deriving from the greater or lesser length of the containers and the greater or lesser depth of the waters they contain. These have no effect on the motion of the waters: this is the product only of the primary cause, without which there would be no tides at all. But they have a very powerful effect in determining the frequency of their oscillation, so powerful that the primary cause has to submit to them. So a period of six hours is no more inevitable or natural than any other time interval; but it may well be the one that has been most observed, since this is the period that pertains in the Mediterranean, which for many centuries was the only sea that could be explored. As a matter of fact, it is not universal even in the Mediterranean; in some of the narrower parts of the sea, such as the Aegean and the Hellespont, the tides have a much shorter period, and even vary a great deal among themselves. Indeed, some say that after Aristotle had observed the tides at length from some cliffs in Euboea, their unpredictability and his inability to find an explanation for it drove him to such desperation that he threw himself into the sea and drowned.*

Why some seas have no tides, despite being very long.

Thirdly, we can readily explain why it is that some seas which are very long, such as the Red Sea, are almost impervious to the effect of tides. The reason for this is that the length of the Red Sea extends not from east to west but from south-east to north-west, and since the Earth's motion is from west to east the impulses moving the water are always across the lines of longitude rather than of latitude. Therefore seas which extend lengthwise towards the poles, lying across the lines of latitude, and are narrow in the other direction, have no reason to be subject to tides, apart from any effect they may share with other seas which flow into them and which are subject to large tidal variations.

Fourthly, it should be easy for us to see why the rise and fall of the tide is most marked at the extremities of a gulf, and most limited in its central parts. We see this clearly in everyday experience here in Venice, at the extremity of the Adriatic, where the tidal range is five or six feet, whereas in parts of the Mediterranean further from the extremities it is not more than half a foot—on the islands of Corsica and Sardinia, for instance, or on the beaches of Rome or Livorno. We can also see how, in contrast, the lateral movement back and forth is extensive where the rise and fall are limited. It is easy, as I say, to understand the reason for these phenomena, since we have seen them clearly replicated in various kinds of artificial containers, where the same effects follow naturally when we subject the container to uneven motion, i.e. to acceleration and slowing down. *Why tides are most pronounced at the extremities of a gulf, and most limited in its central parts.*

Fifthly, we know that a given quantity of water that is moved, albeit slowly, in a wide space, necessarily flows with greater impetus if it has to pass through a narrow opening. So it is not difficult to understand the reason for the strong currents which occur in the narrow channel separating Calabria and Sicily. The Ionian Gulf and the coast of Sicily hold back water in the eastern part of the sea, which flows down towards the west, although only slowly because its area is so wide. But when it is channelled into the strait between Scylla and Charybdis* it falls rapidly and becomes very turbulent. A similar but much greater effect must occur between Africa and the large island of Madagascar, where the waters of the Indian and Atlantic* oceans, which the island divides, are channelled into an even narrower strait between *Why waters flow more rapidly through narrow spaces than open ones.*

it and the coast of Africa. There must also be very strong currents in the Magellan Strait, which links the vast Atlantic and Pacific oceans.

An account of some of the more obscure phenomena which can be observed concerning the tides.

Let us go on now, as our sixth point, to give an account of some of the more obscure and controversial phenomena which can be observed in this matter of the tides; and here we need to consider another important point about their two principal causes, and how they combine and mingle together. As we have already noted more than once, the first and simplest cause is the regular acceleration and deceleration of the different parts of the Earth, which should cause the seas to flow eastwards and westwards at regular intervals in the space of twenty-four hours. The second is the motion deriving from the weight of the water itself which, having once been set in motion by the primary cause, seeks to find its own level again with repeated oscillations. These do not have any one fixed frequency, but are as varied as the varying length and depth of the different basins and inlets of the seas, so that this second principle can cause the seas to flow in one direction and back again in an interval of one hour, or of two, four, six, eight, or ten hours, etc. If we now start to combine the primary cause, which has a fixed period of twelve hours, with one of the secondary causes with a period of, say, five hours, it will sometimes happen that the two causes coincide and both impel the water to move in the same direction; and this conjunction or, as it were, conspiring together of the two motions will mean that the tidal range will be large. At other times it will happen that the primary cause works against the effect of the secondary cause; then, since one of the principles cancels out the effect which the other should have, the motion of the seas will be weakened, and the sea will be reduced to a state where it is tranquil and barely moving. There will be other times again when these two principles neither wholly counteract nor wholly reinforce each other, and then other factors will affect the growth and diminution of the tides. It can also happen that the mingling of the two principles of motion can give rise to contrary motions in two large seas linked by a narrow channel, so that as one is rising the other is flowing in the opposite direction; in such a case the channel between them can become exceptionally rough, with opposing currents and very dangerous whirlpools and turbulence. We continually hear reports from those who have experienced such phenomena. These conflicting motions, arising as they do not only from the different orientation and length of the seas but also to a large extent from differences in their depth, will sometimes produce disturbances in the water which are unpredictable and unobservable. The reasons for these have been and remain a cause of great anxiety to sailors, who encounter them without any apparent impetus from the wind or any other major disturbance in the air which might explain them. We should, indeed, recognize that disturbances in the air play a large part in other phenomena, and indeed we should acknowledge them as a third incidental cause which is powerful enough to bring about major changes in the effects we observe from the two primary, more essential causes. It is quite clear, for example, that a strong wind blowing continuously from the east can hold back the water and prevent the tide from going out. This means that when the second phase of the tide comes in, and then the third, their volume is much increased; so that, if the wind holds the water back for several days, the tides rise higher than usual and cause exceptional floods.

We should also note—and this will be our seventh question—another cause of motion, which arises when a large volume of water from rivers flows into a sea which is not very large. In the channels or straits which communicate with such a sea the water will always flow in the same direction, as happens in the Bosphorus at Constantinople, where the water always flows out of the Black Sea towards the Sea of Marmara. The reason for this is that in the Black Sea, since it is not very long, the principal causes of the tides have little effect; but some very large rivers flow into it, and all this volume of water has to pass through the strait, so that there is a notable current in the strait which always flows towards the south. We should note, too, that although this strait or channel is quite narrow, it is not subject to the same perturbations as the Strait of Messina. This is because the Bosphorus has the Black Sea to the north, and the Sea of Marmara, the Aegean, and the Mediterranean to the south, although these extend over a long distance; and

The reason why in some narrow straits the sea's water always flows in the same direction.

we have seen that seas which run north–south, however long they are, are not subject to tides. The strait of Messina, on the other hand, lies between the parts of the Mediterranean which extend for a great distance from east to west, i.e. in the direction of the ebb and flow of the tides, and this is why the turbulence is so great there. It would be even greater between the Pillars of Hercules, if the Strait if Gibraltar were any narrower; and it is said that in the Magellan Strait the water is very rough indeed.

This is all that comes to my mind to say now concerning the causes of this first, daily motion of the tides and the phenomena associated with it. If there are any points to be raised we can consider them now, before going on to discuss the other two motions, the monthly and the annual.

SIMPLICIO. I don't think it can be denied that your reasoning is very plausible if you argue *ex suppositione*, as we say— that is, assuming that the Earth moves with the two motions attributed to it by Copernicus. But take these two motions away, and everything else falls to the ground; and your own argument clearly shows that such a hypothesis cannot be maintained. You presuppose the twofold motion of the Earth to explain the ebb and flow of the tides, and then you argue in a circle, citing the tides as evidence to confirm these same motions. More specifically, you say that as water is a liquid and so is not firmly fixed to the Earth, it is not obliged to follow the Earth's motion in every detail, and from this you deduce the motion of the tides. I shall argue the opposite by following in your footsteps, as follows. The air is much thinner and more fluid than water, and even less attached to the surface of the Earth; water adheres to the Earth if only because its own weight presses down on it, unlike the air which is very light. Therefore the air should be much less obliged to follow the motion of the Earth; and hence if the Earth moved in the way you claim, we who inhabit the Earth and are carried along by it at the same velocity ought to feel a constant unbearably strong wind from the east. Our everyday experience alerts us to this. If we ride post-haste at only eight or ten miles an hour when the air is still, we feel it against our face like a moderately strong wind; so what would it feel like if we were moving at

Against the argument that the hypothesis of the Earth's motion is supported by the ebb and flow of the tides.

800 or 1,000 miles an hour, in the face of the air which did not share this motion? And yet we feel no such effect.

SALVIATI. To this objection, which appears very convincing, my response is as follows. It is true that air is much thinner and lighter than water, and that its lightness means that it adheres less to the Earth than water, which is so much heavier and more corporeal. But you draw an erroneous conclusion from this when you say that the air's lightness, thinness, and lesser degree of attachment to the Earth make it less constrained than water to follow the Earth's motions, and that therefore we who share fully in these motions ought to feel the effect of this resistance. In fact, what happens is precisely the opposite. If you remember, I said that the tides were caused by the fact that the water does not follow the uneven motion of its container, but retains the impetus which it had received earlier, and that the impetus does not diminish or increase in the water at the same rate as it does in the container. Since, therefore, conserving and maintaining an impetus received earlier means resisting a new increase or decrease in motion, a body which is more able to conserve an impetus will also be better placed to demonstrate the effect which follows from conserving it. Now we can clearly see how water tends to maintain a disturbance imparted to it even when the cause of the disturbance has ceased, because when the sea is whipped up by strong winds it still remains in motion long after the wind has dropped; the divine Poet described this elegantly when he wrote 'Even as the Aegean Sea', etc.* It maintains its motion in this way because of the weight of the water; because as we have said on another occasion,* light bodies are more easily moved than heavy ones, but are correspondingly less apt to maintain the motion imparted to them once the cause of the motion has ceased. So the air, because it is so thin and light, is easily set in motion by the slightest force, but it has very little aptitude for conserving its motion when the moving force ceases. As regards the air surrounding the globe, my view is that it adheres to the Earth and is carried along with it no less than water is; this is especially true of that part of the air which is enclosed in a confined space, such as a plain surrounded by mountains. In fact, I think it is much more reasonable to say that this air is swept around by the Earth's rough surface than it is to claim, as you Peripatetics do, that the higher levels of air are swept around by the motion of the heavens.

What I have said so far is, I think, a very adequate response to the objection raised by signor Simplicio. But I intend to go further in satisfying him, and to confirm the motion of the Earth to signor Sagredo, by identifying a new objection and providing a new response to it, based on a remarkable experience. I have said that the air is carried along by the roughness of the Earth's surface, especially those parts of the air which are below the summits of the highest mountains. It would seem to follow from this that if the Earth's surface were not uneven, but smooth and polished, there would no longer be a reason for it to pull the air along with it, or at least it would do so less uniformly. Now the surface of our globe is not all rough and jagged; there are very large areas which are quite smooth, namely the surface of the open seas. As these areas are a long way from any mountain ranges which might surround them, there seems to be no reason why they should be able to carry the air above them along with them; and if they don't carry it along with them, then the consequence of their not doing so should be felt in those places.

The rotation of the Earth confirmed by a new argument derived from the air.

SIMPLICIO. I was going to raise just this objection, which seems to me a very effective one.

SALVIATI. You're quite right, signor Simplicio; and so, if we do not feel the effect in the air which should follow as a consequence of our globe's rotation, you deduce from this that it does not move. But if this effect, which you consider a necessary consequence of the globe's rotation, were in fact felt and experienced, would you accept this as strong evidence in favour of its motion?

SIMPLICIO. In that case you would have to speak to others and not just to me. If such a thing happened, others might know its cause even though I might not.

SALVIATI. It's impossible to win against you, and it would really be better not to play at all rather than always being on the losing side. But I'll press on, so as not to let down our third companion.

Let me repeat and add some details to what I said earlier. There appears to be no reason why the air, as a thin, fluid body not firmly fixed to the Earth, should be obliged to follow the Earth's motion, with the exception of a part of the air close to the Earth's surface, not extending far above the top of the highest mountains, which is carried along by the roughness of the Earth's surface. This part of the air ought to be all the less resistant to following the Earth's rotation because it is full of vapours, fumes, and exhalations, all of which partake of the qualities of earth, and so naturally follow its intrinsic motions. But where the causes of this motion are absent—where there are wide, smooth spaces on the surface of the globe, and where there are fewer vapours from the earth mixed in with the air— then the cause which makes the surrounding air submit entirely to the Earth's rotation should be partly removed. In such places, if the Earth rotates towards the east, one ought to feel a constant wind blowing from east to west; and this wind should be most strongly felt in the places where the Earth's rotation moves fastest, i.e. in places which are furthest from the poles and nearest to the largest circle of the daily rotation. Now experience does indeed strongly endorse this philosophical reasoning. In the open seas furthest from land in the torrid zone—i.e. in the tropics—where also there are no evaporations from the Earth, one does feel a permanent breeze blowing from the east. It is so constant that ships have an easy passage to the West Indies, and the same wind favours ships which sail across the Pacific Ocean from the coast of Mexico to what we call the East Indies, although for them they are in the west. Sailing east from these parts, on the other hand, is difficult and uncertain; ships cannot follow the same routes, but must stay closer to land so as to find other, more fortuitous and unpredictable winds which have other causes, such as we land-dwellers constantly experience. There are many different causes giving rise to these winds, which there is no need to go into now; these are the fortuitous winds which affect every part of the Earth without distinction, causing storms in seas far removed from the equator and which are surrounded by the rough surface of the land. In other words, they affect the parts of the Earth subject to those disturbances in the air

The vaporous air close to the Earth's surface shares in its motions.

Perpetual breeze felt in the tropics, blowing towards the west.

Ships sail easily towards the West Indies, but with difficulty when they return.

The seas are disturbed by winds coming from the land.

which work against the primary wind, which would be felt perpetually, especially over the sea, if it were not for these fortuitous winds blowing against it. So you can see how the effects we see in the water and the air marvellously accord with our observations of the heavens to confirm that our terrestrial globe is in motion.

SAGREDO. I would like to set the seal on this question by telling you about another circumstance which I think you may not be aware of, which also confirms the same conclusion. Signor Salviati, you have cited the phenomenon encountered by sailors in the tropics, of the constant unvarying wind from the east, which I have heard about from those who have made this voyage several times. I understand, moreover—and this is a significant point—that sailors don't refer to this as a wind, but use some other term which I can't recall, perhaps referring to its steadiness and constancy. Apparently, once they have encountered it they set their sails and then, without having to touch them again, make steady progress even when they are asleep. Now this steady breeze could be recognized as such because it blows continually without being disturbed by other winds; if other winds had blown to disturb it, it would not have been recognized as a distinctive effect quite different from any other wind. This makes me suspect that our Mediterranean sea may also be subject to the same effect, but that we do not notice it because it is often deflected by other winds which override it. I don't have any firm foundations for saying this, but I put it forward as a likely conjecture, based on what I had occasion to note when I was our country's ambassador in Syria. I kept a log and record of the dates of departure and arrival of ships between the ports of Alexandria and Alexandretta and here in Venice. For curiosity's sake, I compared a large number of these and found that on average the homeward voyage to Venice, in other words voyages from east to west in the Mediterranean, were completed in about 25 per cent less time than those in the opposite direction. So it appears that in general winds blowing from the east are stronger than those coming from the west.*

SALVIATI. I'm very glad to have this information, which is not insignificant as confirmation of the motion of the Earth. It

Another observation derived from the air confirming the motion of the Earth.

Voyages in the Mediterranean from east to west are made in a shorter time than those from west to east.

could be argued that all the water of the Mediterranean flows constantly towards the Strait of Gibraltar because all the rivers which flow into it must be discharged into the Atlantic; but I don't think this is enough on its own to account for such a notable difference. This is clear too from the fact that in the Strait of Messina, water flows no less towards the east than towards the west.

SAGREDO. Unlike signor Simplicio, I don't have any need to satisfy anyone other than myself, and I'm convinced by all that has been said in this first part of our discussion. So when you're ready to proceed, signor Salviati, I am here to listen to you.

SALVIATI. I'll do as you ask, but first I would like to hear what signor Simplicio thinks, since his verdict will give me an idea of the reception I can expect my exposition to receive from the Peripatetic schools, if it should reach their ears.

SIMPLICIO. I wouldn't want my opinion to be taken as representing the views of others, or as a basis for speculating on what they might say. As I have said more than once, I have no expertise in these kinds of study; those who have penetrated the innermost secrets of philosophy will be able to answer in ways which would not occur to me, as I have only a nodding acquaintance with the subject. Still, to show that I'm not entirely devoid of ideas, I will reply that the effects which you describe, and especially the last, can equally well be explained by the motion of the heavens, with the Earth being at rest. There is no need to bring in any new idea, but only the opposite of what you yourself have introduced. It has been the received teaching in the Peripatetic schools that the element of fire and also a large part of the air are carried along with the daily motion, from east to west, by their contact with the inner surface of the sphere of the Moon, within which they are contained. So now, still following in your footsteps, I would like us to establish that the part of the air which shares in this motion comes down almost to the top of the highest mountains, and that it would reach down to the surface of the Earth if the mountains themselves did not impede it. This is the counterpart to what you say: you affirm that the air which is confined between mountain ranges is carried along by the

Reversing the argument, the constant motion of the air from east to west is shown to derive from the motion of the heavens.

rough surface of the Earth as it moves; we say, conversely, that the element of air is all carried along by the motion of the heavens, except for the part confined between mountain ranges, which is prevented from moving by the rough surface of the Earth which is at rest. And whereas you say that if it were not for this roughness of the surface the air would not be carried along, we can reply that if it were not for this roughness the air would all continue in motion. Hence, because the surface of the open sea is smooth and flat, it is exposed to the constant motion of the breeze blowing from the east; and this motion is felt most strongly in the tropics, near the equator, where the motion of the heavens is fastest. And if the motion of the heavens is powerful enough to carry along with it all the air that is unimpeded, it is reasonable to say that this motion also influences the motion of the seas, since water is a liquid and is not bound by the immobility of the Earth. We can be all the more confident in affirming this because, as you have acknowledged yourself, this motion is very small in comparison to its efficient cause: the motion of the heavens, which encircles the whole terrestrial globe in a natural day, must exceed many hundreds of miles per hour, especially at the equator, whereas the currents in the open sea flow at only a few miles per hour. This would mean that ships sail quickly and easily westwards not only because of the constant breeze from the east, but also because of the current. This same current may also be the cause of the tides, combined with the varying alignment of the seashore, as the water strikes the shore and rebounds in contrary motion; we can see this effect in the flow of rivers, where the water forms eddies and rebounds if it encounters a projection in the river bank or if there is a hollow under the surface. Hence, it seems to me that the same effects which you cite as evidence for, and which you attribute to, the Earth's motion, can be quite satisfactorily explained assuming that the Earth is fixed and that motion belongs to the heavens.

The motion of water derives from the motion of the heavens.

The ebb and flow of the tides may also derive from the daily motion of the heavens.

SALVIATI. There's no denying that your argument is ingenious and appears very plausible—but plausible in appearance, not in reality. What you say is in two parts: in the first part you give an explanation for the constant motion of a breeze from the east, and for a similar motion in water; in the second you suggest that the cause of the tides may also originate from the same source. The first part has, as I have said, some semblance of probability, although much less than my explanation based on the motion of the Earth; the second is not just wholly improbable, but absolutely impossible and mistaken. Coming to the first, you say that the inner surface of the sphere of the Moon carries along the element of fire and the air as far down as the tops of the highest mountains. To this I reply, first, that it is doubtful whether the element of fire exists; and that even if it does, it is even more doubtful whether the sphere of the Moon, and for that matter all the other spheres, exist as vast solid bodies. Rather, a good number of philosophers are now beginning to believe that beyond the outer limit of the air is an uninterrupted expanse of a substance much thinner and purer than our air, through which the planets follow their courses. But whichever view is correct, there is no reason to suppose that mere contact with a sphere, which you yourself say is absolutely smooth and polished, should be able to carry the whole element of fire with it in a circular motion alien to its natural inclination. That has been proved and demonstrated at length, citing the evidence of the senses, in *The Assayer*. This is quite apart from the further improbability that this motion could be transmitted from the very refined element of fire to the much denser element of air, much less from air to water. Whereas it is not only probable but necessary that a body with a rough and mountainous surface, as it rotates, should carry along with it the air which is in contact with it and which collides with its projecting parts. The evidence of this is plainly visible, although even without seeing it I don't think anyone with intelligence could doubt it.

The constant motion of air and water can be more plausibly explained by assuming that the Earth moves than by assuming it to be at rest.

It is not plausible that the element of fire could be carried along by the inner surface of the sphere of the Moon.

As for the second part of your argument, even if the motion of the heavens were imparted to both the air and the water, this motion would not have anything to do with the tides. A single uniform cause can only produce a single uniform effect, and therefore the only effect which this motion could produce would be a constant uniform motion from east to west. Moreover, this could only occur in a sea which encircled the whole globe; such a motion could not happen in an enclosed sea like

The tides cannot derive from the motion of the heavens.

the Mediterranean, which is blocked at its eastern end, because if its water could be driven westwards by the course of the heavens it would have dried up centuries ago. And in any case, its waters do not just flow towards the west; they also flow back towards the east, and at regular intervals. You cite the example of rivers to argue that even if the sea originally flowed only from east to west, the varying alignment of the shore can make some of the water flow back in the other direction, and I grant that this can happen. But, my dear signor Simplicio, you must realize that if water rebounds in this way it does so perpetually, and if it flows forwards, it flows always in the same way, as the example of rivers demonstrates. But to explain the ebb and flow of the tides, you have to find a reason which will explain how they flow now in one direction and now the opposite, both in the same place; and as these are diverse and contrary effects, they cannot be deduced from a constant and uniform cause. This fact, which destroys the case for saying that the daily motion of the heavens contributes to the motion of the sea, also invalidates the argument of those who would attribute it only to the daily motion of the Earth, thinking that this alone could explain the motion of the tides. This diverse effect must be the product of a diverse and changeable cause.

SIMPLICIO. I have no reply to make, either on my own account, because of the limitations of my wit, or on behalf of others, because the theory is such a novel one; but I don't doubt that if it becomes generally known in the schools, there will be no lack of philosophers who will be able to challenge it.

SAGREDO. Well, we shall wait for that to happen; in the meantime, signor Salviati, if you please, let us proceed.

SALVIATI. All that has been said up to now relates to the daily motion of the tides. We began with a general demonstration of their primary universal cause, without which nothing of this whole effect would come about. Then, moving on to particular varied and in some ways irregular phenomena which occur in relation to the tides, we have discussed the secondary and concomitant causes which bring these about. We must go on now to consider the other two periods, the monthly and the annual. These do not introduce any new or different phenomena from those we have already considered in connection with the daily period, but they have an effect on the daily motions, making them greater or smaller at different times in the lunar month and the solar year. It is almost as if the Moon and the Sun were playing a part in bringing about these effects, an idea which my intellect totally rejects: I can see that the tides are a local, physical movement involving an immense volume of water, and I cannot bring myself to subscribe to such vain ideas as the effect of light, temperate warmth, the prevalence of hidden qualities, or other such fancies. In fact, so far are these from even possibly being causes of the tides, that the very opposite is true: the tides gave rise to these ideas, putting them into the heads of those who are more suited to talking and showing off than to speculating and reflecting on the secret works of nature. They would rather hold forth, and sometimes even write, about the most absurd ideas, than pronounce those wise, simple, and modest words, 'I do not know'. Anyone can see that the Moon and the Sun cannot produce any effect on even the smallest container of water by means of their light, their motion, or their great or temperate heat; in fact, to make water rise at all by means of heat it has to be brought almost to boiling point. In short, there is nothing we can do to replicate artificially the motions of the tides, apart from moving the vessel containing the water. Surely this is enough to convince anyone that any other cause that is put forward to explain this effect is a vain fantasy that has nothing whatever to do with the truth?

I say, therefore, that if it is true that an effect has only one primary cause, and that there is a fixed and constant relationship between cause and effect, it must follow that any fixed and constant change in the effect must come from a fixed and constant change in the cause. The changes which affect the tides at different times of the year and the month follow a fixed and constant time period; therefore there must be regular changes in the same time intervals affecting the primary cause of the tides. Now the changes which occur in the tides at these times only affect their size—the extent to which the water rises or falls, and the greater or lesser impetus with which it flows; so it follows that the primary cause of the tides must grow or

Changes in effects imply changes in their causes.

Causes of the monthly and annual periods of the tides explained at length.

diminish in strength at these determined times. But we have concluded that the primary cause of the tides is the inequality and disparity in the motion of the basins containing the water; therefore this disparity must vary at the corresponding times, sometimes being greater and sometimes less. At this point we must recall that this disparity—that is, the difference in the velocity of motion of the basins, i.e. the different parts of the surface of the globe—derives from their composite motion, the product of the combined annual and daily motions of the globe as a whole. Of these it is the daily rotation, sometimes adding to and sometimes subtracting from the annual motion, that produces the disparity in the composite motion; so the additions to and subtractions from the annual motion as a result of the daily rotation are the original cause of the uneven motion of the basins, and hence of the tides. It follows that if these additions and subtractions always affected the annual motion to the same extent, the tides would continue to exist, but they would always be constant in their effect. But we need to find the cause which makes the ebb and flow of the tides

Monthly and annual changes in the tides can only derive from changes in the additions to and subtractions from the annual motion as a result of the daily motion.

greater and less at different times, and therefore (assuming we want to maintain the identity of the cause) we must find some change in these additions and subtractions, so that they are more or less powerful in producing the effects which follow from them. And I cannot see how this power or lack of it can derive from anything other than greater or lesser additions and subtractions, producing greater or lesser degrees of acceleration and deceleration in the composite motion.

SAGREDO. I feel as if I am being led gently by the hand, and although I haven't encountered any obstacles in the way, I feel like a blind man who can't see where his guide is taking him. I can't imagine where this journey is going to end.

SALVIATI. I know there is a great difference between your quick reasoning and the slow pace of my philosophy; but in this matter we are discussing now, I'm not surprised that your piercing insight is still defeated by the dense fog which conceals the goal to which we are travelling. Any possible surprise vanishes when I recall how many hours, days, and even more nights I have spent speculating over this matter, and how often I despaired of ever getting to the bottom of it. I was like the wretched Orlando,* trying to find consolation by persuading myself that the evidence before my eyes from so many trust-worthy witnesses was not true. So don't be surprised if, for this once, you aren't able to anticipate our conclusion; and if you are still surprised, I don't think you will remain so when you see the outcome, unexpected though I think it will be.

SAGREDO. Thank God your despair didn't lead you to the fate we read of poor Orlando, or the perhaps no less fictitious fate which is told of Aristotle.* If it had, I and everyone else would have been deprived of the revelation of something as hidden as it is sought after; so please satisfy my thirst for it as quickly as you can.

SALVIATI. I will. What we have to discover is how the additions and subtractions to the annual motion as a result of the Earth's daily rotation can vary in their proportions, since this variation is the only possible explanation for the monthly and annual changes which we see in the extent of the tides. I come now to consider three ways in which these additions and subtractions to the annual motion produced by the daily rotation can vary in their extent.

Additions to the annual motion as a result of the daily motion can vary in three ways.

First, this can happen because of an increase or reduction in the velocity of the annual motion, while the additions and subtractions of the daily motion remain unchanged. This is because the annual motion is roughly three times greater, i.e. faster, than the daily motion, even at the equator; so if it is increased even further, the additions or subtractions of the daily motion will produce less of a change. Conversely, if the annual motion is slowed down, the changes resulting from the same daily motion will be proportionately greater. In the same way, an increase or decrease of four degrees of velocity is proportionately less for something moving with a velocity of twenty degrees than it is for a velocity of only ten degrees. The second way would be for the additions and subtractions to be increased and decreased while the velocity of the annual motion remains the same; this is easily understood, since a velocity of, say, twenty degrees is clearly changed more by the addition or subtraction of ten degrees than by the addition or subtraction of four. The third way would be if these two changes were combined, with the annual motion decreasing and the daily additions and subtractions increasing. This much, as you can see, is not difficult to follow. What has been much harder for me has been to find out how this can happen in nature; yet I have found that nature does indeed make marvellous use of such changes, in ways which could hardly be foreseen. I say they are marvellous and unforeseeable for us, but not for nature herself, who can bring about things which are infinitely astonishing to us with consummate ease and simplicity; and what is supremely difficult for us to understand is supremely easy for her to achieve. I have shown, then, that the proportions between the additions and subtractions of the daily rotation and the annual motion can increase and decrease in two ways; I say two, because the third is simply a combination of these two. Now I will go on to add that nature makes use of both these ways; and I will add further that if she made use of only one of them, then one of the two periodic changes would have to be eliminated. The monthly variation would cease if there were no change in the annual motion, and there would be no annual variation if the additions and subtractions caused by the daily rotation were constantly the same.

What is supremely difficult for us to understand is supremely easy for nature to achieve.

If the annual motion did not change, there would be no monthly variation. If the daily motion did not change, there would be no annual variation.

SAGREDO. Do you mean to say that the monthly variation in the tides depends on the Earth's annual motion, and the annual variation in the tides depends on the additions and subtractions of the daily motion? Now I'm more confused than ever, and have even less hope of solving this conundrum, which strikes me as more tangled than the Gordian knot. I envy signor Simplicio, who I assume from his silence understands everything, and doesn't share any of the confusion which is perplexing me.

SIMPLICIO. Signor Sagredo, I can well believe that you are confused, and I think I know the source of your confusion: I imagine the reason is that you understand some parts of what signor Salviati has just been saying but not others. And you're right to say that I'm not confused, but not for the reason you think; rather the opposite. It's not that I understand everything, but I don't understand anything of it at all; and confusion is caused by a multiplicity of things, not by nothing at all.

SAGREDO. Look, signor Salviati, how some of the tugs on the reins that you've given to signor Simplicio over the last few days have tamed him, and transformed him from a steeple-chaser into a docile nag. But please don't keep us both in suspense any longer.

SALVIATI. I shall do my best to explain myself more clearly, and your sharp wits will compensate for any obscurity in my expression. There are, then, two phenomena whose causes we have to discover: the first concerns the variations in the tides which occur in a monthly cycle, and the other relates to the annual cycle. We shall discuss the monthly variation first, and then the annual one; and we shall resolve them both by referring to the principles and suppositions we have already established, without introducing any new concepts, either in astronomy or in general, in order to account for the tides. Rather, we shall show that all the various phenomena that are observable in the tides have their cause in facts which are already known and accepted as true and beyond doubt.

It is, then, naturally and necessarily true that the same moving object, impelled in a circular motion by the same motive force, will take a longer time to complete a revolution of a larger circle than of a smaller one. This is a truth which all accept and which is confirmed by every experience, of which we shall cite some examples. In rotary clocks, especially large ones, clockmakers make it possible to adjust their time-keeping by means of a stalk pivoted horizontally, with a lead weight attached to each end. If the clock runs slowly, they have only to bring these weights a little nearer to the centre of the stalk, and so make it oscillate more frequently; and conversely, to make the clock run more slowly, they move the weights more towards the extremities, making its oscillations more infrequent, and hence lengthening its measurement of the hours. In this example the motive force is the counterweight, and so does not change; the moving bodies are the lead weights, which also do not change; and their oscillations are more frequent when they are nearer the centre, in other words when they are moving in smaller circles. Or again, take equal weights and attach them to cords of different lengths, and then move them away from the perpendicular and let them go. We shall see that the weights attached to the shorter cords complete their oscillations in a shorter time, since they are

It can be confidently supposed that the revolution of a small circle is completed in a shorter time than of a larger circle. Two examples show this.

First example.

Second example.

moving in smaller circles. More than this, attach one of these weights to a cord which runs over a nail driven in to a bench, and hold the other end of the cord in your hand. Release the weight and, while it is swinging, pull on the end of the cord in your hand, so that the weight rises. You will see that as the weight rises, the frequency of its oscillations increases, as it is continually moving in smaller circles.

Here there are two particular points which should be noted.

Two particular phenomena which should be noted concerning pendulums and their oscillations.

The first is that the oscillations of a pendulum necessarily follow a fixed time interval, which is impossible to change except by lengthening or shortening the cord. You can confirm this by experience straight away, if you tie a piece of string to a stone and hold the other end of the string in your hand; then see if there is anything you can do to alter the time it takes to complete its swing, apart from lengthening or shortening the string, and you will find it is quite impossible. The other point which is truly marvellous is that the same pendulum will oscillate with the same frequency, or with only minimal and almost imperceptible variations, whether it is swinging in very large or very small arcs of the same circumference. For whether you move the pendulum from the perpendicular by only one, two, or three degrees, or whether you move it by seventy, eighty, or even ninety degrees, and then let it go, it will oscillate with the same frequency, even though in the first case it moves through an arc of only four or six degrees, and in the other through an arc of 160 degrees or more.

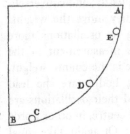

This can be demonstrated by taking two equal weights and suspending them from two cords of equal length. If they are moved from the perpendicular, the first by only a small amount and the other by a much larger amount, and then released, they will both swing to and fro in the same time interval, the first

Remarkable problems concerning objects falling along a quadrant of a circle, and along any chord of a whole circle.

in a very small arc and the second in a much larger one. This allows us to solve a most intriguing problem, which is this. Take a quadrant of a circle—call it AB in this sketch which I'll draw here on the ground—and set it up vertically so that it rests on a horizontal plane, touching it at point B. Make an arc with a piece of wood which is smooth and polished on its concave side, following the curve of a circumference ADB, so that a smooth round ball can run freely around it; the frame of a sieve is very suitable for this purpose. Now if you place the ball at any point, as near or far as you like from the lowest point B—point C, D, or E, for example—and let go of it, it will always arrive at B in the same space of time, or with only imperceptible differences. It makes no difference whether it starts from C, D, or E, or any other point—a truly remarkable phenomenon. And there is another phenomenon which is no less intriguing: if you draw a chord from B to C, D, E, or any other point, not only on the quadrant BA but anywhere on the circumference of the complete circle, a moving object will fall along these chords in exactly the same time; so it would fall along the whole diameter of the circle vertically above B in the same time as it would along the arc BC, even if that was an arc of only one degree or less. Add to this a further remarkable fact, which is that an object falls along any arc of the quadrant AB in a shorter time than it would along the chord of that same arc; so the most rapid motion and the shortest time for an object to travel from A to B will not be along the straight line AB—even though that is the shortest distance from A to B— but along the circumference ADB. And if you take any other point on this same arc, such as point D, and draw two chords AD and DB, the object will travel from A to B in a shorter time along the chords AD and DB than it would simply along AB— but the shortest time of all would be if it travelled along the arc ADB. And the same would apply to any other smaller arc above the lowest point B.

SAGREDO. Please stop—I can't take in any more. You're so overloading me with wonder, and distracting my mind in so many different directions, that I'm afraid I won't have enough mental capacity left to concentrate on the main subject that we're discussing, which is quite obscure and difficult enough in itself. But I do hope you will favour me by staying a few more days after we've finished our investigation of the tides, and will honour this house of mine which is also yours for a little longer. There are so many other problems to discuss which we have left in suspense, which I think are no less fascinating and intriguing than this question we have been discussing in these past days, and which we must conclude today.

SALVIATI. I shall be glad to; but we shall need more than just one or two sessions if we are to deal, not just with the other questions which we have set aside to discuss separately, but also with the many matters relating to local motion—both natural motion and that of projectiles—which our friend the Lincean Academician has dealt with at length. But to come back to our original subject, we were saying that objects moved in a circle by a motive force which remains constant have a period for their circulation which is fixed and determined, and cannot be made longer or shorter; and we gave examples and cited experiments which we could replicate to confirm this. Now experience also confirms that this same truth applies to the celestial motion of the planets, where we can see that the same rule is maintained; the planets which move in larger orbits take a longer time to complete them. This can be very clearly observed in the Medicean planets, which take only a short time to complete their revolutions around Jupiter. So there is no reason to doubt, in fact we can confidently affirm, that if the Moon, for example, were to be drawn gradually into a smaller orbit while still continuing with its own intrinsic motion, it would acquire a tendency for its orbital period to become shorter. This would be just like the pendulum where we reduced the length of the cord in the course of its oscillations, so shortening the radius of the circumference on which it was moving.

Now what I have just cited as an example with the Moon does in fact happen in reality. You'll remember that we concluded, in agreement with Copernicus, that the Moon cannot be separated from the Earth, and that it moves around the Earth in a month; this is generally agreed. We recall, too, that the Earth, always accompanied by the Moon, revolves in its great orbit around the Sun in a year, in which time the Moon revolves around the Earth almost thirteen times. This revolution of the Moon means that it is sometimes closer to the Sun and sometimes much further away: closer when it is between the Sun and the Earth, and further away when the Earth is between it and the Sun. In short, it is closer to the Sun when it is in conjunction, at the new Moon, and further away when it is in opposition, at full Moon. The difference between its greatest and its shortest distance from the Sun is the same as the diameter of the lunar orbit. Now if it is true that the force moving the Earth and the Moon around the Sun is constant; and if it is true, further, that the same object, impelled by the same motive force but in unequal circles, moves through a similar arc in a shorter time if the circle is smaller; we must conclude that when the Moon is at a shorter distance from the Sun, i.e. when it is in conjunction, it moves through a larger arc of the Earth's great orbit than when it is further away, i.e when it is in opposition, at full Moon. This inequality in the Moon must also be shared by the Earth. If we imagine a straight line from the centre of the Sun, through the centre of the Earth, and extended as far as the orbit of the Moon, this will be the radius of the great orbit which the Earth would follow uniformly if it were unaccompanied. But now place another body, carried along by the Earth, on this radius, locating it sometimes between the Earth and the Sun and at other times beyond the Earth, at a greater distance from the Sun. It must follow that the shared motion of both bodies along the circumference of the great orbit will be somewhat slower in the latter case, when the Moon is further away from the Sun, than in the former, when the Moon is between the Earth and the Sun and so closer to the Sun. The effect is just the same as when the clockmaker adjusts the clock's time-keeping, with the Moon functioning like the lead weight which is placed sometimes further from the centre, to make the stalk oscillate more slowly, and sometimes closer, to make its oscillations more frequent. Hence it is clear that the Earth's annual motion in its great orbit along the ecliptic is not uniform, and that its irregularities derive from the Moon, varying in a monthly cycle.

The Earth's annual motion along the ecliptic is unequal, because of the motion of the Moon.

Now we concluded earlier that the monthly and annual variations in the tides could only derive from the varying proportions of the additions and subtractions caused by the daily rotation in relation to the annual motion. We concluded, further, that these varying proportions could arise in two ways: either by a change in the annual motion, with the size of the additions remaining constant, or by a change in the latter with the annual motion remaining constant. We have now identified the first of these two reasons, arising from the irregularity of the annual motion caused by the Moon with its monthly cycle. Hence, for this reason the tides must increase and decrease in a monthly cycle; so now you can see how the cause of the monthly cycle lies in the annual motion, and also how the Moon plays a part in this process even though it has nothing to do with the sea and the tides.

SAGREDO. If you were to show a very high tower to someone who had no knowledge of any kind of staircase, and asked them if they thought they could reach its highest point, I'm quite sure they would say no, as they wouldn't be able to see any way of reaching it other than by flying. But if you then showed them a stone not more than half a *braccio* high and asked them if they thought they could step onto it, they would certainly say yes, and equally they wouldn't deny that they could easily climb onto it not just once but ten, twenty, or a hundred times. So when they were shown a staircase, by which they could by their own admission easily climb to the point which they had just said was impossible to reach, I'm sure they would laugh at themselves and acknowledge their own lack of insight. Well, signor Salviati, you have led me so gently step by step that, to my astonishment, I have reached a height which I never thought I could get to, and with minimal effort. The staircase, indeed, was so dark that I didn't realize I was approaching the top, or that I had got there until I came out into the daylight and discovered a great expanse of sea and countryside. And just as there is no effort in climbing a single step, so each of your propositions seemed so clear that I thought I was making little or no progress, as there was little or nothing in it that was new to me. So I am all the more amazed at the unexpected outcome of this argument, which has led me to the point where I can understand something which I thought was inexplicable. I am left with just one difficulty which I would very much like to have resolved for me, and it is this. If the Earth's motion around the zodiac, together with the

Moon's, is irregular, surely this irregularity should have been observed and noted by astronomers; and yet as far as I know this has not happened. I know you are informed about such things, so please resolve my difficulty for me, and tell me how the matter stands.

SALVIATI. That's a very reasonable question. My reply is that, although astronomy has made great strides over many centuries in investigating the structure and motions of the heavenly bodies, it is still at a stage where many things remain in doubt, and I dare say there are many more still which are as yet undiscovered. I think it likely that the first people who observed the heavens recognized only the motion common to all the stars, that is, the daily motion. It may well have taken them only a few days to realize that the Moon was inconstant in its motion in relation to the other stars, but then it was probably many years before they identified all the planets. In particular, I think that Saturn and Mercury were probably the last to be recognized as planets or wandering stars—Saturn because it moves so slowly, and Mercury because it is seen so rarely. Then I expect many more years passed before they observed the stations and retrogradations of the three outer planets, and likewise their variable distance from the Earth, which makes it necessary to introduce eccentrics and epicycles; these were unknown even to Aristotle, since he makes no mention of them. How long did Mercury and Venus keep astronomers in suspense with their baffling appearances before they were able to establish their location, never mind anything else? So just the order of the celestial bodies and the overall structure of the parts of the universe which are known to us were in doubt until the time of Copernicus; it was he who finally showed us the true structure and system according to which these various parts are ordered, so that now we know for certain that Mercury, Venus, and the other planets revolve around the Sun, and that the Moon revolves around the Earth. But we still have not established beyond doubt how each planet behaves in its individual revolution, and what exactly is the structure of its orbit—what is commonly called its theory. For evidence of this you have only to look at Mars, which causes so many problems for modern astronomers; and even the Moon*

There may still be many things in astronomy which have not yet been observed.

Saturn and Mercury were among the last planets to be observed, Saturn because it moves so slowly and Mercury because it is seen so rarely.

The individual structures of the planetary orbits are still not definitively resolved.

has had various theories applied to it since Copernicus so notably changed the Ptolemaic theory.

But let's come to the particular question we are discussing, namely the apparent motions of the Sun and the Moon. A great irregularity has certainly been observed in the motion of the Sun, since there is a marked difference in the time it takes to pass through the two halves of the ecliptic, divided by the equinoctial points:* it takes around nine days longer to move through one half than through the other, which as you can see is a very significant difference. But no one has yet observed— perhaps no one has even investigated—whether its motion through a small arc, such as each of the twelve signs of the zodiac, is entirely regular, or whether it moves more rapidly at some points and more slowly at others, as it must if the annual motion is really that of the Earth accompanied by the Moon, and only apparently of the Sun. As for the Moon, its cycle has been studied mainly in relation to eclipses, for which it is enough to have an accurate knowledge of its motion around the Earth; its progression through particular arcs of the zodiac has not been investigated so closely. So there is no reason to doubt that as the Earth and the Moon move around the zodiac, they accelerate somewhat at the new moon and slow down at the full moon, simply because this inequality has not been observed. There are two reasons for this: first, because it has not been thoroughly studied, and secondly, because it may not be very great.

It doesn't need to be very great to produce the effect which we see in the variation of the range of the tides, because not only these variations but the tides themselves are very small compared to the size of the bodies which they affect, even though they may seem large to us from our small perspective. One degree more or less in a natural velocity of 700 or 1,000 degrees can hardly be called a large alteration, either for the force which imparts it or for the body which receives it. The water in the Mediterranean travels at about 700 miles per hour as a result of the daily rotation, although this motion is imperceptible to us because it is the common motion which it shares with the Earth. The tidal motion which is apparent to us in the currents of the sea is less than one mile an hour (in the open sea, that is, not in

The Sun moves through one half of the zodiac in nine days less than through the other.

The motion of the Moon has been investigated principally in relation to eclipses.

The ebb and flow of the tides is very small compared to the vastness of the seas and the rapidity of the earth's motion.

narrow straits), and this is the effect of the greater primary, natural motion. This is a significant alteration for us and for ships; for a vessel which can be rowed at, say, three miles an hour in still water, the difference between having such a current following it or against it is the equivalent of doubling its speed—a very significant difference in the motion of the boat, but tiny in relation to the motion of the sea, which is modified by just one seven-hundredth of its speed. The same can be said of a rise and fall of one, two, or three feet, or perhaps four or five feet at the end of a gulf more than two thousand miles long where the water is hundreds of feet deep: an equivalent variation would be if the water in one of the barges carrying fresh water to Venice were to rise at the bow of the barge by the thickness of a leaf when the boat stops. So we can conclude that tiny variations in the sea, in proportion to its immense size and great speed, are enough to produce great changes in relation to the smallness of ourselves and our affairs.

SAGREDO. I am fully satisfied on this point. It remains for you to explain to us how the additions and subtractions produced by the daily rotation can vary in size; you said that the annual increase and decrease in the range of the tides depended on these variations.

SALVIATI. I shall make every effort to make myself understood, but I am daunted by the difficulty of the phenomenon, and the capacity for abstract thought which is needed to understand it. The variations in the additions and subtractions which the daily rotation produces in the annual motion derive from the inclination of the axis of the daily motion in relation to the plane of the Earth's orbit, or ecliptic. This inclination means that the equator intersects the ecliptic and is inclined and oblique to it by the same amount as the inclination of the axis. The additions are equal to the whole diameter of the equator when the centre of the Earth is at the solstitial points, and become less and less as the centre approaches the equinoctial points, which are the points where the additions are the smallest.* This is the whole explanation, but couched in rather obscure terms, as you can see.

Causes for the variations in the additions and subtractions to the annual motion produced by the daily rotation.

SAGREDO. Or rather as I can't see, since so far I understand nothing at all.

SALVIATI. I feared as much. Let's see if we can shed some more light by means of a drawing, although it would be better if we could represent it with solid bodies rather than just a drawing; but we'll use perspective and foreshortening to help us out. So I shall draw here, as we did before, the circumference of the Earth's great orbit, with point A as one of the solstitial points, and the diameter AP as the

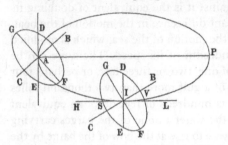

section of the solstitial colure* which it has in common with the plane of the Earth's orbit, or ecliptic. If we take the centre of the terrestrial globe to be at point A, the globe's axis CAB, which is inclined above the plane of the Earth's orbit, falls in the plane of this colure, which passes through the axes of both the equator and the ecliptic. For simplicity's sake we shall just draw the circle of the equator and mark it with these letters, DGEF. The line DE is the section which this circle has in common with the plane of the Earth's orbit, so that half of the equatorial circle, DFE, is below this plane, and the other half, DGE, is above it. Now let us assume that the equator is turning in the sequence D, G, E, F, and that the centre, A, is moving towards E. Since the centre of the Earth is at A, and its axis CB (which is perpendicular to the equatorial diameter DE) falls, as we have said, in the solstitial colure, of which the diameter PA is the section which it has in common with the Earth's orbit, this line PA must also be perpendicular to DE, because the colure is perpendicular to the Earth's orbit. Therefore DE is the tangent of the Earth's orbit at point A. So when the Earth is in this position, the motion of its centre along the arc AE, which amounts to one degree per day, varies hardly at all; in fact it is the same as it would be along the tangent DAE. Finally, since the daily rotation carrying point D through G to E adds to the motion of the centre, which has moved along virtually the whole length of the diameter DE, while the motion of the other semicircle EFD falls short of the motion of the centre by the same amount, it follows that in this position, i.e. at the solstice, the additions and subtractions will be equivalent to the whole of the diameter DE.

Now let us go on to see whether they would still be the same at the equinoxes. We shall move the centre of the Earth to point I, ninety degrees from point A, and we shall assume the same equator, GEFD, the same line DE as the section which it has in common with the Earth's orbit, and the same axial inclination CB. But the tangent of the Earth's orbit at point I is no longer DE, but another line intersecting it at a right angle which we can mark here as HIL; this is the line along which the centre, I, will move as it progresses around the circumference of its great orbit. Now in this position the additions and subtractions are no longer equivalent, as they were before, to the diameter DE, because this no longer corresponds to the line followed by the annual motion, HL, but intersects it at a right angle. Therefore points D and E have no bearing on the additions and subtractions; these will now be based on the diameter which falls on the plane perpendicular to the plane of the Earth's orbit and intersecting it with the line HL. Let us draw this diameter here, GF: now what we might call the additional motion will be the motion of point G around the semicircle GEF, and the rest will be the subtracted motion around the other semicircle, FDG. But the additions and subtractions are not determined by the whole length of this diameter GF, because it is not aligned with the annual motion HL but rather intersects it, as we can see, at point I; so the extent of the additions and subtractions will be taken from that part of the line HL which lies between two perpendiculars drawn to it from points G and F. We can call these GS and FV. The extent of the additions is therefore the line SV, which is less than either GF or DE, which was the extent of the additions at the solstice, A.

Hence we can establish the extent of the additions and subtractions for any location of the centre of the Earth along the quadrant AI. If a tangent is drawn from a given point, and perpendiculars are dropped to it from each end of the equatorial diameter defined by the plane through this tangent perpendicular to the plane of the Earth's orbit, the part of the tangent between these perpendiculars will give the extent of the additions and subtractions; and this will always be smaller at a point nearer to the equinoxes and greater when it is nearer to the solstices. As for the difference between the maximum and minimum extent of the additions, this is easy to find, because it is the same as the difference between the whole axis or diameter of the sphere and the part of it that lies between the polar circles. This is approximately a twelfth less than the whole diameter, assuming the additions and subtractions as they occur at the equator; they will be less at other latitudes in proportion to their diminishing diameters.

This is as much as I am able to tell you on this matter, and I dare say it is as much as it is possible for us to know with any certainty, since as you know, certain knowledge can come only from conclusions which are fixed and constant. Such are the three general periods of the tides, deriving as they do from causes which are one, invariable, and eternal. But these primary and universal causes are intermingled with secondary detailed causes which are capable of producing many changes. Some of these, such as the changing winds, are inconstant and un-predictable; others, such as the length of the sea's inlets, their different geographical orientations, and the great variations in the depth of the water, are fixed and established, but cannot be observed because they are so many and diverse. It would take very prolonged observations and absolutely reliable reports to compile an account of them which could provide a basis for firm suppositions about how they combine to produce all the appearances, not to say peculiarities and anomalies, which can be found in the motion of the tides. So I shall content myself with pointing out that such accidental effects occur in nature and are capable of producing extensive changes; I shall leave it to those who have practical knowledge of the various seas to observe them in detail.

To conclude our discussion I shall add just one further point which needs to be considered in relation to the exact timing of the tides. It seems to me that these are affected, not only by the varying length and depth of the sea's inlets, but also to a great extent by the confluence of different stretches of sea, all differing in their length and also in their position or orientation. This diversity is apparent here in the Adriatic, a gulf which is much smaller than the rest of the Mediterranean and quite different in its orientation: the Mediterranean is closed at its eastern end by the coast of Syria, whereas the Adriatic is closed at its western end. And since the tidal range is much greater at the extremities—indeed these are the only places where there is a very large rise and fall—it is very likely that the times of high tide in Venice coincide with low tide in the rest of the Mediterranean, which in a sense dominates the Adriatic because it is so much bigger and because it extends directly from west to east. So it would not be surprising if the effects due to the primary causes did not occur at the times and following the periods we would expect in the Adriatic, but rather at those prevailing in the rest of the Mediterranean. But such details require prolonged observations, such as I have not so far been able to undertake, nor do I expect to be able to do so in future.

SAGREDO. I think you have already done a great deal by pointing out the way for us to follow in investigating this profound question. Even if you had only expounded your first general proposition, it seems to me so far superior to the foolish ideas advanced by so many others that just thinking about them again makes me feel ill. I don't see how there can be any possible counter-argument to your demonstration that, if the basins containing the waters of the sea are fixed, it would be impossible in the ordinary course of nature for the motions which we observe in the tides to come about; and that on the contrary, if we assume the motions which Copernicus for other reasons attributes to the terrestrial globe, such alterations in the seas must necessarily follow. I find it quite astonishing that among so many men of profound understanding, not one recognized the incompatibility between the immobility of the vessel containing the water and the motion of the water which it contains; for this incompatibility now seems to me to be self-evident.

SALVIATI. What's more astonishing is that some of them, who did show more understanding than most by attributing the cause of the tides to the motion of the Earth, were not able

A simple motion of the terrestrial globe is not enough to explain the motion of the tides.

The opinion of the mathematician Seleucus criticized.

to follow their insight through to a conclusion. They didn't realize that it was not enough to identify a simple uniform motion such as the daily rotation of the Earth; the explanation required an unequal motion, faster at some times and slower at others, because if the motion of the containers is uniform the waters they contain adapt to it and do not move at all. An ancient mathematician* is reported to have said that the tides were caused by the opposition between the motion of the Earth and the motion of the lunar sphere; this is totally unfounded, not only because no explanation is offered for how this might happen, but also because there is no opposition between the rotation of the Earth and the motion of the Moon, since they both move in the same direction. So this is manifestly false. And any other views which have so far been put forward or imagined are, in my view, completely invalid. But of all the great men who have speculated on this marvellous effect of nature, the one who most astonishes me is

Kepler respectfully corrected.

Kepler: enlightened and acute thinker as he was, he had grasped the motions attributed to the Earth, and yet he still listened and assented to the notion of the Moon's influence on the water, and occult properties, and similar childish ideas.*

SAGREDO. In my view these speculative thinkers were in much the same situation as I now find myself in, of not being able to grasp how these three periods, the annual, monthly, and daily motions, interact with each other, or how they appear to derive from the Sun and Moon even though the Sun and the Moon have nothing to do with water. I need more time to apply my mind at length to the whole question before I can understand it fully; at the moment it still eludes me because of its novelty and difficulty. But I don't despair of being able to master it if I spend some time reflecting on it in solitude and silence, to digest what for the moment is simply stored up in my imagination.

In our discussions over these last four days we have, then, seen strong evidence in favour of the Copernican system. Three arguments in particular seem to be quite conclusive: the first based on the planetary stations and retrogradations, and the varying distances of the planets from the Earth; the second, on the rotation of the Sun and the observations concerning the sunspots; and the third on the ebb and flow of the tides.

SALVIATI. We could perhaps add a fourth, and maybe even a fifth. The fourth argument, based on the fixed stars, could be proved if accurate observation should confirm the minimal mutations which Copernicus posits but says are imperceptible.* And a fifth new discovery has just arisen which could also provide evidence for the mobility of the terrestrial globe, thanks to the very refined observations which are being made by signor Cesare, of the noble family of Marsili in Bologna, who is also a member of the Lincean Academy.* He has written a most learned account of how he has observed a continual, although very gradual, change in the line of the meridian, which I have recently read with amazement, and which I hope he will make available to all those who study the marvels of nature.

The motion of the meridian observed by signor Cesare Marsili.

SAGREDO. This is not the first time I have heard mention of this gentleman's great learning, and of his assiduous support of all men of letters; and if this or any other work of his is published, we can be sure that it will be something of great note.

SALVIATI. Now that the time has come for us to bring our discussions to an end, it remains for me to beg your indulgence if, as you reflect at more leisure on the points I have made, you should come across difficulties or doubts which I have not adequately resolved. Please excuse my shortcomings, partly because of the novelty of the ideas, partly because of the weakness of my own intellect, and partly because of the sheer magnitude of the subject matter. Excuse me, finally, because I do not expect and never have expected others to give the assent which I myself do not give to this fantasy, which I could readily admit to be a vain illusion and a tremendous paradox. In our discussions you, signor Sagredo, have often applauded and shown yourself convinced by the thoughts I have put forward; but I believe this is, in part, more because of their novelty than their certainty, but even more because, courteous as you are, you wanted to give me that gratification which we all feel when our efforts are approved and praised. And as I am obliged to you for your kindness, so I have appreciated the

ingenuity of signor Simplicio; his constancy in so strongly
and resolutely defending his master's teaching has greatly
endeared him to me. So I thank you, signor Sagredo, for your
kindness and consideration, and I beg signor Simplicio's
pardon if I have sometimes offended him with my excessively
bold and assertive talk. I assure him that I was not motivated
by any ill will; I simply wanted to give him more opportunity
to put forward his own thoughts so as to have a greater understanding myself.

SIMPLICIO. There is no need for you to apologize, least of
all to me; I am used to academic and public debates, so I have
often heard the disputants become heated and angry, and
indeed trade insults and sometimes come close to blows. As
regards the discussions we have had, and in particular this last
on the causes of the tides, the fact is that I am not entirely
convinced; but I confess that, from the rather vague idea of it
that I have formed in my mind, your theory strikes me as much
more ingenious than the others I have heard, although I don't
regard it as conclusively proved. Rather, I always keep in mind
a very sound principle that I was taught by a most learned and
eminent person, to which one must perforce submit.* I know
that both of you, if you were asked whether God in his infinite
power and wisdom could endow the element of water with its
observable oscillating motion by some means other than that of
moving the basin in which it is contained, you would reply that
he could have done such a thing in many different ways, which
might indeed be inconceivable to us. So I conclude that, this
being the case, it would be the height of presumption to try
to limit or restrict the divine power and wisdom to any one
particular fantasy.

SALVIATI. A fine and truly angelic principle, to which I will
add another, also divine, which is entirely in keeping with it. It
is this: that while we are free to debate about the structure of
the universe, we shall never discover all God's handiwork—
lest perhaps we cease to exercise our human minds and become
idle. So let us use the exercise which God allows and ordains
for us to recognize and wonder all the more at his greatness,
the more we find the profound depths of his infinite wisdom
outstrip our ability to penetrate them.

SAGREDO. Let this, then, be the conclusion of our four days of discussions. We must now curb our curiosity and allow signor Salviati an interval of rest, if he chooses to take it, but on the condition that he returns at a time convenient to him to satisfy our wish—mine especially—to resolve the topics we put to one side. I wrote these down as questions to put to him in one or two further sessions, as we agreed; and I am anxious above all to hear the elements of the new science which our friend the Academician has set out concerning local motion, both natural and enforced. Meanwhile let us go, as is our wont, to enjoy the cool of the evening for an hour in the gondola which awaits us.

THE TRIAL

REPORT OF THE COMMISSION OF ENQUIRY
(SEPTEMBER 1632)

In conformity with your Holiness's order, the whole sequence of events concerning the printing of Galileo's book, as it was eventually printed in Florence,* is set out below.

In substance, the affair progressed as follows.

In the year 1630 Galileo brought his book to the Revd Master of the Sacred Palace in Rome so that it could be reviewed in preparation for printing. The Revd Master gave it for revision to Father Raffaele Visconti, his fellow Dominican and a professor of mathematics. He amended it in several places, and was minded to give it his approval in the usual way, provided the book was printed in Rome.

The said Father has been written to and asked to send a copy of his decision, and this is awaited. The original text of the book has also been requested, so that the corrections made can be checked.

The Master of the Sacred Palace also wanted to review the book, and to save time agreed that it should be sent to him page by page; he gave his *imprimatur* for it to be printed in Rome, so that he could deal directly with the printer.

The author went to Florence and requested the Revd Master for permission to print the book there. This was refused, and the Revd Master referred the matter to the Inquisitor in Florence, withdrawing from the case himself. He advised the Inquisitor as to the conditions to be observed in the printing, while leaving it to him to decide whether it should be printed or not.

The Revd Master of the Sacred Palace has provided a copy of the letter which he wrote to the Inquisitor on this matter, and also a copy of the Inquisitor's reply to him, in which the Inquisitor stated that he had given the book to Father Stefani, Consultant to the Holy Office, for correction.

Thereafter the Master of the Sacred Palace had no further knowledge of the matter, but had seen the book, which was printed in Florence and published with the Inquisitor's *imprimatur*, and also bearing the *imprimatur* for Rome.

It is alleged that Galileo has acted contrary to the orders given to him, by asserting absolutely and not as a hypothesis that the Earth moves and the Sun is at rest;

that he has wrongly identified the motion of the tides, which is a fact, with the claim that the Earth moves and the Sun is at rest, which is not. These are the principal charges; in addition:

that he has fraudulently failed to disclose an injunction given to him by the Holy Office in the year 1616, to the effect that 'he should relinquish altogether the above opinion that the Sun is the centre of the universe and the Earth moves, and that he should not henceforth hold, teach, or defend it in any way, verbally or in writing. Otherwise, proceedings would be taken against him by the Holy Office. He acquiesced in this ruling and promised to obey it.'

It has now to be decided how to proceed, both against the person and against the book which has already been printed, the facts being as follows:

CONCERNING THE LICENCE

1. Galileo came to Rome in the year 1630, bringing his original manuscript, which he produced so that it could be reviewed in preparation for printing. The case being notified, an order was received that no element of the Copernican system was to be approved except as a pure mathematical hypothesis. It was immediately found that the book did not conform to this requirement, but that the arguments for and against the absolute truth of the system were put forward, although without any decision being reached. The Master of the Sacred Palace resolved that the book should be revised to make it purely hypothetical, and a new preface and conclusion were to be added, with which the body of the book should conform. These would explain how the book was constructed, and would preface the whole debate with a statement that the criticisms of the Ptolemaic system were made purely *ad hominem*, and in order to show that the Sacred Congregation was aware of all the arguments when it rejected Copernicus.

2. To execute this decision the book was given, with these instructions, to the Revd Brother Raffaello Visconti, a fellow Dominican of the Master of the Sacred Palace, as he was a professor of mathematics. He revised the book and amended it in many places,

and also informed the Master of other places where he took issue with the author, which the Master corrected without hearing further submissions. The rest of the book being approved, he was minded to issue his approval to be printed at the beginning of the book in the usual way, provided it was printed in Rome, as was the declared intention at that time.

The Inquisitor has been written to with a request that he should send a copy of his decision, which is expected by the first courier. The original text of the book has also been requested, so that the corrections made can be checked.

3. The Master of the Sacred Palace wished also to revise the book himself. When the author complained that it was not usual for there to be a second revision, and that this would cause delay, it was agreed that to expedite matters the Master would read the text page by page as it was ready to be sent to the press. In the meantime the *imprimatur* was granted for the book to be printed in Rome, so that he could deal directly with the printer; the preliminaries of the book were prepared, and printing was expected to begin imminently.

4. The author then went to Florence, and after some time had elapsed made a request for the book to be printed there. The Master of the Sacred Palace rejected this absolutely. When the request was renewed, he said that the original should be returned to him so that he could make the final revision as had been agreed, and that if this was not done he would never give permission for the book to be printed on his authority. It was replied that the original could not be sent because of the danger of loss or infection. The request being pressed further, with the intervention of his Highness, the Master of the Sacred Palace sought to find a solution by withdrawing from the case himself and referring it to the Inquisitor in Florence. He informed the Inquisitor of the conditions to be observed in the correction of the book, leaving him to decide on his own authority whether the book should be printed or not, without any commitment on the part of the Master's Office. Accordingly, he wrote a letter to the Inquisitor dated 24 May 1631, receipt of which was acknowledged in a letter from the Inquisitor, where he reports that he had referred the book for correction to Father Stefani, Consultant to the Holy Office in Florence.

A brief outline of the preface to be added to the work was subsequently sent so that the author could embellish it and incorporate it as he chose, and so that he could also provide a conclusion to the Dialogue in conformity with it.

5. The Master of the Sacred Palace had no further involvement in the affair until the first copies of the book arrived, having been printed and published without his knowledge. He impounded them, seeing that the conditions had not been observed; on subsequently receiving the Holy Father's command he had all copies of the work recalled, as far as time and due diligence allowed.

6. The substantial offences to be considered in the book are as follows:

(a) Having included the Rome *imprimatur* without approval, and without notifying the person over whose signature it appeared.

(b) Having printed the Preface in a different typeface and separate from the rest of the work, making it ineffective. Further, the correction at the end is placed in the mouth of a foolish character, at a point where it is not easily found, and moreover it is only coldly accepted by the other interlocutor, with just a brief acknowledgment which suggests that he accepts it only grudgingly.

(c) There are many places in the work where the motion of the Earth and the stability of the Sun are not treated hypothetically but asserted absolutely, or the arguments on which they are based are presented as necessarily demonstrated, while the counter-arguments are dismissed as impossible.

(d) The matter is treated as undecided, as if a resolution is awaited without any presupposition as to the outcome.

(e) Opposing authors on whom the Church most relies are misrepresented.

(f) A degree of equivalence is wrongly claimed between human and divine understanding in matters of geometry.

(g) It is cited as an argument for the truth that adherents of the Ptolemaic system are won over to the Copernicans, but not the other way round.

(h) The motion of the tides, which is a fact, is wrongly identified with the claim that the Earth moves and the Sun is at rest, which is not.

All the above points could be amended if it were considered that the book had some utility which would justify this concession.

7. The author received an injunction from the Holy Office in 1616 to the effect that 'he should relinquish altogether the above

opinion that the Sun is the centre of the universe and the Earth moves, and that he should not henceforth hold, teach, or defend it in any way, verbally or in writing. Otherwise, proceedings would be taken against him by the Holy Office. He acquiesced in this ruling and promised to obey it.'

GALILEO'S FIRST DEPOSITION*

Tuesday, 12 April 1633.

Galileo, son of Vincenzo Galilei, of Florence, in the 70th year of his age, having been summoned to the palace of the Holy Office in Rome, appeared in person in the lodgings of the Father Commissioner before the Revd Brother Vincenzo Maculano of Firenzuola, Commissioner-General, and the Revd Carlo Sinceri, Procurator Fiscal of the Holy Office.* Having taken the oath to tell the truth, he was questioned as follows.*

How and at what time had he come to Rome?
I arrived in Rome on the first Sunday in Lent, travelling in a litter.
Had he come on his own account or had he been summoned, or had he received any injunction that he should come to Rome, and from whom?
The Father Inquisitor in Florence ordered me to come to Rome and present myself to the Holy Office, this being commanded by the ministers of the Holy Office.
Did he know or could he imagine any reason why he had been enjoined to come to Rome?
I imagine that I was ordered to present myself to the Holy Office in Rome so as to give an account of my book which has recently been printed. I imagined that this was the case because, a few days before I was ordered to come to Rome, I and the bookseller were instructed not to distribute any more copies of the book, and also because the bookseller was ordered by the Father Inquisitor to send the original of my book to the Holy Office in Rome.
That he should explain why he imagined the book might be the reason for his having been enjoined to come to Rome.
It is a book written in the form of a dialogue, which deals with the structure of the universe, that is, the two principal systems, concerning the disposition of the heavens and the elements.

If he were shown a copy of the said book, would he be prepared to recognize it as his?

I hope so, and that if I were shown the book I would recognize it.

He was shown a copy of the book printed in Florence in 1632, entitled *Dialogo di Galileo Galilei Linceo, etc.*, which deals with the two world systems; and having examined it carefully, he said: 'I know this book very well; it is one of those that were printed in Florence, and I recognize it as mine, and composed by me.'

Would he equally acknowledge each and every part of its contents as his?

I know this book that has been shown to me; it is one of those that were printed in Florence, and I acknowledge everything it contains as composed by me.

When and over how long a time did he compose this book, and where?

As regards the place, I composed it in Florence ten or twelve years ago; and I must have been occupied with it for around seven or eight years, but not continuously.

Had he been in Rome on other occasions; specifically, had he been in the year 1616, and for what reason?

I was in Rome in 1616, and again in the second year of the pontificate of the Holy Father Urban VIII. I was last here three years ago, the reason being that I wanted to have my book printed. The reason for my visit to Rome in 1616 was that I had heard that doubt had been expressed about the theory of Nicolaus Copernicus concerning the motion of the Earth and the stability of the Sun, and the order of the celestial spheres; I came to learn what it was permissible to believe on this matter, as I wanted to be sure of holding only godly and catholic opinions.

Did he come on his own account or was he summoned? If he was summoned, what was the reason? With what person or persons did he discuss these matters?

I came to Rome in 1616 on my own account, without being summoned, for the reason I have given. I discussed this matter in Rome with several Cardinals who were members of the Holy Office at that time; in particular, Cardinals Bellarmine, Aracoeli, Sant'Eusebio, Bonsi, and d'Ascoli.*

That he should state in detail what he discussed with the abovementioned Cardinals.

The reason for my discussion with the said Cardinals was that they wished to be informed about the teaching of Copernicus, since his book is difficult to understand for those who are not professional mathematicians or astronomers. In particular, they wanted to know the disposition of the heavenly spheres according to Copernicus' hypothesis: that he places the Sun at the centre of the planetary spheres; around the Sun, and closest to it, the sphere of Mercury; around this, the sphere of Venus, and then the Moon around the Earth; and then outside these, Mars, Jupiter, and Saturn. Then concerning their motion, he places the Sun immobile in the centre, and the Earth rotating on its axis and revolving around the Sun; that is, rotating on its axis in its daily motion, and revolving around the Sun in its annual motion.

When he came to Rome seeking the above-mentioned resolution and the truth, what was resolved on this matter?

Concerning the controversy surrounding the above-mentioned theory that the Sun is at rest and the Earth in motion, it was decided by the Congregation of the Index that this theory, taken absolutely, was repugnant to Holy Scripture, and that it was to be admitted only as a hypothesis, as it is taken by Copernicus.

Was this decision notified to him, and by whom?

This decision of the Congregation of the Index was notified to me, by Cardinal Bellarmine.

What did his Eminence Cardinal Bellarmine notify him about this decision? What else, if anything, did he say to him about it?

Cardinal Bellarmine notified me that the said theory of Copernicus could be held as a hypothesis, as Copernicus himself had held it. His Eminence knew that I held this view as a hypothesis, that is in the same way in which Copernicus holds it, as can be seen from the Cardinal's reply to a letter* from Father Paolo Antonio Foscarini, Provincial of the Carmelites, of which I have a copy, and in which he writes: 'It seems to me that both you and signor Galileo are acting prudently in confining yourselves to speaking hypothetically and not in absolute terms.' This letter from the Cardinal is dated 12 April 1615. And he said that it should not be held or defended in any other way, that is, in absolute terms.

What was resolved and notified to him at that time, i.e. in the month of February 1616?

In February 1616 Cardinal Bellarmine told me that since the theory of Copernicus, taken absolutely, was contrary to Holy Scripture, it could not be held or defended, but that it could be adopted and used as a hypothesis. As confirmation of this I have a statement made by Cardinal Bellarmine himself, dated 26 May 1616, in which he says that the theory of Copernicus cannot be held or defended because it is contrary to Holy Scripture; and I submit a copy of this statement here.

He produced a sheet of paper, with twelve lines written on only one side, beginning 'I Robert, Cardinal Bellarmine, having', and ending 'on this 26th day of May 1616', and signed 'the above Robert, Cardinal Bellarmine'. I received this and marked it as Document B.* *Note*: I have the original of this statement in my possession here in Rome, and it is written entirely in the hand of the above-mentioned Cardinal Bellarmine.

> *When the above matters were notified to him, were any others present, and who?*

When Cardinal Bellarmine notified me and told me what I have said about the theory of Copernicus, some Dominican Fathers were present, but I did not know them, and I did not see them again.

> *Was he given any other injunction concerning this matter at that time, when the said Fathers were present, either by them or by others; and if so, what?*

I remember that it happened in this way. One morning Cardinal Bellarmine sent for me, and said something to me which I would rather say in the ear of His Holiness before anyone else; but the conclusion was that he told me that the theory of Copernicus could not be held or defended because it contradicted Holy Scripture. I do not remember whether the Dominican Fathers were present at the beginning or whether they came afterwards, nor do I recall whether they were present when the Cardinal told me that the said theory could not be held. It is possible that I was also given an injunction that I should not hold or defend this theory, but I don't remember, as it is something that happened some years ago.

> *If the things that were said and intimated to him at that time, together with the injunction he was given,* *were read to him, would he remember them?*

I don't remember that anything else was said to me, and I cannot say whether I would remember what was said to me then even if it was read to me. I am saying freely what I remember, because I do not claim ever to have contravened this injunction. That is, I have never held or defended in any way the said theory that the Earth is in motion and the Sun is at rest.

> *He was told that in this injunction, which was given to him in front of witnesses, it was stated that he could not in any way hold, defend, or teach the said theory.* Did he remember how and by whom this was intimated to him?*

I do not remember the injunction being intimated to me by anyone other than Cardinal Bellarmine in person. I remember that it said I could not hold or defend it; it may also have said 'nor teach'. And I don't remember whether it also had the words 'in any way', but it's possible that it did. I didn't think further about it or commit it to memory, because a few months later I had Cardinal Bellarmine's statement of 26 May, which informed me of the order not to hold or defend this theory. I didn't remember the other two details of the injunction which have just been notified to me, 'nor teach' and 'in any way', I think because they were not spelt out in the Cardinal's statement, which was what I relied on and committed to memory.

> *Having received this injunction, had he ever obtained permission to write the book which he has acknowledged, and which he subsequently had printed?*

After receiving this injunction I did not seek permission to write the above book, which I have acknowledged, because I do not believe that in writing this book I was contravening the injunction I was given not to hold, defend, or teach this theory, but rather I was refuting it.

> *Had he received a licence to print this book, and from whom; and was it for himself or for someone else?*

I acted on my own initiative three years ago to obtain a licence to print this book. I had offers of payment for it from France, Germany, and Venice, but I refused them all and came to Rome to submit the book to the chief censor, the Master of the Sacred Palace, who has absolute authority to add, delete, or change anything he wished. He had it very thoroughly checked by his fellow Dominican Father Visconti and, as I had submitted it to him, the Master of the Sacred Palace also revised it himself. He licensed the book, that is, he signed the book off and gave the licence to me, but with the order that it should be printed in Rome. We agreed that I would return to Rome in the autumn, since the summer was approaching and I wanted to return home to avoid the risk of falling ill, and I had already been here for the whole of May and June. Then, while I was in Florence the plague broke out, and traffic was blocked; so seeing that I couldn't come to Rome, I wrote a letter to the Master of the Sacred Palace asking if he would allow the book to be printed in Florence. He told me that he wished to review my original text, and that therefore I should send it to him. I used every possible diligence, and even consulted the Grand Duke's own first secretaries and chief couriers, to see if there was any way that the said original could be sent securely, but it was impossible to ensure that it would be delivered safely, and it would undoubtedly have been lost, soaked, or burnt, so bad were the conditions. I explained the difficulty of sending the book to the Revd Master, and he ordered that the book should be carefully checked once more by a person approved by himself. The person whom he approved was the Revd Master Giacinto Stefani, a Dominican brother, teacher of Sacred Scripture in the University of Florence, preacher to their Highnesses, and Consultant to the Holy Office. The book was delivered by me to the Revd Inquisitor of Florence, who gave it to the said Father Giacinto Stefani, who returned it to the Revd Inquisitor. He sent it to signor Nicolò dell'Antella, reviewer of books to be printed for the Most Serene Highness of Florence. The printer, whose name was Landini, collected it from this signor Nicolò and, having consulted the Revd Inquisitor, printed it, following exactly all the instructions given by the Revd Master of the Sacred Palace.

When he asked the Master of the Sacred Palace for permission to print the above-mentioned book, did he show him the injunction he had been given concerning the command of the Sacred Congregation, mentioned above?

I said nothing to the Master of the Palace about the above injunction when I asked for the licence to print the book, because I saw no need to say anything to him about it. I had a clear conscience, since I had not defended the theory that the Earth moves and the Sun is at rest in the said book; rather, in it I give the opposite of Copernicus' theory, and show that Copernicus' arguments are invalid and inconclusive.

After this he was dismissed, and a room was assigned to him in the official residence in the Palace of the Holy Office, instead of prison, with an injunction not to leave there without special permission, on pain of the judgement of the Sacred Congregation. This was given to him to sign, and he was sworn to silence on oath.

[In Galileo's handwriting]
I, Galileo Galilei, have made this deposition as above.

DOCUMENT B

I, Robert Cardinal Bellarmine, having learnt that signor Galileo Galilei has been calumniously reported to have abjured in our hand, and to have been subjected to salutary penance, and being asked to state the truth, hereby declare that the said signor Galileo has not abjured either in my hand or in that of anyone else here in Rome, or anywhere else as far as I know, of any theory or teaching held by him, nor has he received any salutary or any other kind of penance. He has only been apprised of the declaration made by the Holy Father and published by the Sacred Congregation of the Index, which states that the teaching attributed to Copernicus, that the Earth is in motion round the Sun and that the Sun is at the centre of the universe and does not move from east to west, is contrary to Holy Scripture and therefore may not be held or defended. In witness whereof I have signed this statement in my own hand, on this 26th day of May, 1616.

The above-named Robert Cardinal Bellarmine.

GALILEO'S SECOND DEPOSITION

Saturday, 30 April 1633.

The above-mentioned Galileo Galilei, having requested a hearing,* was interrogated in person in Rome, in the chamber of the Congregation, in the presence of those mentioned above. Having taken the oath to tell the truth, he was questioned as follows.

Let him declare what he wanted to say.
I have reflected continually for some days on the questions put to me on the 16th of this month, in particular on whether sixteen years ago I had been forbidden, by order of the Holy Office, to hold, defend, or teach in any way the theory which had just then been

condemned, that the Earth was in motion and the Sun at rest. I thought I should reread my printed Dialogue, which I had not looked at for three years, to establish carefully whether, contrary to my wholly innocent intention, I had inadvertently written anything which might lead a reader or my superiors to suspect any trace of disobedience on my part, or any other detail which might give the impression that I was someone who goes against the orders of the Church. Since, with the kind permission of my superiors, I was at liberty to send a servant on errands on my behalf, I managed to obtain a copy of my book, and I set about reading it attentively and considering its contents in detail. It is so long since I looked at it that coming to it now is almost like reading a new work by a different author, and I freely confess that in several places it struck me as being set out in such a way that a reader who was not aware of my inner thoughts would have had reason to think that the arguments advanced for the false position, which I was intending to refute, are put forward in a way which makes them appear more effective in reinforcing their position than easy to disprove. Two in particular, one based on the sunspots and the other on the tides, are presented as being so strong and robust that the reader has the impression they are being endorsed more than would be appropriate if the author considered them inconclusive and intended to refute them, as I myself privately and genuinely considered them, and as I still do. To excuse myself in my own eyes for falling into an error which was so remote from my intention, I do not consider it enough to plead that when putting forward the arguments of an opponent which one intends to refute—especially when writing a dialogue—they ought to be set out as cogently as possible, not distorting them so as to disadvantage one's adversary. As I have said, I do not consider this sufficient excuse, and so I plead that natural pride which everyone has in their own ingenuity, in wanting to show themselves cleverer than the common run of men in finding ingenious and plausible arguments even for propositions which are false. But nonetheless, and even though, like Cicero, I am more desirous of glory than is seemly,* if I were to write these same arguments again now I would certainly present them less forcefully, so that they should not appear to have a validity which in reality they do not have. So my error, which I confess, has been one of vain ambition and sheer ignorance and carelessness.

This is all I had to say on this matter, which came to me on rereading my book.

After this, and when he had signed his statement, he was dismissed, and was sworn to silence on oath.

[In Galileo's handwriting]
I, Galileo Galilei, have made this deposition as above.

Returning after a short interval, he said:

As further confirmation that I have not held and do not hold the theory, which has been condemned, that the Earth is in motion and the Sun at rest, I am willing to demonstrate this more clearly, if as I desire I am given permission and time to do so. The opportunity for this is readily available, as in the book which I have published the interlocutors agree that, after a certain interval, they should meet again to discuss a number of scientific questions which they set aside from the material covered in their meetings. This would provide the opportunity for me to add one or two more days, in which I promise I would return to the arguments previously brought forward in support of this false and condemned theory, and would refute it, with God's help, in the most effective way I could. I therefore pray this Tribunal to concur with this good resolution of mine, by allowing me the possibility of putting it into effect.

Here he again added his signature.

[In Galileo's handwriting]
I, Galileo Galilei, affirm as above.

GALILEO'S THIRD DEPOSITION AND WRITTEN DEFENCE

Tuesday 10 May, 1633.

The above Galileo Galilei was summoned and appeared in person in the chamber of the Congregation in the Palace of the Holy Office in Rome, before the Revd Father Vincenzo Maculano of the Order of Preachers, Commissioner-General of the Holy Office. The said Galileo appearing before him, the Revd Commissary set a time limit of eight days for him to prepare his defence, if he should wish or intend to submit one. In reply he said:

I have heard what your Reverence has said, and in reply I present this written text in my defence, not to excuse in any way my having gone too far in some places, as I have already said, but in order to show the sincerity and purity of my intentions. I also submit a statement by his late Eminence Cardinal Bellarmine, written by the Cardinal in his own hand, of which I have already presented a copy in my own hand. For the rest I submit myself entirely to the customary mercy and clemency of this Court.

Having signed this statement, he was dismissed to the residence of the above Ambassador of the Grand Duke, in the manner and form already notified to him.

[In Galileo's handwriting]
I, Galileo Galilei, with my own hand.

In the interrogation recorded above, in which I was asked whether I had informed the Revd Master of the Sacred Palace of the command given to me some sixteen years ago, by order of the Holy Office, that I should not hold, defend, or teach in any way the theory that the Earth moves and the Sun is at rest, I replied that I had not. As I was not asked the reason for my not having informed him, I did not have an opportunity to add anything further. It now seems to me necessary to explain this, to demonstrate my wholly pure intention, which has always rejected the use of dissimulation or fraud in any of my affairs.

I declare therefore that at that time, some who were not well disposed towards me were spreading a rumour that I had been summoned by his Eminence Cardinal Bellarmine to abjure some of my theories and teachings, and that I had had to abjure them and do penance for them. I was obliged, therefore, to appeal to his Eminence and ask him to give me a statement explaining why I had been summoned. He provided this statement written in his own hand, which I submit together with the present declaration. It is clear from this only that I was instructed that the teaching attributed to Copernicus, that the Earth is in motion and the Sun at rest, could not be held or defended; but there is no trace of any other command given specifically to me, over and above this general pronouncement which was applicable to all. Thereafter, since I had this authentic attestation written in the Cardinal's own hand as a reminder, I did not think further about or try to remember the words used orally in notifying me of this injunction, that it could not be defended or held, etc. So the two details, that in addition to holding or defending, it was not to be taught in any way, which I am told were part of the command given to me and were recorded as such, have come to me now as something completely new that I am hearing for the first time. And I do not think I should be doubted when I say that in the course of fourteen or sixteen years I have completely forgotten these words, especially as I had no need to reflect on them, since I already had such an authoritative statement in writing. Now if these two extra details are removed and just the two noted in the present attestation are retained, there is no room for doubt that the command it contains is the same as in the injunction given in the decree of the Sacred Congregation for the Index; and so I believe I can reasonably be excused for not having notified the Revd Master of the Sacred Palace of the injunction given to me privately, since it was the same as the one given by the Congregation for the Index.

Since therefore my book was not subject to any stricter censorship than that required by the ruling of the Index, I think it is clear that I followed the most secure and appropriate way of ensuring that it was free of any suspicion of error, by presenting it to the supreme Inquisitor at the very time when many books dealing with the same topics were being prohibited, solely on the basis of the said ruling.

I believe that what I have said gives me firm grounds for hoping that their Eminences, the most prudent judges, will no longer harbour any suspicion that I knowingly and deliberately transgressed the commands that were given to me. Rather, I hope they will see that such shortcomings as are found to be in my book were not deliberately introduced with any deceptive or insincere intention, but purely out of vain ambition and a desire to appear more perceptive than the common run of popular writers, and that I inadvertently allowed my pen to run away with me, as I have already confessed in my earlier deposition. And I shall be ready to make good and amend any such shortcoming with all possible diligence, at any time that their Eminences may command or allow me to do so.

It remains finally for me to ask them to consider my poor state of physical health, to which I have been brought by ten months of continuous mental torment, and by the discomforts of a long and arduous journey at the most inclement season, at the age of seventy years, which has deprived me of the greater part of such remaining years as my previous state of health might have led me to expect. I am encouraged in this by the faith I have in the clemency and benevolence of their Eminences who are my judges. I hope that, if in their justice they deem that I have not suffered enough as a punishment for my offences, my plea may move them to be lenient to my declining old age, which I humbly commend to them. Equally I commend to them my honour and reputation against the calumnies of those who wish me ill, whose persistence in seeking to harm my good repute is witnessed by my having had to ask for the attestation from Cardinal Bellarmine which I have just submitted with this statement.

GALILEO'S FOURTH DEPOSITION

Tuesday 21 June, 1633.

Galileo Galilei, of Florence, of whom see elsewhere, being arraigned to appear in person in Rome, in the palace of the Congregation of the Holy Office, in the presence of the Revd Father the Commissioner-General of the Holy Office, with the Revd the Procurator Fiscal in attendance,
took the oath to tell the truth, and was questioned as follows.

Was there anything he wished to say?
I have nothing to say.
Did he hold or had he held, and until what time, that the Sun is the centre of the universe, and that the Earth is not the centre of the universe but moves in a daily motion?
A long time ago, before the resolution of the Sacred Congregation and before the injunction was given to me, I was undecided between the two theories, that of Ptolemy and that of Copernicus, considering them to be debatable, as either was possible in nature. But after the above resolution, having confidence in the prudence of my superiors, my uncertainty was resolved, and I held and still do hold the theory of Ptolemy to be undoubtedly true, namely that the Earth is at rest and the Sun is in motion.

It was said to him that the way in which he had treated and defended the said theory in the book which he had had printed after the aforesaid time, and especially the fact that he had written the said book and had it printed, led to the presumption that he had held the said theory after the specified time; and he was asked to declare truthfully whether he holds or had held this theory.

As regards writing the Dialogue which has been published, I did not write it because I hold Copernicus' theory to be true. I thought only that it would serve the common good to set out the scientific and astronomical reasons which can be advanced on both sides, and I tried to make clear that on neither side were the arguments sufficient to prove conclusively in favour of one theory or the other. Therefore the secure way to proceed was to look to more sublime doctrines for a resolution, as many places in the Dialogue make clear. So my inner conviction is that I do not hold the theory which has been condemned, and have not held it since the resolution of my superiors.

It was said to him that from the book, and from the arguments put forward in favour of the view that the Earth moves and the Sun is at rest, it could be presumed that he held Copernicus' view, or at least that he had held it at that time. Therefore, if he did not resolve to confess the truth, recourse would be had against him to the appropriate remedies prescribed by the law.

I do not hold the theory of Copernicus, and I have not held it since I received an injunction intimating to me that I should abandon it. For the rest, I am in your hands, for you to do with me as you will.

It was said to him that he should speak the truth, otherwise recourse would be had to torture.

I am here to obey. I have not held this theory since the resolution was made, as I have said.

Since there was nothing further to be done in the execution of the decree, having signed his statement he was sent back to his place.

[In Galileo's handwriting]
I, Galileo Galilei, have made a deposition as above.

GALILEO'S ABJURATION

I, Galileo, son of the late Vincenzo Galileo of Florence, aged seventy years, am arraigned to appear personally in judgement and kneel before your Eminences, Reverend Cardinals, the general Inquisitors against heretical wickedness throughout the whole Christian Commonwealth.* Having before my eyes the Holy Gospels and laying my hands on the same, I swear that I have always believed, believe now, and with the help of God will believe in future, all that is held, preached, and taught by the Holy Catholic and Apostolic Church. But whereas it was intimated to me with an injunction by this Holy Office that I should wholly abandon the false theory that the Sun is at the centre of the universe and is at rest, and that the Earth is not at the centre of the universe and is in motion, and that I could not hold, defend, or teach this false doctrine in any way, either verbally or in writing; and whereas, after being notified that this doctrine is contrary to Holy Scripture, I wrote and caused to be printed a book in which I expound this same condemned doctrine and put forward many effective arguments in its favour, without offering any counter-arguments to them; I have been judged by this same Holy Office to be vehemently suspect of heresy, that is, of having held and believed that the Sun is the centre of the universe and is at rest, and that the Earth is not the centre of the universe, and is in motion.

Therefore, desiring to remove from the minds of your Eminences and of all faithful Christians this vehement suspicion which is justly held against me, I sincerely and with unfeigned faith abjure, curse, and detest the above-mentioned errors and heresies, and in general any other error, heresy, or sect contrary to Holy Church. And I swear that henceforth I shall never say nor assert, verbally or in writing, anything which may cause a similar suspicion to be held against me; and if I come to know of any heretic or anyone suspected of heresy, I will denounce them to this Holy Office, or else to the Inquisitor or Ordinary of the place where I may be.

I swear and promise, further, to carry out and observe in full any penance which this Holy Office has imposed or may impose on me; and if I should contravene any of these promises and undertakings which I have sworn—which God forbid—I submit to such penalties and punishments as canon law and other general or particular constitutions may impose and promulgate against such offenders. So help me God and these Holy Gospels, on which I now lay my hands.

I, the above-named Galileo Galilei, have abjured, sworn, promised, and undertaken as set out above; in witness whereof I have signed the present copy of my abjuration in my own hand, and have read it word for word, in Rome, in the Convent of S. Minerva, on this 22nd day of June, 1633.

I, Galileo Galilei, have abjured as stated above, in my own hand.

TWO NEW SCIENCES

FROM THE FIRST DAY

Speakers: SALVIATI, SAGREDO, AND SIMPLICIO*

SALVIATI. Regular visits to your famous Arsenal* here in Venice, gentlemen, must provide great scope for speculation to those of a philosophical turn of mind, especially in the field of mechanics. For instruments and mechanisms of every kind are constantly being set up here by a large number of craftsmen, among whom I'm sure there are some who have acquired great expertise and ability to expound their skills, drawing both on the observations made by their predecessors and on the solutions they are continually working out for themselves.

SAGREDO. You're quite right. My natural curiosity often leads me to come here to converse with those who, because of their preeminence over the rest of the work-force, we call master shipwrights; and speaking to them has often helped me to understand the causes of effects which were not just extraordinary, but obscure and indeed quite baffling. It's true that their conversation has sometimes confused me so much that I have despaired of being able to understand how something could come about which was quite foreign to my way of thinking, but which my senses showed me to be true. What that old man said to us just now is a very well-known saying; but I thought it was completely foolish, like many things which those of limited intelligence say when they want to sound as if they know something about a matter which they don't understand at all.

SALVIATI. I imagine you mean his last reply to us, when we were trying to understand why they were constructing such a great panoply of supports, props, and other defences around the large galley they were about to launch, which they don't do for smaller vessels. He replied that it was to avoid the danger of the ship breaking up under the pressure of its great weight and size, a problem which doesn't arise with smaller ships.

SAGREDO. Yes, that was what I had in mind; and especially his last remark, when he added that you can't extrapolate from small mechanisms of this kind to large ones, because there are many kinds of mechanism which work on a small scale but not when they are enlarged. I've always considered this to be a foolish popular idea, because the principles of mechanics all have their foundation in geometry, and I don't see that large or small size makes any difference to the properties of circles, triangles, cylinders, cones, or any other solid figures. If every element of a large mechanism is built conforming to the proportions of a smaller one, and the smaller one is robust and resistant enough to serve the purpose for which it was intended, then I see no reason why the larger one too should not be able to withstand any harmful or destructive forces it might encounter.

SALVIATI. The popular saying is completely groundless—so much so that one could equally truthfully assert the opposite, and say that many mechanisms can be more perfectly realized on a large scale than on a small one. A clock, for instance, can be made to show the time and strike the hours more accurately if it is a given size rather than a smaller one. The same saying has been taken over with more justification by rather more intelligent people, who explain that if large mechanisms of this kind do not work in the way pure abstract geometrical proofs suggest they should, this is the result of the imperfection of matter, which is subject to all kinds of changes and imperfections. My response to this may sound a little arrogant, but I will say it nonetheless: reference to the imperfections of matter, and their contaminating effect on the purity of mathematical proofs, is inadequate as an explanation of why an actual mechanism fails to obey abstract ideal laws. Even disregarding any imperfections of matter and assuming it to be perfect, immutable, and immune to any accidental change, the mere fact of being a material object means that a larger mechanism, made of the same material and with the same proportions as a smaller one, will correspond perfectly to the smaller one in every respect but one, namely its robustness and resistance to violent force. In fact, the bigger it is the weaker, proportionately, it will be. Now since I assume matter to be immutable, that is, always the same, it is clear that it has eternal and necessary effects which are susceptible to proof just as much as those of pure mathematics. So, signor Sagredo, you will have to abandon the opinion which you have held, I dare say in common with many others who have studied mechanics, to the effect that mechanisms and constructions made of the same materials, and carefully observing the same proportions between their different parts, must be equally—or rather, proportionately—able to resist and to yield to external forces and impacts; because it can be geometrically proved that larger structures are proportionately less resistant than smaller ones. Therefore, not only for artificial mechanisms and constructions but for natural ones as well, there is a necessary limit set which neither art nor nature can go beyond—as long, that is, as they maintain always the same proportions and the same material.

SAGREDO. This all leads to a very apt answer to the question, which is much debated among philosophers, about the cause of the acceleration of natural motion in falling objects. When a heavy object is projected upwards, the force impressed upon it by the projector continually diminishes. As long as this force was stronger than the contrary force of the object's heaviness, it continued to propel the object upwards. When the point is reached where these two forces are in equilibrium, the object ceases to rise and passes through a state of rest, in which the impressed impetus is not eliminated, but the amount by which it exceeded the object's heaviness and hence propelled it upwards is absorbed. As this extraneous impetus continues to diminish, and consequently the object's heaviness begins to prevail, the object begins to fall, slowly at first because it still retains a considerable part of the impressed force. But as the diminution of this force continues, it is increasingly exceeded by the object's heaviness, and hence the falling object continually accelerates.

SIMPLICIO. That's a clever thought, but I think it is more subtle than sound. It may be conclusive, but it only applies to natural motion which is preceded by a violent motion, where part of the extraneous force is still effective. Where there is no such residual force, and the object starts from a long-standing state of rest, your whole argument is ineffective.

SAGREDO. I think you are mistaken, and this distinction you make between the two cases is redundant, or rather non-existent. Tell me: is it possible for the projector to impart varying amounts of force to the projectile, sometimes more and sometimes less, so that it can be thrown a hundred *braccia* into the air, or twenty, or four, or one?

SIMPLICIO. Undoubtedly, yes.

SAGREDO. In that case it is equally possible for the force impressed to exceed the resistance of the object's heaviness by enough to raise it by only an inch; and finally, for the force of the projector to be just enough to equal the resistance of heaviness, so that the object is not projected upwards but simply supported where it is. When you hold a stone in your hand, are you not simply exerting enough upward force on it to equal the effect of its heaviness which is pulling it downwards? And do you not continue to exert this force on it for as long as you hold it in your hand? Surely it does not diminish simply because you hold it for a long time; and surely it makes no difference whether this support which prevents the stone from falling comes from your hand, or from a table, or from a cord from which it is suspended? So you must conclude, signor Simplicio, that it makes no difference at all whether the stone falls after a period of rest which is long, or short, or only momentary; for in every case, as it starts to fall it is subject to exactly that amount of force opposing its heaviness which was needed to keep it in a state of rest.*

SALVIATI. I don't think this is the time to embark on an investigation of the cause of the acceleration of natural motion, on which different philosophers have put forward a variety of different ideas. Some attribute it to the fact that the object is coming closer to the centre; others to the object having to make its way through progressively less of the intervening medium; and others again to an extruding effect of the surrounding medium, which continually impels the object forward as it closes behind it. We would have to examine these fantastic ideas, and others besides, and try to resolve them, to little effect. For now, it suffices for our Author's purpose* if we understand that his intention is to investigate and demonstrate some properties of accelerated motion—whatever the cause of such acceleration might be—which mean that an object's velocity when it starts from a state of rest increases in simple proportion to the passing of time, in other words that equal increases in velocity occur in equal intervals of time. If we find that the phenomena which we demonstrate occur in the naturally accelerated motion of falling objects, then we can take it that the principle we have postulated applies to the motion of falling objects, and that it is correct to say that their acceleration increases as the time and duration of their motion increase.

SAGREDO. As far as I understand it at the moment, it seems to me that it might be clearer to define the principle as follows, without changing its substance: that in uniformly accelerated motion, the velocity increases at the same rate as the distance travelled. So, for example, the degree of velocity acquired by an object falling four *braccia* is twice that which it had acquired when it had fallen only two *braccia*, which it turn is twice what it acquired in falling the first *braccio*. I think there can be no doubt that an object falling from a height of six *braccia* strikes with twice the impact of one which had fallen three *braccia*, three times that of one which had fallen two *braccia*, and six times the impact it would have had if it had fallen just one *braccio*.*

SALVIATI. It's a consolation to be in such good company in my error. What you say seems so plausible that when I said the same thing to our Author, he didn't deny that for some time he had fallen into the same fallacy. But what astonished me more than anything else was that just four simple words sufficed to show me not just the falsehood but the impossibility of two propositions which seemed so convincing that no one, among the many people to whom I expounded them, had ever refused to accept them.

SIMPLICIO. I must admit that I would have been one of those who would have accepted them. To say that a falling object acquires force as it travels—*vires acquirat eundo**—so that its velocity increases at the same rate as the distance travelled, and that its momentum on impact is twice as great if it falls from twice the height, are propositions which seem to me to be quite straightforward and uncontroversial.

SALVIATI. And yet both propositions are false and impossible, just as it is impossible for motion to be instantaneous. This can be demonstrated very clearly. If velocity increases in the same proportions as the distance already travelled and the distance still to be travelled, then these distances are travelled in an equal period of time; so if a falling object travels four *braccia* at twice the velocity at which it travelled the preceding two *braccia* (since it is twice the distance), it must travel both distances in the same time. But for the same falling body to travel four *braccia* in the same time as it takes to travel two *braccia*, motion would have to be instantaneous; whereas we can see that a falling object travels in time, and that it travels two *braccia* in less time than it takes to travel four. So it is false to say that velocity increases as the distance increases. The falseness of the second proposition can be demonstrated equally clearly. If the falling object is the same, any difference in its momentum on impact must be due to the difference in velocity; so for the impact to be twice as great when the object falls from twice the height, it would have to strike at twice the velocity. But if it travelled at twice the velocity, it would travel twice the distance in the same time, whereas we can see that it takes a longer time to fall from a greater height.

SAGREDO. You make these hidden conclusions appear too clearly and too easily; by making them so straightforward you make them less highly prized than when they were concealed under appearances which seemed to suggest the contrary. I think that people in general would place a lower value on knowledge they acquire with so little effort, than on matters which are the subject of long and incomprehensible debates.

SALVIATI. Those who demonstrate briefly and clearly the falsehood of propositions which have been generally believed to be true can live with the disappointment of being greeted only with scorn rather than appreciation. What is really unpleasant and harmful is the reaction of some who claim to be anyone's equal in studies of this kind, when conclusions which they have passively accepted as true are then easily and briefly shown by someone else to be false. Their reaction is not what I would call envy, which easily turns to hatred and anger against those who expose such fallacies, but rather a desire to uphold long-established errors rather than allowing newly discovered truths to be accepted. This desire sometimes drives them to write contradicting these truths even though they themselves recognize the truth in their hearts, simply so as to damage the reputation of others in the eyes of the undiscerning common people. Our friend the Academician* has told me about quite a number of such false conclusions, which are accepted as true even though they are very easily refuted, and I have kept a note of some of them.

SAGREDO. I hope you will not deprive us of the chance of hearing about them, and that you will share them with us in due course, even if we need a separate meeting for the purpose. But for now, taking up the thread of our conversation, it seems to me that we have so far established the definition of uniformly accelerated motion, which will be the subject of our forthcoming discussions. It is as follows: *Equally or uniformly accelerated motion is that which, starting from a state of rest, acquires equal degrees of velocity in equal intervals of time.*

SALVIATI. This definition being established, there is just one principle which our Author requires and presupposes, namely: *The degrees of velocity acquired by the same moving object on differently inclined planes are equal, if the elevations of the planes are equal.*

By the elevation of an inclined plane, he means the perpendicular which falls from the upper end of the plane to the horizontal drawn from the lower end of the same plane. Thus, if the line AB is horizontal and the two planes CA and CD are inclined above it, the perpendicular CB, which falls to the horizontal BA, is what he calls the elevation of the planes CA and CD. His supposition is that

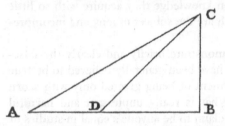

the same moving object descending along the inclined planes CA and CD will acquire the same degree of velocity when it reaches their respective ends, A and D, because both planes have the same elevation, CB. Moreover, this will be the same velocity as would be acquired by the same object falling vertically from C when it reached point B.

SAGREDO. This supposition certainly seems sufficiently probable to be accepted without argument, always assuming that any accidental external impediments are removed, that the planes are smooth and solid and that the moving object is perfectly round, so that neither the plane nor the moving object has any unevenness. Provided all such obstacles and impediments are removed, my natural reason has no difficulty in telling me that a heavy, perfectly round ball descending along the lines CA, CD, and CB would reach points A, D, and B with the same impetus in each case.

SALVIATI. Your reasoning is highly probable. But I would like us to increase the probability by means of an experiment, to the point where it is little less than conclusively demonstrated. So let us imagine that this sheet of paper is a vertical wall, with a nail fixed in it from which a thin thread, AB, two or three *braccia* long, hangs vertically with a lead weight of one or two ounces attached to it. Draw a horizontal line, DC, on the wall, intersecting at a right angle the vertical line AB, which should be about two inches clear of the wall. Then move the thread AB, with the lead weight attached to it, to AC, and let go of the weight. You will see that the weight first falls through the arc CBD, going so far beyond point B along the arc BD that it rises almost to the horizontal line CD, from which it will fall short only by a very small amount due to the resistance of the air and

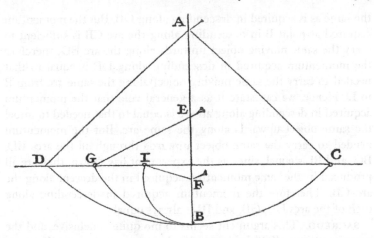

the thread. This allows us to conclude that the impetus which the weight has acquired at point B, in the course of descending along the arc CB, is sufficient to impel it to the same height along a similar arc BD. When we have repeated this experiment several times, let us then fix another nail to the wall, touching the vertical line AB, as for instance at point E or point F. The nail should project five or six inches from the wall, so that when we let the thread AC fall as before, with the weight C falling through the arc CB, at the point where it reaches B the thread will catch on the nail, E. The weight will then have to move along the circumference BG around the centre E, and we shall be able to observe the effect of the impetus it has acquired at point B, which previously impelled it along the arc BD to the height of the horizontal line CD. Now, gentlemen, I am sure you will be gratified to see that the weight reaches the horizontal at point G; and the same would happen if the nail were fixed at a lower point such as F, in which case the weight would follow the arc BI, still rising precisely to the level of the line CD. If the nail were fixed at a point so low that there was not enough thread below it to reach up to the level of CD (as would happen if it were closer to B than to the point where AB and the horizontal CD intersect), then the thread would rise above the nail and would wrap itself around it.

This experiment leaves no room for doubt that our supposition is correct. Since the two arcs CB and DB are equal and similarly placed, the momentum acquired in descending along the arc CB is the same as is acquired in descending along DB. But the momentum acquired at point B in descending along the arc CB is sufficient to carry the same moving object upwards along the arc BD; therefore the momentum acquired in descending along DB is equal to that needed to carry the same moving object along the same arc from B to D. Hence, we can state it as a general rule that the momentum acquired in descending along an arc is equal to that needed to impel the same object upwards along the same arc. But the momentum needed to carry the same object upwards through all the arcs BD, BG, and BI is equal, since as the experiment has shown, they are all produced by the same momentum acquired in the descent along the arc CB. Therefore the momentum acquired in descending along each of the arcs DB, GB, and IB is always equal.

SAGREDO. This argument seems to me quite conclusive, and the experiment so well devised to verify the supposition we made that it can be considered proven.

SALVIATI. I wouldn't want us to assume more than is warranted, signor Sagredo, especially since we have to use this supposition chiefly in relation to motion on flat surfaces rather than on curved ones, where the effect of acceleration is quite different from what we are assuming for planes. So even though the experiment we have used shows us that descending motion along the arc CB imparts sufficient momentum to the moving object to carry it upwards to the same height along any of the arcs BD, BG, BI, we cannot show with the same clarity that the same would happen if a perfectly round ball were to descend along plane surfaces inclined along the chords of these same arcs. Indeed, since such plane surfaces would form an angle at point B, it seems likely that a ball descending along the chord CB would be impeded by the ascending planes along the chords BD, BG, BI, and so would lose some of its impetus and would not be able to reach the height of the line CD. But if one could eliminate this impediment, which vitiates the experiment, then I think we can accept the principle that the impetus (which after all derives its force from the height of the descent) would be sufficient to carry the object back up to the same height. So let us take this as our supposition, the absolute truth of which remains to be established when other conclusions resting on this hypothesis are shown to be experimentally correct.

[. . .]

SAGREDO. Please suspend the reading* for a while, so that I can puzzle out an idea which has just occurred to me. It will make it easier for me and for you if I make a drawing: The line AI represents time elapsed starting from an initial moment A. I draw a straight line from A, AF, at any angle, and I join the ends of the two lines, I and F. Dividing the time AI in half at C, I draw a line CB parallel to IF. This line CB represents the maximum velocity reached by an object starting from a state of rest at A, and increasing in proportion to the length of lines parallel to BC intercepting the triangle ABC— which is the same as saying that velocity increases in proportion to time elapsed. It is clear from what has been said so far that the distance travelled by a falling object accelerating in this way

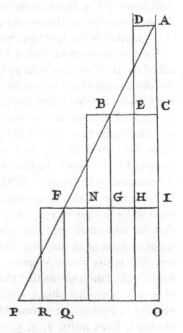

will be equal to the distance it would have travelled if, in the same time AC, it had been moving in uniform motion at a velocity equal to EC, this being half of BC.

Now, taking a step further, I deduce that if an object falling with accelerated motion had a velocity of BC at the moment C, and if it then continued to travel at the same speed BC without accelerating further, in the following interval of time CI it would travel twice the distance as in the equal interval of time AC, when it was travelling at a uniform velocity of EC, or half of BC. However, since it falls with uniformly accelerated motion, gaining equal degrees of velocity in equal intervals of time, this velocity CB must be augmented by the same amount again in the following interval of time CI, increasing in proportion to the parallel lines intercepting the triangle BFG, which is equal to the triangle ABC. So, to the velocity GI we must add half of FG, this being the maximum velocity reached by the object as a result of its acceleration in proportion to the parallel lines in the triangle BFG; this gives the velocity IN, which is the speed at which it would have travelled in the time interval CI in uniform motion. And since IN is three times the velocity EC, it follows that the distance travelled in the second time interval CI must be three times that travelled in the first time interval, CA.

Now let us suppose that a further equal interval of time, IO, is added to AI. If we now enlarge the triangle to APO, it is clear that if the object continued in motion throughout the time interval IO at the velocity IF, this being the speed it had reached as a result of its accelerated motion in the time interval AI, the distance which it travelled in the time IO would be four times that travelled in the first equal time interval AC, because the velocity IF is four times the velocity EC. But since the uniform acceleration continues to grow, as represented by the triangle FPQ which is equal to the triangle ABC, by an amount which in uniform motion would be equal to EC, we must add a further increment QR which is equal to EC. This means that the equivalent uniform velocity in the time interval IO is five times as great as in the time interval AC, and therefore the distance travelled is five times as great as in the first interval AC. So this simple calculation shows that the distance travelled in equal intervals of time by a moving object which, starting from a state of rest, gains velocity in proportion to time elapsed, increases in the ratio of odd numbers, from unity, 1, 3, 5, etc.* The total distance travelled in twice the time will be four times the amount travelled in one unit of time; in three times the time the distance will be nine times the amount travelled in one unit of time. In short, distances travelled are in double proportion to the time elapsed; that is, they are the squares of the lengths of time elapsed.

SIMPLICIO. I must say that I have appreciated this clear and straightforward exposition of signor Sagredo's more than our Author's proof, which I found somewhat obscure. I can readily see that, given the definition of uniform accelerated motion, the effect must be as described. But whether this is the acceleration actually used by nature in the motion of falling heavy objects, is something about which I am still in doubt. So I think that, to help my understanding and that of others like me, it would have been helpful at this point to cite some of the experiments, of which I understand there have been many, which confirm the conclusions that have been demonstrated in various ways.

SALVIATI. That's a very reasonable request, as befits a true scientist. This is indeed the usual and proper way to proceed in those sciences which apply mathematical proofs to conclusions about the natural world; perspective painters, astronomers, mechanics, musicians, and others all use sense experiments to confirm their principles, which then provide the foundation for the whole structure which they go on to build. So I hope no one will think we have spent too long discussing this first and most important foundation, on which rests a whole immense structure with an infinite number of conclusions, only a few of which are included in this book. The Author has, indeed, done a great deal to open up a door which has long been locked against speculative enquiry. He has certainly not neglected experiments, and I have been with him on many occasions when he has undertaken to prove that the acceleration of naturally falling objects does indeed happen in the proportions described above. He has done this in the following way.

We took a wooden rule or beam, about twelve *braccia* long, half a *braccio* wide, and three inches thick, and hollowed out a channel just over an inch wide along its edge. The channel was cut absolutely straight, and lined with parchment which had been burnished and polished to make it as smooth as possible. Then, inclining the rule by raising one end one or two *braccia* from the horizontal, we rolled a hard, perfectly round and polished bronze ball down the channel. Using a method which I shall describe below, we measured the time taken for the ball to travel down the whole length of the channel, repeating this numerous times to ensure that the time measurement was correct; we found that the measurement never varied by even a tenth of a second. Once this had been established, we rolled the same ball down just a quarter of the length of the channel, and measured the time taken for its descent; this was consistently found to be exactly half of the previous measurement. We repeated the experiment for other proportions, comparing the time taken for the full length of the channel with that for a half, two-thirds, three-quarters, or for any other fraction. In more than a hundred instances altogether, the distances travelled were always found to be the square of the time elapsed, regardless of the degree of inclination of the plane, i.e. of the channel down which the ball was rolled. Moreover, the time taken for the descent at different inclinations maintained exactly the relative proportions which the Author assigns to them, as we shall see demonstrated below. As for how the time was measured, we had a large bucket full of water fixed in a high position, with a thin tube soldered into its base, through which a thin stream of water ran into a small beaker for the duration of the time in which the ball was rolling down the channel or part of the channel. The water collected in this way was then weighed with a precision balance on each occasion, so that the proportions between the different weights corresponded to the proportions between the different intervals of time. This was done to such a high degree of accuracy that, as I have said, there was no significant variation in the results over the very many times that the experiment was repeated.

EXPLANATORY NOTES

For details of the standard edition of Galileo's works see the Note on the Text and Translation, p. xxx. This edition is referred to in the notes below as *Opere*, followed by volume and page number.

A SIDEREAL MESSAGE

1 *[Title-page]*: the Latin that Galileo uses is *nuncius*, that can stand for either messenger or message, but Galileo clearly intended the latter. See Edward Rosen, 'The Title of Galileo's *Sidereus Nuncius*', *Isis*, 41 (1950), 287–9.

patrician: used here in the sense of member of the nobility.

recently discovered by him: Galileo recognized the priority of the Dutch but he insisted that he had not been informed of the way they made their telescope. See the Introduction, p. xii.

planets: Galileo uses the word *planetae* to describe the four celestial bodies that we call satellites.

Medicean Stars: Galileo had toyed with the idea of calling the satellites *Cosmici*, which would have enabled him to dedicate them exclusively to Cosimo II, and to pun on the semantic ambiguity between *Cosmo* (the name of the Grand Duke Cosimo II in Latin) and 'cosmic'. But since the Grand Duke had three brothers, he had also considered calling them 'Medicean' to honour the whole family. On 13 February 1610 he wrote to the Grand Duke's Secretary of State, Belisario Vinta, to ask his advice (*Opere*, 10, p. 283). Vinta answered on 20 February that the second title was more appropriate since *cosmici* could be interpreted in different ways while 'Medicean' was unequivocal (*Opere*, 10, pp. 284–5). But without waiting for Vinta's reply, Galileo had already rushed into print with a title-page bearing 'Cosmica Sidera'. When Vinta's letter arrived and he realized that he had made the wrong choice, Galileo had slips of paper with the word *Medicea* pasted over the word *Cosmica*. The correction has not been found in all copies, however, and this confirms that the substitution was only made when the book was in press.

3 *Fourth Grand Duke of Tuscany*: the first Medici to bear the title of Grand Duke was Cosimo I, who reigned from 1570 to 1574. He was succeeded by his eldest son, Francesco I, who had no heir and was followed by his brother Ferdinando I, who died on 3 February 1609, when Cosimo II (12 May 1590–28 February 1621) became the Fourth Grand Duke.

columns and pyramids ... as the poet says: Galileo has in mind, 'The pyramids reared to the stars, at such expense', a line from the *Elegies* of Sextus Propertius, a Roman poet who lived in the latter half of the first century BC (*Elegies*, 3.2, verse 19). The verse recalls a passage from the *Odes* of his contemporary, Horace, quoted in the next note below.

3 *imperishable monuments of literature*: the Latin poet Horace wrote in his *Odes*, 3.30, verses 1–5: 'I have finished a monument more lasting than bronze and loftier than the Pyramids' royal pile, one that cannot be destroyed by wasting rain, furious north wind, or the endless passing of years and the swift flight of time.'

inscribed on the . . . orbs: in traditional astronomy the stars and the planets were said to be carried round by the crystalline spheres in which they were embedded like diamonds in a ring. As a name can be inscribed on a ring, so poets imagined that the names of gods and heroes could be engraved on the surface of the spheres.

introduce Julius Caesar into their company: Suetonius, the second-century AD historian, writes in his *Lives of the Caesars* ('The Deified Julius', ch. 88): 'at the first of the games that Augustus gave in honour of his apotheosis, a comet (*stella crinita*) shone for seven successive days . . . and was believed to be the soul of Caesar, who had been taken to heaven'. Ovid in his *Metamorphoses*, 15.843–50, had described how Venus caught up the passing soul of Caesar and bore it towards the stars of heaven. When she released it, it rose higher than the Moon and, leaving behind a hairy train, gleamed like a star. Both the Greek *cometes* and the Latin *crinitus* mean 'hairy'.

4 *noblest of the planets*: in traditional mythology Jupiter is the king of the heavens (Ovid, *Metamorphoses*, 15. 858–9). Placed between Mars, a 'hot-tempered' and warlike planet, and Saturn, a 'cold' and contemplative one, Jupiter represented for astrologers the source of just and temperate behaviour, and had been linked to the fortunes of the Medici family by Cosimo I, the grandfather of Cosimo II.

the Sun itself: this is Galileo's first public commitment to the view that the centre of the world is not the Earth, as Aristotle and Ptolemy believed, but the Sun.

twelve years: the time it takes for Jupiter to complete one revolution around the Sun is 11 years and 315 days.

the Creator of the stars . . . himself: Galileo not only repeatedly declared that his celestial discoveries had been inspired, he went as far as claiming that God had personally chosen him to be 'the first and only one' to make all those that were rendered possible by the telescope, as he wrote to Belisario Vinta on 30 January 1610 (*Opere*, 10, p. 280).

in preference to all others: princes elsewhere soon became interested in having their names hoisted into the heavens. King Henry IV of France had married Maria de' Medici, a cousin of the Grand Duke Cosimo II, and from Paris came, with the promise of a generous reward, the request that Galileo call 'Henry' the next celestial body he discovered (letter of 20 April 1610 quoted in Galileo's letter of 25 June to Vincenzo Giugni,

Opere, 10, p. 381). The French astronomer, Jean Tarde, in his haste to ingratiate himself with the king of France, dedicated the sunspots (which he assumed were satellites) to the reigning Bourbon family. Not to be outdone, his competitor, the Jesuit Karl Malapert, dedicated the same sunspots to the reigning house of Austria.

Jupiter: the basic textbook of astrology in use in Galileo's time was written by the astronomer Claudius Ptolemy who lived in Alexandria in the second century AD. Here is what he says about the influence of Jupiter: 'He makes those who are subjected to his influence magnanimous, generous, god-fearing, honorable, agreeable, kind, magnificent, liberal, just, high-minded, dignified, mindful of their own business, compassionate, good conversationalists, beneficent, affectionate, and with qualities of leadership' (translation, slightly emended, by W. G. Waddell and F. E. Robbins in *Manetho, Ptolemy*, Loeb Classical Library (Cambridge, Mass.: Harvard University Press, 1971), 347).

mid-heaven ... eastern angle: the mid-heaven lies at the intersection of the ecliptic and the meridian. The eastern angle is situated at the intersection of the ecliptic and the eastern horizon. When drawing a horoscope it is important to know what celestial body was rising in the east at the moment of birth. Cosimo II was born, according to his birth record in the Archives in Florence, 'on the first hour of the night of 12 May 1590, and was baptized by Cardinal Alessandro de' Medici, Archbishop of Florence', the future Pope Leo XI. It was customary in Galileo's day to indicate hours beginning with sunset. On 12 May 1590 the Sun set over Florence at 19.13 hours local mean time. Given that the birth occurred at one hour after sunset this means that he was born around 20.13 hours.

to teach your Highness mathematics: in 1605 Galileo was invited to teach Cosimo, who was then 15 years old, on the use of the geometrical and military compass that he had recently perfected. He spent the summer at the Tuscan court that year and he returned again to coach the young prince in 1606 and 1608.

6 *Council of Ten*: in Venice the Council of Ten, which was responsible for internal security including the police force and censorship, was elected for a period of one year by the *Great Council*, which comprised all male patricians who were over 25 years old.

Overseers: the Governing Board of the University of Padua was composed of three Overseers (*Riformatori* in Italian), who were chosen among the members of the Venetian Great Council and appointed for a period of two years.

Sidereal Message, etc.: the original of the report, dated 26 February 1610, that was sent by the Overseers of the University to the Council of Ten, bore the title 'An Astronomical Announcement to Astronomers' (*Astronomica Denuntiatio ad Astrologos*, in *Opere*, 19, pp. 227–8). In Latin *astrologus* was often used in the broad sense of *astronomus* probably because professional astronomers saw casting horoscopes as part of their job. Galileo substituted *Astronomica Denuntiatio*, first by 'Astronomical Message' (*Astronomicus Nuncius*), and then by 'Sidereal Message' (*Sidereus Nuncius*).

6 *Lunardo Marcello*: the original of the report of the Overseers has Lunardo Mocenigo not Lunardo Marcello (*Opere*, 19, p. 228).

7 *cosmic stars*: this title of the work, as it appears here on the inside, bears the early form *Astronomical Message*, which was replaced by *Sidereal Message* on the cover. We also find 'cosmic stars' instead of 'Medicean stars' for the reason explained above in the note to p. 1.

more than ten times as numerous . . . familiar: before Galileo the standard number was the one given in Ptolemy's *Almagest* written in the second century AD, where we find the position and magnitude of 1,022 stars. In Galileo's own day the Danish astronomer, Tycho Brahe (1546–1601), had published a list of 777 stars in 1592, which he later increased in some haste in order to offer to his patron, the emperor Rudolph II, a catalogue of 1,000 stars.

sixty terrestrial diameters: Galileo wrote *diameters* when what is meant are *radii*, namely *semi-diameters*. That this is not a mathematical error but a peculiar linguistic usage is clear from the way he consistently uses *diameter* for *radius* in a letter dated 7 January 1610 (*Opere*, 10, p. 273, line 4, and p. 277, line 124). When Galileo intended *diameter*, he used the expression 'the whole diameter', as we find when he speaks of the Moon. In the *Dialogue on the Two Chief World Systems*, however, he gave the distance of the Moon from the Earth as '56 times the semi-diameter of the Earth' (see p. 287 below).

two diameters away: in August 1609 Galileo had an instrument that magnified nine times (letter to Doge Leonardo Donato, 24 August 1609, *Opere*, 10, p. 250, lines 10–11). By January 1610 he had one that magnified twenty times, and he claimed that he would soon be able to increase that power to thirty (letter to Antonio de' Medici, 7 January 1610, *Opere*, 10, pp. 273, 277).

thirty times . . . with the naked eye: when the linear or transverse magnification is thirty, an object appears thirty times closer, its surface is 30^2, namely 900 times greater, and its volume is 30^3, namely 27,000 times bigger. Galileo gives the increment of magnification not only for the distance but also for the surface and the volume, something that may not be very useful but it allowed him to mention the sensational increases from thirty to 900 to 27,000.

8 *wandering stars*: the planets were called wandering stars to distinguish them from the 'fixed' stars that rotate every twenty-four hours but do not change position relative to each other.

never stray beyond certain limits: in Ptolemy's system, the orbits or spheres (called *deferents*) of Mercury and Venus revolve around the Sun, but the two planets themselves revolve on *epicycles* that are attached to their *deferents* in such a way that they always rise or set a little ahead or a little after the Sun. An alternative system was constructed by Heraclides of Pontus (fourth century BC) in which Mercury and Venus travel around the Sun, and it is probably this model that Galileo had in mind. By the end of January 1610 Galileo had convinced himself that the satellites of Jupiter were not carried by *epicycles* but had 'their own proper and particular movements' around Jupiter in the same way as Venus and Mercury, 'and perhaps the other known planets, go around the Sun' (letter of 30 January 1610 to Vinta, *Opere*, 10, p. 280).

About ten months ago: in the manuscript of the *Sidereal Message* that he had started writing halfway through January 1610, Galileo had written 'about eight months ago' (*Opere*, 3, folio 18). Taken literally this would mean that he heard about the telescope in May 1609, but in an earlier letter dated 29 August 1609, Galileo had told his brother-in-law, Benedetto Landucci, that 'about two months ago the rumour was spread here that in Flanders a spyglass (*occhiale*) had been given to Count Maurice' (*Opere*, 10, p. 253). This would indicate June or early July as the time when the news reached him.

Jacques Badouère: (also written Badouer, Badovere, or Badoire), the son of a wealthy French Protestant, who studied at the University of Padua between 1598 and 1599 and was a paying guest in Galileo's house during most of that time.

sixty times bigger: if the telescope enlarged the area more than sixty times, it had a magnification power of about eight times, and it must be the one that Galileo presented to the Doge on 24 August 1609 (*Opere*, 10, p. 250).

9 *two terrestrial diameters away*: two diameters stand here for two semi-diameters, i.e. two earth-radii. See the note to p. 7 above.

the rays are carried . . . ECF and EDG: Galileo considers the rays as coming from the eye.

only a few minutes of an arc: 'object FG' is the field of vision of the tube without the lenses, and 'object HI' the field of vision that is observed when the lenses are inserted. The visual angle (CED) is the same in both cases since HI is seen with the telescope under the same angle as was FG with the naked eye. If the procedure with the sheets of different sizes is carried out, the ratio of these two fields of vision will be known. But how can we determine the distance HI? The difficulty is that the tube, as it is drawn, does not give us the focal length of the objective lens or the eyepiece. Galileo realized that varying the size of the aperture of the objective lens produced a change in the field of view but he did not work out the ratio.

10 *we shall provide . . . instrument*: Galileo never got around to it because he could not work out the optical properties of his telescope in spite of urgent requests from friends, notably Sagredo (e.g. letters of Sagredo to Galileo of 2 June, 30 June, and 7 July 1612, *Opere*, 11, pp. 315, 349–50,

356). Writing from Rimini on 13 September 1616, six years after the publication of the *Sidereal Message*, Malatesta Porta voiced his disappointment: 'How long will you keep us on tenterhooks? You promised in your *Sidereal Message* to let us know how to make a telescope so that we could see all the things that are invisible to the naked eye, and you haven't done it to the present day' (*Opere*, 12, p. 281).

13 *a certain spot . . . region of the Moon*: the large spot meant here is not the circular one discussed below but an irregular area near the top that crosses the boundary between light and darkness at first and last quarter as shown in the figure.

17 *they present themselves . . . unbroken line*: Galileo's argument is cogent for a telescope that only magnifies twenty times, but with modern instruments the unevenness of the Moon's rim can easily be observed. Most lunar mountains are less than 6,000 metres (4 Italian miles; see note to p. 19), but the highest ranges happen to be close to the rim. Mt Leibniz on the southern limb rises to 9,000 metres.

aether: aether for the ancients and up to the time of Galileo was the material out of which celestial bodies were made. It was considered perfect and unchangeable in contrast to the four terrestrial elements of earth, water, air, and fire. Hence the belief, mentioned below on p. 23, that the Earth is the refuse of the world.

19 *7,000 Italian miles*: the Italian mile, which is also called the Roman mile, is approximately 1.5 km or 0.93 English mile. So 7,000 such miles give 10,500 km for the diameter of the Earth, which is roughly 18% off the correct value of 12,756 km for the diameter at the equator.

the sum of the squares . . . 1,010,000: Galileo is applying the well-known Pythagorean theorem to the right-angle triangle DCE so that we have $(DE)^2 = (DC)^2 + (CE)^2$.

even 1 mile high: at 8,850 metres Mt Everest is approximately six times higher than Galileo's estimate for the highest mountain on the Moon. In his day the heights of mountains on Earth could not be determined accurately, and widely diverging values were produced.

21 *The System of the World*: Galileo wanted to call the book *On the System of the World* as he told Belisario Vinta, the Grand Duke's Secretary of State, in a letter of 7 May 1610 (*Opere*, 10, p. 351), but he later changed this to *Dialogue on the Tides*, to emphasize his physical proof for the motion of the Earth. When the censors objected to this title, he altered it to *Dialogue on the Two Chief World Systems*.

so childish . . . answer: this is a harsh indictment of the idea that Venus might be responsible for the Moon's secondary light. In the *Dialogue on the Two Chief World Systems* Galileo rejects the suggestion with equally strong language and calls it a 'vain idea' (*vanità*) (*Opere*, 7, p. 116; see p. 216 in this volume).

permeating . . . the solid body of the Moon: the theory that the secondary light of the Moon is caused by sunlight rested on the assumption that the lunar globe is partly translucent, and that when it is exposed to the rays of the Sun it soaks up this light, so to speak.

22 *ecliptic*: the apparent path that the Sun traces out in the sky during the year. The Moon crosses it about twice per month. If this happens during new moon a solar eclipse occurs; if during full moon a lunar eclipse.

23 *will be considered more fully*: in the First Day of the *Dialogue on the Two Chief World Systems* Galileo discusses the Moon at length (*Opere*, 7, pp. 85–126; pp. 183–226 in this volume).

round of stars: Galileo's phrase *stellarum corea* (literally 'dance of stars') is rendered by 'round of stars' to convey his conviction that the Earth formed part of the harmonious dance of the stars and the planets that go round and round.

surpasses the Moon in brightness: 'The Earth', writes Galileo in the *Dialogue on the Two Chief World Systems*, 'repays the debt when the Moon needs it most, reflecting the Sun's rays to give it a very strong light, which I reckon must exceed the light which the Earth receives from the Moon by the same amount as the Earth's surface is larger than the Moon's' (*Opere*, 7, p. 92; pp. 189–90 below).

a thousand physical arguments: where we would use a vaguer expression such as 'a very large number', Galileo writes 'six hundred physical arguments' which I have rendered by 'a thousand'.

four or five times greater: as Galileo had mentioned earlier, when a telescope (in this case his best) has a linear magnification of thirty, objects appear thirty times closer, their area is increased by 30^2, and their volume by 30^3. See the note to p. 7 above. Here he refers to a telescope that made objects appear 100 times bigger but stars merely four or five times greater. This means that the magnification decreased from ten to about two, but his numbers are probably meant as suggestive and should not be taken as rigorous.

the brightness that surrounds it: in the *Optics* of Ptolemy, which Galileo follows, what is taken into consideration is not how the rays travel but the triangle that has the object as its base and the eye as its summit. Here the object, which is assumed to be real, is not only the body of the star but the brightness that surrounds it.

24 *the Dog Star*: Sirius, the brightest star in the constellation Canis Major.

six new orders of magnitude: the magnitude scale dates back to the second century BC, when the Greek astronomer Hipparchus ranked the stars visible to the naked eye into six groups. The brightest stars were characterized as first magnitude. The next-brightest stars were labelled second magnitude, and so on, down to the faintest stars visible, which were classified as sixth magnitude. The human eye is such that a change of one magnitude corresponds to a factor of 2.5 in a pattern of brightness, so that a first-magnitude star is roughly 2.5 brighter than a second-magnitude star, which is roughly 2.5 brighter than a third magnitude star, and so on. Galileo introduces a new sequence of six magnitudes for stars that are invisible to the naked eye. A star of the seventh magnitude is seen with the telescope as if it were a star of the second magnitude observed with the naked eye, and so on. The idea is interesting, but Galileo did not have a telescope that enabled him to carry it out properly.

24 *five hundred*: to be interpreted as 'very many' and not as the result of a careful counting of the number of stars by Galileo.

25 *Pleiades . . . hardly ever visible*: in Greek mythology the seven daughters of Atlas and Pleione were changed into stars. Six are brighter than the fifth magnitude. The seventh (and two other ones) are brighter than the sixth magnitude but can only be seen by people with especially good eyesight.

26 *Milky Way*: Aristotle took the Milky Way to be a sort of high fog in the elemental region, formed, like comets, from hot vapours that rise from the uppermost part of the terrestrial sphere (*Meteorology*, book 1, chapter 8, 345 a 11–346 b 15). At the end of the sixteenth century Tycho Brahe maintained that the Milky Way was made of the same matter as the stars but more 'diffused' in character. Christopher Clavius, who taught at the Roman College during this period, considered it more likely that the Milky Way was a denser part of the aether and could absorb light from the Sun, the view that the telescope was just about to discredit. On the history of the Milky Way, see Stanley L. Jaki, *The Milky Way: An Elusive Road to Science* (New York: Science History Publications, 1972).

Galaxy: *galaxia* is the Greek word for the Milky Way, from *gala* = milk.

nebulous: not used here in the modern sense of interstellar cloud of dust or gas but in the broader sense of 'cloudlike' (from the Latin *nebula* which means cloud) as we find it in Ptolemy's *Almagest*, who mentions seven such formations including the star clusters in Orion and Praesepe that Galileo describes.

27 *Aselli*: the two large stars in the engraving of the nebula Praesepe are the Aselli or ass-colts, which are now referred to as γ and δ Cancri. The 'cloudlike' area between them is what Ptolemy listed as nebulous.

28 *7th of January . . . at the first hour of night*: the observation is recorded as follows in Galileo's letter of 7 January 1610: 'Many fixed stars are seen with the spyglass but not without it. This very evening I saw Jupiter accompanied by three fixed stars that are completely invisible because they are so small' (*Opere*, 10, p. 277). The 'first hour of the night' does not refer to one hour after midnight as it does for us. In Galileo's day time was recorded 'from sunset', and the civil day began and ended with the setting of the Sun. The first hour in Padua on 7 January 1610 began around 16.30 hours our time. On this day Galileo began keeping an Astronomical Logbook of his observations (see Jean Meeus, 'Galileo's First Records of Jupiter's Satellites', *Sky and Telescope*, 24 (1962), 137–9).

three little stars: on account of the very narrow visual field of his telescope Galileo was not able to see the fourth satellite until 13 November when it had come closer to the other three.

parallel to the ecliptic: if the three celestial bodies had been stars, the fact that they were lying on a straight line parallel to the ecliptic would have enabled Galileo to measure the movement of Jupiter exactly. 'Ori.' and 'Occ.' (for 'oriens' and 'occidens') in this and the following illustrations indicate east and west respectively.

30 *15th of January*: on this day Galileo began making his entry in his Astronomical Logbook in Italian as he had done until then, but halfway through he switched to Latin (*Opere*, 3, pp. 427–8). The realization that the celestial bodies near Jupiter were satellites probably made him think of publishing this extraordinary news in Latin, the language of the scientific community.

32 *the narrow circle along which they travel ... effect*: the orbits of the satellites are very nearly circular and too small to account for the considerable change in brightness of the satellites at different positions.

an oval motion ... appearances: Galileo is thinking of a stretching out of the line, known as the apsis, which connects the point on the orbit of the satellite that is nearest to Jupiter to the point on the other side that is furthest from the planet. He makes no reference to Kepler, who had published his discovery that the orbits of the planets are elliptical in his *Astronomia Nova* of 1609.

the great lights: this expression for the Sun and the Moon is borrowed from the Bible: 'And God made the two great lights, the greater light to rule the day, and the lesser light to rule the night' (Genesis 1: 16).

the Moon does not: the largest angle subtended by the Moon amounts to about 32 seconds of an arc. It seems much larger on the horizon than high in the sky, but this is an optical illusion. A simple experiment will prove this: take a slim pencil of about 5 mm across, hold it at arm's length between your eye and the Moon. It will be completely covered whether it is near the horizon or high above in the sky.

System: in the First Day of the *Dialogue on the Two Chief World Systems* (the name under which his *System* finally appeared in 1632) Galileo points out the reasons why there is no atmosphere on the Moon: 'if clouds ever formed over any part of [the Moon's] surface as they do around the Earth, they would block out some of the things which we can see through a telescope, and we would see some change in its visual appearance; but in all my long and diligent observations I have never seen any such change' (*Opere*, 7, p. 126; pp. 225–6 below).

sphere of elements: according to Aristotelian cosmology the Earth is surrounded by successive layers of the four elements: earth, water, air, and fire.

LETTERS ON THE SUNSPOTS

33 *Mark Welser*: a member of a prominent family of bankers in Augsburg. He had studied in Rome and was a friend of the Jesuits.

Apelles: the writer was the Jesuit Christoph Scheiner; for this pseudonym see the Introduction, n. 17.

34 *for the past eighteen months*: this would imply that Galileo first observed the sunspots around October 1610 which is the time Christoph Scheiner also saw them, as he mentions in his first letter to Mark Welser dated 12 November (*Opere*, 5, p. 25); see the Introduction, p. xvi above. Galileo insisted to his dying day that he was the first to have seen them but both he and Scheiner were probably preceded by the Dutch astronomer Johann Fabricius, who was the first to publish information about them.

35 *parallax*: the technical word that is used to indicate the apparent difference in the position of an object in the sky when it is viewed from different positions on Earth.

37 *as the Pythagoreans and Copernicus maintain*: Nicolaus Copernicus (1473–1543) was a Polish astronomer who had studied in Italy. His book, *On the Revolutions of the Heavenly Spheres*, was published in the year he died. In it he defended the belief, which he attributed to Pythagoras (sixth century BC), that the Sun is at rest and the Earth is in motion, against the view of Claudius Ptolemy (flourished at Alexandria *c.* AD 150) that the Earth was at rest.

bodily conjunctions: these refer to both the 'transits' of Venus, when it passes directly across the face of the Sun, and the 'occultations', when it passes directly behind the Sun.

40 *eccentrics, deferents, equants, epicycles, etc.*: in the Ptolemaic system the planets are assumed to rotate on a small circle, called an epicycle, whose centre is attached to the rim of another circle, called the deferent, which in turn moves around the centre of the entire system.

45 *none will be found in the future*: Galileo later changed his mind. See his third letter, pp. 53–4 below.

47 *method discovered by a pupil of mine*: this pupil is Benedetto Castelli, a Benedictine monk who studied with Galileo in Padua, and followed him to Florence in 1611. At Galileo's request he was appointed professor of mathematics at the University of Pisa in 1613. His method consisted in drawing a circle on paper and then fitting the telescopic image of the sunspots on this circle in order to trace the exact placement of the spots.

48 *thus forming . . . varying shapes*: Prince Federico Cesi had written from Rome to Galileo on 14 September 1612 to inform him that a Dominican had supported his position in a debate at the Roman College while the Jesuits had sided with Scheiner in calling them small stars. When the Dominican pointed out that stars were round and not irregular in shape,

the Jesuits retorted that clusters of stars might have a variety of shapes (*Opere*, 12, p. 395).

49 *have their own individual motion*: in other words, the 'wandering stars' are planets.

51 *Change is not alien to the heavens*: Aristotelians considered the heavens immutable. See below, First Day of the *Dialogue*, p. 158 ff.

53 *I have already written*: see Galileo's first letter, pp. 44–5 above.

Has Saturn devoured its children?: in classical mythology, Saturn devoured his newborn children to forestall a prophecy that he would be overthrown by one of his sons.

plausible conjectures: since the Earth has one satellite (the Moon) and Jupiter had been found to have four, Galileo thought it very likely that Saturn had two. See the Introduction, p. xv above.

SCIENCE AND RELIGION

LETTER TO DON BENEDETTO CASTELLI

55 *Signore*: for Benedetto Castelli see the note to p. 47 above, and the Introduction, p. xviii.

Niccolò Arrighetti: a mutual friend of Castelli and Galileo; he called on Galileo in Florence on 20 December 1613, the day before Galileo wrote this letter. Castelli, in a letter of 14 December, had already told Galileo that he had been the guest of the Grand Duke two days earlier, on 12 December.

the whole university: Castelli began teaching mathematics at the University of Pisa in November 1613.

Madame: mother of the Grand Duke Cosimo II, Christina of Lorraine, the widow of Grand Duke Ferdinand I. Her title was Grand Duchess.

Archduchess: Maria Maddalena of Austria, the wife of Cosimo II.

Don Antonio: Antonio de' Medici, the natural son of the Grand Duke Francesco I and Bianca Cappello.

the passage in Joshua: the reference is to Joshua 10: 12–13, where it is narrated that the Sun stood still in the sky in response to the prayers of Joshua, so that his army was able to complete the rout of their enemies in battle.

56 *experience of our senses . . . necessary demonstrations*: Galileo stresses that his arguments are based on rigorous mathematical reasoning ('necessary demonstrations') as well as careful observation and experimentation ('the experience of our senses').

58 *matters of faith*: those assertions that a Catholic cannot doubt. Galileo uses the technical Latin phrase 'de Fide'.

59 *Joshua's prayers*: the reference is again to Joshua 10: 12–13.

59 *two motions*: in the geocentric system used prior to Copernicus the Sun
has two motions. The first carries it in twelve months across the twelve
signs of the zodiac. The second is the daily motion around the Earth that
it shares with all the planets and the stars.

Primum Mobile: the tenth and outermost concentric sphere of the universe in Ptolemaic astronomy. It was thought to revolve around the Earth
from east to west in twenty-four hours and to cause the other nine
spheres to revolve with it.

60 *all the more quickly*: Galileo maintains that arresting the Sun's annual
motion would not prolong but shorten the length of the day.

faster than the Sun's: the time the Moon takes to complete its daily
rotation from east to west is greater than that that the Sun requires for its
daily rotation. This is why the Moon rises fifty minutes later every day.
The reason is that the Earth revolves around its axis in twenty-four hours
but the Moon takes about 29.5 days to revolve around the Sun while
making one rotation around its own centre. The relation between the two
motions is 24 divided by 29.5 which gives about fifty minutes.

turns on its own axis: the important discovery that the Sun rotates around
itself was communicated by Galileo in his *Letters on the Sunspots*
published in 1613 (not included in the extracts in this volume).

imparts . . . planets which revolve around it: Galileo conjectures that the
revolutions of the planets around the Sun in the Copernican system
are caused by the rotation of the Sun on its axis. This entails a better
explanation of the miracle of Joshua because stopping the Sun would
freeze all the heavenly motions and not just the daily revolution of the
Sun around the Earth, as would be the case in the Ptolemaic system.

LETTER TO THE GRAND DUCHESS CHRISTINA

61 *Augustine writes*: Galileo gives as reference, 'Saint Augustine, *On the
Literal Meaning of Genesis*, book 2, at the end'. The translations of
Galileo's quotations from Augustine are adapted from *De Genesi ad
Literam: The Literal Meaning of Genesis*, translated and annotated by
John Hammond Taylor, SJ (New York: Paulist Press, 1982).

62 *my first announcement*: Galileo is referring to his *Sidereal Message*
published in 1610.

63 *arguments . . . this alternative position*: the phases of Venus that Galileo
discovered with his telescope in December 1610 could not be explained
on the Ptolemaic system.

the injury . . . mathematicians everywhere: Galileo is referring to an attack
that was made by a Dominican, Tommaso Caccini, in the Florentine
church of Santa Maria Novella on 21 December 1614.

a priest and a canon: although Copernicus was a canon he was never
ordained a priest.

Bishop of Fossombrone: the Dutch priest and mathematician, Paul van
Middelburg, was appointed to the Italian see of Fossombrone in 1594. In

the preface to his *On the Revolutions of the Heavenly Spheres* Copernicus states that it was Middelburg who got him interested in the problem of the reform of the calendar.

64 *tables ... movements of all the planets*: this is a reference to the *Prutenic Tables* published by Erasmus Reinhold in 1551 and based on the Copernican system.

Cardinal ... Kulm: the Cardinal of Capua is Nikolaus von Schönberg (1472–1537), a Dominican friar who was made Cardinal in 1535. He wrote to Copernicus in 1536 to urge him to publish his new astronomy as soon as possible, but the work only appeared in 1543. Tiedeman Giese (1480–1550) was the Bishop of Kulm and a friend of Copernicus, whom he encouraged.

65 *Lactantius*: Christian apologist of the third–fourth centuries, widely admired for the elegance of his Latin style, for which he was called the 'Christian Cicero'. He considered it ridiculous that anyone should believe that there are people at the south pole who have their feet opposite to ours.

67 *Tertullian ... wrote*: Galileo gives as reference, 'Tertullian, *Against Marcion*, book 1, chapter 18'.

69 *in the words of St Augustine*: Galileo gives as reference, 'Saint Augustine, *On the Literal Meaning of Genesis*, book 2, chapter 9', adding, 'This can also be found in Peter Lombard, the master of the *Sentences*'. A text that is close to the one of St Augustine that Galileo quotes can be found in Distinction 14 of book 2 of the *Sentences*.

70 *a very eminent churchman*: Galileo gives the name of Cardinal Baronio in the margin. Cesare Baronio (1538–1607) was a famous Church historian whom Galileo could have met in Padua.

'In discussing ... through experience and reason': Galileo gives as his reference, 'Pereira, *On Genesis*, near the beginning'. Benedetto Pereira was a Jesuit who taught at the Roman College and he had published his book in Venice in 1607.

in Saint Augustine we read: Galileo gives as reference, 'In his seventh letter, to Marcellinus'.

71 *'What we know ... what we do not know?'*: this was a commonplace in Galileo's day. The source is probably Themistius, a fourth-century AD philosopher, who made this observation when commenting on Aristotle's *On the Soul*, book 3, chapter 3, 427b 1–13.

'He has given up ... to the end': Galileo gives as reference, 'Ecclesiasticus chapter 3 [verse 11]'.

as Aristotle tells us: Aristotle, *On the Heavens*, book 2, chapter 13 (293a 30–2).

Plutarch ... life of Numa: Plutarch, *Life of Numa Pompilius*, 11.3, in Plutarch, *Lives* (Dryden's translation revised by Arthur Hugh Clough) (New York: Random House, reprint of the 1864 edition), 83.

71 *as we learn from Archimedes*: Archimedes, *The Sand Reckoner*, in T. L. Heath (ed.), *Works of Archimedes* (New York: Dover [no date]), 222.

Seleucus the mathematician: lived in the second century BC.

Nicetas, according to Cicero: Cicero wrote Hicetas (*Academica*, 2.39) but Galileo misspells the name, repeating the error 'Nicetas' that had been made in the dedication of Copernicus' *On the Revolutions of the Heavenly Spheres*, which must have been his source.

Seneca . . . in his book On Comets: Seneca, *Natural Questions*, 7.2. Galileo quotes the passage in his *Observations on the Copernican Theory* (p. 97 below).

72 *Scripture . . . could not exist*: Galileo is referring to writers who used Scripture to refute the Copernican theory. These include Martin Horky's *A Very Short Excursion Against the Sidereal Message*, Ludovico delle Colombe's *Against the Motion of the Earth* (1610 or 1611), and Francesco Sizzi's *Dianoia astronomica, optica, physica* (1611).

writer . . . recently published a book: Galileo has in mind a recent book by Ulisse Albergotti who claimed that the Moon does not shine because it reflects light coming from the Sun but on account of some kind of internal light.

73 *St Jerome*: Galileo gives as reference, 'Letter to Paulinus' and indicates the number as 103 but modern scholars and editors designate it as number 53.

74 *Archimedes, Ptolemy, Boethius, and Galen*: the classical authorities in geometry, astronomy, music, and medicine, respectively.

76 *the following words of St Augustine*: Galileo gives as reference, '*On the Literal Meaning of Genesis*, book 1, chapter 21' but the text is given in the version in which it appears in the book by Pereira that Galileo had already mentioned (see note to p. 70).

the former mathematician . . . Pisa: Galileo alludes to Antonio Santucci who had been professor of mathematics at the University of Pisa from 1594 until his death in 1613.

77 *other mathematicians*: in the margin Galileo gives the name of Clavius, the most famous mathematician of the Roman College, who had recently died in 1612.

the glory and greatness of God . . . the heavens: Galileo's phrase recalls specifically Psalm 19: 1, 'The heavens are telling the glory of God, and the firmament proclaims his handiwork'.

79 *the heavens are stretched out like a skin*: Psalm 104: 2; modern translations render as 'stretched out like a tent'.

But someone may ask . . . conclusions: Galileo gives as reference, '*On the Literal Meaning of Genesis*, chapter 9'.

80 *Although this problem . . . Old or the New Testament*: the quotation is from *On the Literal Meaning of Genesis*, book 2, chapter 18.

81 *St Jerome writes*: Galileo gives as reference, '*On Jeremiah*, chapter 28'.

and *elsewhere*: Galileo gives as reference, '*On Matthew*, chapter 13.'

Job chapter 27: the reference should be to Job 26: 7.

82 *In St Thomas's own words*: Thomas Aquinas (1225–74) was considered one of the greatest theological authorities of the Catholic Church. The reference is to Aquinas' *Commentary on the Book of Job*, translated by Brian Mullady, now available as E-text, www.opwest.org/Archive/2002/Book_of_Job/tajob.html.

mean motion . . . prosthaphaeresis of the Sun: when discussing the technical details of his new heliocentric theory Copernicus continued to use the terminology of the geocentric theory. For instance, in *On the Revolutions of the Heavenly Spheres*, book 3, chapter 14, he published tables on 'the simple uniform motion of the Sun' (namely the apparent motion of the Sun against the background of the fixed stars) and the 'regular motion of the anomaly of the Sun' (namely the angular distance of the Sun from its apogee, the place where it is furthest from the Earth). Copernicus uses 'prosthaphaeresis' for the shift in the apparent position of the Sun that is caused by the Earth's annual motion.

83 *Didachus of Stunica*: Diego de Zuñiga was a Spanish theologian. His *Commentaries on Job* appeared in Toledo in 1584 and were reprinted in Rome in 1591. On 6 March 1616 the Congregation of the Index decreed that the book, along with Copernicus' *On the Revolutions of the Heavenly Spheres*, was to be 'suspended until corrected'.

84 *Council of Trent . . . fourth Session*: special authority was given to the pronouncements of formally constituted Councils of the Church. The Council of Trent (1545–63) reasserted the traditional teaching of the Catholic Church in response to the challenge of the Protestant Reformation. The Council held its fourth session on 8 April 1546.

My reply . . . useful in the Church: Galileo gives as reference, '*On the Literal Meaning of Genesis*, book 2, chapter 10'.

Dionysius the Areopagite: mentioned in Acts 17: 34 as being converted by St Paul. The writings ascribed to him and held to be genuine at the time of Galileo are now thought to have been written in the fifth century AD.

Bishop of Avila: Alfonso Tostado, Bishop of Avila in the fifteenth century, wrote a commentary of the book of Joshua.

Paul of Burgos: the name taken by the Spanish Jew Selemch-Ha-Levi (*c.*1351–1435) after his conversion to Christianity. He became archbishop of Burgos and was appointed Lord Chancellor by King Henry in 1416.

miracle at the time of Hezekiah . . . the sundial: the reference is to Isaiah 38: 8, where the shadow on a sundial is said to have miraculously moved back as a sign that Hezekiah's life would be prolonged.

86 *In matters that are obscure . . . meaning of Scripture to be ours*: Galileo gives as reference, 'St Augustine, *On the Literal Meaning of Genesis*, book 1,

chapters 18 and 19'. This first quotation comes from chapter 18 but the next six are all from chapter 19, which is reproduced in almost its entirety.

90 *motion of the Primum Mobile*: see the note to p. 59 above.

which prompted Dionysius . . . to say: Galileo gives as reference, 'in his letter to Polycarp'. On Dionysius see the note to p. 84 above.

St Augustine . . . is of the same opinion: Galileo gives as reference, '*On the Miracles of Holy Scripture*, book 2'. This work was believed to be by Saint Augustine but it was written by an Irish monk in the seventh century.

the Bishop of Avila . . . at length: Galileo gives as reference, 'Questions 22 and 24 of chapter 10 of his commentary on Joshua'. On the Bishop of Avila see the note to p. 84 above.

91 *as I believe I have demonstrated . . . Sunspots*: see the note to p. 60 above.

Dionysius . . . On the Divine Names: on Dionysius see the note to p. 84 above. The Latin text that Galileo is using was translated from the Greek by Joachim Périon and published in Paris in 1598.

93 *Cajetan*: Thomas de Vio (1468–1534), author of a commentary on Thomas Aquinas's *Summa Theologica*.

Magalhães: Cosme de Magalhães: (1533–1624), the author of a *Commentary on Joshua* that was published in 1612 and is one of the sources of Galileo's patristic quotations.

the hymn: the first two of five stanzas of the hymn 'Caeli Deus sanctissime' that is sung at Vespers on Wednesdays.

94 *'Before he had made . . . hinges of the earth'*: the reference is to Proverbs 8: 26, but the meaning of the word translated 'hinges' (*cardines* in the Vulgate) is obscure and modern translations vary.

LETTER FROM ROBERTO BELLARMINE TO PAOLO ANTONIO FOSCARINI, 12 APRIL 1615

Paolo Antonio Foscarini: In February or March 1615 Foscarini published in Naples a 64-page *Letter Concerning the Opinion of the Pythagoreans and Copernicus about the Mobility of the Earth and the Stability of the Sun and the New Pythagorean System of the World*, in which he argued that it is not at variance with the Bible. A copy was immediately sent to Galileo by Prince Federico Cesi, his Roman friend and patron. Foscarini himself sent a copy to Cardinal Bellarmine and this elicited the present letter. See the English translation in Richard J. Blackwell, *Galileo, Bellarmine and the Bible* (Notre Dame, Ind.: University of Notre Dame Press, 1991), 217–51.

all the appearances are saved: see the Introduction, n. 25.

95 *'The Sun also . . . where it rises'*: Ecclesiastes 1: 5, traditionally attributed to Solomon.

96 *it appears . . . moving away from the ship*: Bellarmine is replying to Foscarini who had quoted a verse from Virgil, 'When we set out from

harbour the shore and town recede' (*Aeneid*, 3.72), that had already been used by Copernicus in his *On the Revolutions of the Heavenly Spheres*, book 1, chapter 8.

OBSERVATIONS ON THE COPERNICAN THEORY

The manuscripts of these essays bear no date but were written in 1615 and 1616.

97   *calculations . . . of astrologers*: the Copernican hypothesis made it easier to work out some of the calculations related to horoscopes, something that Galileo did regularly, like all the astronomers of his day.

Plato's teacher . . . and Seleucus the mathematician: to the 'precursors' of Copernicus mentioned in the *Letter to the Grand Duchess* (see p. 71 above) Galileo now adds the name of Ecphantus of Syracuse (between 500 and 300 BC).

Seneca . . . On Comets: Seneca, *Natural Questions*, book 7, chapter 2.

98   *On the Magnet*: William Gilbert (1544–1603) published his *On the Magnet* in 1600. The work was closely studied by Galileo, but although Gilbert says that the Earth rotates on its axis in twenty-four hours he does not say that it revolves around the Sun.

Johannes Kepler: Kepler (1571–1630) not only championed Copernican-ism but discovered that the orbit of the Earth around the Sun is not a circle but an ellipse.

past and present emperor: Kepler was the court mathematician of the emperor Rudolph II (1552–1612) and his brother Matthias (1557–1619) who succeeded him.

Ephemerides: in his *Ephemerides* published in 1609, the German astronomer David Tost (known as Origanus) states that the Earth turns on its axis but he does not come out in favour of the Copernican hypothesis but rather in favour of the one of Tycho Brahe for whom the planets go around the Sun, which, in turn, makes an annual revolution around the Earth.

prince of present-day philosophy: namely Aristotle.

101 *Cardinal of Capua and the Bishop of Kulm*: see the *Letter to the Grand Duchess*, p. 64 above and note.

showing them to be invalid: Galileo is referring to chapters 7 and 8 of book 1 of Copernicus' *On the Motions of the Heavenly Spheres*.

103 *properties . . . declinations of the ecliptic*: the stars on the celestial sphere are like cities on a globe where they are located using latitude and longitude. Longitude says how far a city is east or west *along* the Earth's equator; latitude says how far a city is north or south *of* the Earth's equator. Right ascension (RA) is like longitude. It locates where a star is along the celestial equator. The declination corresponds to latitude, and the ecliptic is the apparent path that the Sun traces in the sky during the year.

103 *"To us who are being carried along . . . move away"*: these two lines are two verses at the end of the Preface to book 2 of *On the Revolutions of the Heavenly Spheres*, and are almost certainly by Copernicus himself.

104 *Ptolemy . . . simpler than the second*: In book 3, chapter 3 of his *Almagest*, Ptolemy explains how we can account for the motion of the Sun by using one of two models: the first uses an epicycle revolving on a deferent circle, and the second a circle that is eccentric, i.e. rotates around a point that is not the centre of the system. In chapter 4 Ptolemy goes on to say that he prefers the second model.

The four Medicean planets: see Galileo's account in the *Sidereal Message* and the first of the *Letters on the Sunspots*, pp. 27–32 and 44 above.

the three superior planets: Mars, Jupiter, and Saturn.

105 *second anomaly*: the principal motion of the planets is a regular west-to-east motion, and ancient astronomers had good values for the average speed of each planet. But the planets were observed to have two *anomalies* to their otherwise regular motion. The speed of each planet is not constant, and there is a point of minimum speed (the apogee) and a point of maximum speed (the perigee). This irregularity was called the first anomaly. The planets were also observed to stop their forward (west-to-east) motion, reverse direction, stop again, and finally resume forward motion. This was called the second anomaly.

a preface . . . unsigned: Johann Kepler had revealed at the beginning of his *Astronomia Nova* published in 1609 that in his own copy of Copernicus' book there was a note, written by Jerome Schreiber of Nuremberg, stating that this preface had been inserted by Andreas Osiander, a Protestant theologian who had supervised the printing, for reasons similar to those set forth here by Galileo.

106 *not just sixteen times as he says*: Galileo stresses the error of Osiander who compared the size of Venus to the area of a circle when he should have considered the area of a sphere.

107 *let it be recognized as true and correct*: Galileo describes here the implications of the discovery of the phases of Venus that he discovered with his telescope.

110 *the Council*: the Council of Trent (1545–63); see note to p. 84 above.

111 *because Scripture says they did*: Bellarmine had written, 'It would be just as heretical to deny that Abraham had two sons and that Jacob had twelve, as it would be to deny that Christ was born of a virgin' (see Letter of Bellarmine to Foscarini, p. 95 above). Here Galileo adds Tubal's dog (*Tobit* 11: 9 as mentioned in the Latin Vulgate but not in other editions).

FROM THE ASSAYER

115 *Sarsi*: the pen name of Orazio Grassi, a Jesuit professor of the Roman College, who published in 1619 a book entitled *The Astronomical and Philosophical Balance* to which Galileo replied in *The Assayer* that appeared in 1623. He stressed that 'assayer' (*saggiatore* in Italian) is a delicate weighing instrument employed in the assay of pure gold, not a crude balance or steelyard like the one Grassi used.

Orlando furioso: a romantic poem by Ludovico Ariosto, first published in 1532 and a highly popular work of fiction in Galileo's day. Galileo will refer to Ariosto's poem again in his *Dialogue on the Two Chief World Systems*; see below, pp. 245 and 341 and notes.

this great book: The metaphor, 'book of nature', which is often used in the seventeenth century, might come from the Bible. In his early notebooks, the *Juvenilia* (*Opere*, 1, p. 64), Galileo quotes Isaiah 34: 4: 'The skies will be rolled up like a scroll' ('book' in the Latin Vulgate edition that Galileo was using), and Revelation 6: 14: 'The sky will vanish like a scroll that is rolled up' (again 'book' in the Latin Vulgate edition).

signor Mario: Grassi had criticized a booklet written by Galileo but published under the name of one of his young friends, the Florentine Mario Guiducci.

116 *And yet I have shown ... anyone who cares to look*: in the first of the *Letters on the Sunspots*; see pp. 36–7 above.

118 *Suidas*: the *Suda* ('Fortress') is the title of a Greek lexicon-encyclopedia, but before the twentieth century the name, in the form 'Suidas', was thought to refer to its compiler, who seems to have lived in the tenth century AD. Included in the lexicon are texts from classical Greek works and commentaries. Though mostly derived from late and corrupt sources, the *Suda* preserves much information about Greek literature that would otherwise be lost.

121 *sonority or transonority*: Galileo stresses that sound is a physical phenomenon caused by the air impinging on the eardrums and that there is no need to postulate an occult qualilty of 'sonority' or 'transonority'.

DIALOGUE ON THE TWO CHIEF WORLD SYSTEMS

122 *Most Serene Grand Duke*: Ferdinando II (1610–70), Grand Duke of Tuscany.

124 *To the discerning reader*: this letter is the preface that Riccardi required to be added as a condition of granting his *imprimatur*. For the circumstances of the printing of the *Dialogue* see the Introduction, pp. xxiv–xxvi.

A salutary edict: Copernicus' *On the Revolutions of the Heavenly Spheres* was placed on the Index of proscribed books on 3 March 1616.

124 *the opinion of Pythagoras . . . Earth*: Copernicus believed that Pythagoras (sixth century BC) had taught that the Sun is at rest and that the Earth is in motion.

Peripatetics only in name: 'Peripatetic' is a name given to the followers of Aristotle, originally derived from the colonnades (*peripatoi* in Greek) of the school in Athens where the Aristotelians would walk up and down during their discussions.

125 *cause of the tides*: in 1616 Galileo wrote a *Discourse on the Tides* that became the Fourth Day of the present *Dialogue*.

Sagredo: Giovan Francesco Sagredo (1571–1620) was a Venetian patrician and a talented amateur of science. He had studied with Galileo at Padua and became his close friend. In the *Dialogue*, he speaks for the intelligent layman who is already half-converted to the new astronomy.

Salviati: Filippo Salviati (1538–1614) was a noble Florentine who often had Galileo as his guest in his villa near Florence. Galileo wished to perpetuate his memory by making him his spokesman in the *Dialogue*.

126 *a Peripatetic philosopher*: this philosopher could have been Cesare Cremonino (1550–1631) who was Galileo's colleague at the University of Padua and was famous for his commentaries on Aristotle.

Simplicius: a sixth-century interpreter of Aristotle. It is generally agreed that the Simplicio of the Dialogue is a composite of Galileo's Peripatetic opponents.

FIRST DAY

127 *Aristotelian and Ptolemaic system . . . Copernican system*: see note to p. 37 above.

128 *splendid proofs . . . 'continuity'*: see Aristotle, *On the Heavens*, book 1, chapter 1, section 268a 1–268b 10.

129 *necessary demonstrations*: see the note to p. 56 above.

130 *laughing-stock . . . Senate itself*: pestered by his mother about what had been debated at the Senate, Papirius told her that it concerned the question whether it would be better to allow one man two wives, or one woman two men. The result was that a large and noisy delegation of townswomen appeared before the Senate to argue for the latter alternative. The anecdote is told by Macrobius, *Saturnalia*, 1.6.18–26.

which you know already: this is the first of a number of allusions to the Platonic theory of recollection.

132 *I will follow Aristotle*: Aristotelians insisted that qualitative properties disclosed the nature of things whereas Galileo maintained that quantitative relations provided the genuine clues to an understanding of reality. Mathematics, for him, is the grammar of science. See his description of the language of nature in *The Assayer*, p. 115 above.

133 *simple bodies . . . principle of motion*: the four Aristotelian elements naturally moved either downwards (earth and water) or upwards (air and fire).

138 *integral bodies*: the four elements; what Galileo earlier called 'simple bodies' (see previous note).

139 *worthy of Plato*: Galileo seems to have in mind a passage in Plato's *Timaeus*, 38–9, but he has taken great liberties with the text.

   the Lincean Academician: Galileo is referring to himself. In 1611, as a recognition for his telescopic discoveries, he had been made a member of the Lincean Academy by its founder, Prince Federico Cesi.

140 *If a body . . . impact*: this and later passages enclosed in brackets were added by Galileo in his own copy of the first edition.

141 *braccia*: the unit of length that Galileo uses is the *braccio* (plural *braccia*) which is 58.4 cm or about an inch less than 2 feet.

146 *proofs concerning local motion*: see the Second Day of the *Dialogue*, pp. 284 ff., and the selection from the *Two New Sciences*, pp. 382 ff. below.

148 *maintaining its allotted velocity*: from the viewpoint of Newtonian physics, which was developed half-a-century later, the 'Platonic' cosmology produces no gain. Rather it implies two major miracles. First, it involves changing instantaneously the direction of the movement of the falling planets, which is as difficult as conferring instantaneously the determined velocity to a body. Indeed, in the natural order of things it is impossible. Secondly, it implies that the force of attraction of the Sun has doubled at the very moment when the circular motion is substituted for a downward one. But neither of these considerations can be said to hold for Galileo, who considers the operations of conferring motion on a body at rest and that of changing its direction as altogether different. In the first case something new has to be produced, but in the other the changes are merely accidental. As to the doubling of the force of attraction, Galileo has no need for it whatsoever since, for him, circular motion is inertial and does not engender centrifugal forces, so that no force of attraction from the Sun is necessary to make the planets describe their particular orbits and stay in them. Furthermore, Galileo does not work on the assumption that the Sun attracts the planets; they move towards the Sun by virtue of an inclination that has its origin in their bodies. For a fuller discussion, see William R. Shea, *Galileo's Intellectual Revolution*, 2nd edn. (New York: Science History Publications, 1977), 121–9.

152 *'the same reasoning . . . parts'*: this axiom is quoted by Aristotle, *On the Heavens*, book 1, chapter 3, section 270a, 11.

154 *'there is no arguing . . . first principles'*: see Aristotle, *Physics*, book 1, chapter 2, section 185a, 3.

   even though he proved it . . . contrary motion: see *On the Heavens*, book 4, chapters 4–5, section 311a, 15–312b.

158 *'But the movements of contraries . . . exempt from contraries'*: Galileo has paraphrased a passage in *On the Heavens*, book 1, chapter 3, section 270a, 14–17.

161 *two whole books*: Aristotle's *On Generation and Corruption* consists of two 'books'.

162 *horned arguments known as sorites*: the source of the 'liar's paradox' is the apostle Paul's letter to Titus, where he warns him not to rely on the Cretans for, as one of themselves said, 'All Cretans are liars' (Titus 1: 12). This was called a 'horned argument' or a 'forked question' by medieval logicians. It is not a sorites, whose name derives from the Greek word *soros*, meaning 'pile' or 'heap'. The classic example of a sorites is the paradox that arises when one considers a heap of sand from which grains are individually removed. Is it still a heap when only one grain remains? If not, when did it change from a heap to a non-heap? Note that it is the Aristotelian, Simplicio, who makes the mistake of confusing 'a horned argument' with a sorites.

164 *Cremonino*: see the note on 'A Peripatetic philosopher', p. 126 above.

168 *a priori . . . a posteriori*: the terms *a priori* and *a posteriori* are used to distinguish two types of knowledge or kinds of argument. *A priori* knowledge or justification is independent of experience (e.g. mathematial proofs); *a posteriori* knowledge or justification is dependent on experience or experimentation (e.g. the statement 'some students are bilingual').

170 *Abyla and Calpe*: Calpe is the Rock of Gibraltar, and Abyla is a hill on the African side of the Strait. Abyla and Calpe were called the Pillars of Hercules by the ancients.

172 *Pythagoras sacrificed a hundred oxen*: Pythagoras is said by the ancient Greek writers, Porphyry and Plutarch, to have sacrificed an ox, but Cicero already doubted the veracity of the tale (*On the Nature of the Gods*, 3.88).

173 *two new stars*: a very bright star, now known as a supernova, suddenly appeared in 1572 and was studied by the Danish astronomer Tycho Brahe. A second supernova appeared in 1604 and was the subject of well-attended lectures that Galileo delivered at the University of Padua.

Anti-Tycho: the title of a book published in 1621 by the Italian astronomer Scipione Chiaramonti.

parallax: see the note on p. 35 above.

174 *sunspots*: see the *Letters on the Sunspots*, pp. 33–54 above.

175 *the Sun's eccentric sphere*: in the Ptolemaic system the Sun's centre of revolution is not the centre of the Earth but is slightly off centre, hence the word 'excentric'.

Demosthenes: a prominent statesman and orator of ancient Athens, 384–322 BC.

176 *Mark Welser*: see the note to p. 33 above.

178 *Prytaneum*: the town hall of a Greek city-state, normally housing the chief magistrate and the common altar or hearth of the community.

182 *natura nihil frustra facit*: an ancient aphorism that was often quoted by the Scholastic philosophers but is not found as such in Aristotle.

188 *if the Moon had an epicycle*: Copernicus never abandoned the notion that perfect motion had to be circular and in this he was followed by Galileo. In the case of the Moon, Copernicus uses not only one but two epicycles, the second rotating on the first that is attached, in turn, to a third and larger circle called the deferent (*On the Motion of the Heavenly Spheres*, book 4, chapter 2).

Antichthons: the Greek word *Antichthon* means Counter-Earth, a hypothetical body of the solar system that the Pythagoreans are said to have postulated to support their heliocentric cosmology (see Aristotle, *On the Heavens*, book 2, chapter 13, section 293a, 15–30).

the ray from their eye: on Galileo's theory of optics see the note to p. 205 below.

192 *distinguished professor in Padua*: perhaps Cesare Cremonini. See the note to p. 126 above.

194 *in the Assayer and his Letters on the Sunspots*: these passages, which cover the same topic, are not translated in this anthology. They can be found for *The Assayer* in *Opere*, 6, pp. 283 ff., and for the *Letters* see Galileo Galilei and Christoph Scheiner, *On Sunspots*, trans. Eileen Reeves and Albert Van Helden (Chicago: Chicago University Press, 2010), 284.

201 *'we should not expect . . . luminous body'*: the text that Simplicio is made to quote may be by an author who has not been identified or it could be a literary device that Galileo uses to show how obscure is the material upon which Simplicio bases his arguments.

205 *darkness is the absence of light*: Galileo mocks Simplicio by having him solemnly introduce a trivial and obvious definition of darkness. Aristotle uses the same words but in a less naive context (*On the Soul*, book 2, chapter 7, section 418b ff., and *On Sensation*, section 439a 20).

his visual rays: Galileo was both influenced and hampered by the traditional description of how light-rays travel. We are all familiar today with the correct theory, which is that of *intromission*, meaning that vision is caused by rays of light entering the pupil, whereas the rival theory of *extromission*, which accounted for vision by rays streaming from our eyes, was more commonly accepted in Galileo's day. Whether the rays originate from the object or from the eye, the geometrical description of the situation is the same because the direction of the rays does not alter the way they are traced. We still speak of 'eye contact', of hard stares, and of gazing as 'looking outward'. In the *Sidereal Message* Galileo considers the rays as being carried from the eye to the object when lenses are placed between the eye and what is being observed (see above, p. 9), and in a letter to the Jesuit Christopher Grienberger, he writes, 'our visual rays leave our eye as from the vertex [of a triangle] and stretch out spherically until they reach the perimeter of the Moon' (letter of 1 September 1611, *Opere*, 11, p. 118). In a comment on comets written in 1619 Galileo still speaks of 'visual rays proceeding from the eye' (Note on Orazio Grassi's *De tribus cometis anni MDCXVIII*, *Opere*, 6, p. 107). As late as 1632 we find in the Third Day of the *Dialogue on the Two Chief World Systems* that the pupil is 'that hole from which the visual rays come out' (*Opere*, 7, p. 391, not in the present selection).

212 *pillar of cloud . . . bright by night*: the reference is to Exodus 13: 21–2.

214 *forty times more*: Galileo is comparing the surfaces of the Earth and the Moon when he should have compared their volumes, which are as 14 to 1.

216 *book of conclusions*: Galileo refers to a booklet, *A Mathematical Discourse on Controversies over Astronomical Novelties* published in 1614. It was written by Johann Georg Locher at the instigation of his teacher, the Jesuit Christoph Scheiner.

Cleomedes, Vitellio, Macrobius: Cleomedes is a first-century Greek writer known only through his book *On the Circular Motions of the Celestial Bodies*, which is a compendium of Greek sources (see Alan C. Bowen and Robert B. Todd, *Cleomedes' Lectures on Astronomy: A Translation of The Heavens, with an Introduction and Commentary* (Berkeley and Los Angeles: University of California Press, 2004). Vitellio (or Witelo) was a Polish friar who lived in Italy in the thirteenth century and wrote an important work on perspective. Macrobius was a Roman grammarian and Neoplatonist philosopher who lived at the end of the fourth and the beginning of the fifth century AD. His Commentary on Cicero's *Dream of Scipio* contains the idea mentioned here.

some other modern author: this author could be the Jesuit François de Aguilon, who published a book on optics in 1613.

220 *a man . . . secret device*: this is perhaps the Bohemian Martin Horky, an opponent of Galileo's discoveries. In a letter to Kepler on 24 May 1610, Horky bragged that he had a better telescope than Galileo and that he intended to make an instrument that would enable people to communicate over a distance of 15 miles (*Opere*, 10, p. 359).

221 *ancient spots*: the dark areas that can be seen on the Moon without the aid of the telescope. See the *Sidereal Message*, p. 11 above.

227 *Michelangelo*: Michelangelo Buonarroti (1475–1564), whose reputation as a supremely gifted artist was already established in his lifetime, thanks especially to the status given to him in Vasari's *Lives of the Artists*.

228 *Archytas*: a Pythagorean philosopher (428–350 BC) and a contemporary of Plato. He was born in Taranto in southern Italy, and is said to have experimented with toy flying-machines, including a wooden dove that flew by 'the secret blowing of air enclosed inside', perhaps a primitive compressed-air mechanism.

SECOND DAY

239 *hoops*: these *ruzzole*, as they are called in Italian, are wooden discs about 15 cm in diameter and 2 cm thick. There is a groove in the rim around which a cord is wound and then pulled to set the disc in motion.

Socrates' demon: Socrates called the source of his inspiration his 'demon'. Sagredo pokes fun at Simplicio and offers to become his source of inspiration by using the Socratic method of questioning.

240 *the one explaining . . . square ones*: Galileo is referring to the *Mechanical Questions*, chapter 8, section 851b, 15–852a, 14. The work was considered to be by Aristotle but is now believed to have been written later.

245 *our narrative poem*: Galileo is alluding to the long-standing debate on the relative merits of Tasso's *Gerusalemme liberata* and Ariosto's *Orlando furioso*, to which Galileo contributed forcefully in favour of Ariosto's freewheeling, episodic poem in preference to Tasso's attempt (not entirely successful) at writing a monothematic epic. Sagredo continues the allusion on p. 247 when he says, 'I am content to excuse you from telling this story now.'

246 *a problem . . . yet been able to solve*: the unsolved problem is the law governing the distance covered by freely falling bodies. Galileo provides the solution below; see pp. 284 ff.

247 *the Academician . . . on motion*: the Academician is Galileo himself (see note to p. 139), and the treatise is *On Naturally Accelerated Motion* that was eventually published in 1638 at the end of the Third Day of Galileo's *Discourses on Two New Sciences*. See the extract on pp. 382 ff. in this volume.

spirals: *On the Spirals* is one of the works of Archimedes (287–212 BC), the ancient mathematician that Galileo most admired.

as we have already discussed and established at length: in the First Day; see pp. 139–48 above.

248 *must terminate . . . Earth*: the erroneous belief that falling bodies must reach the centre of the Earth made it impossible to work out the correct path.

249 *this fancy of mine*: Galileo later reached the correct solution that the path of projectiles is a parabola, and he tried to pass off as a jest the explanation given here that the mixture of the straight motion of the falling body and the uniform diurnal motion of the Earth would give rise to a semicircle that ended at the centre of the Earth (letter to Pierre Carcavy, 5 June 1637, *Opere*, 17, p. 89).

250 *which you allowed for it at the outset*: see p. 138 above.

251 *with his great genius*: Galileo's phrase here, 'per altezza d'ingegno', is quoted from Dante's well-known praise of his late friend Guido Cavalcanti (*Inferno*, 10.59).

255 *Alexandretta*: port at the north-east extremity of the Mediterranean (modern Iskenderun), through which Sagredo travelled en route to the Syrian city of Aleppo.

257 *handbook of assertions*: the 'handbook' by the Jesuit Clemente Clementi is entitled *An Encyclopedia Explained and Defended with One Hundred Philosophical Assertions* (Rome, 1624).

266 *in Copernicus' book*: Galileo is referring to *On the Revolutions of the Heavenly Spheres*, book 1, chapter 12 where the tables are printed at the end. Whereas the ancients used chords we now use sines (a chord is equal to double the sine of half the angle).

270 *all humans naturally desire knowledge*: this is the first sentence of Aristotle's *Metaphysics*.

273 *including Ptolemy*: Galileo's source is Copernicus' *On the Revolutions of the Heavenly Spheres*, book 1, chapter 7, where Copernicus paraphrases, rather loosely, Ptolemy's *Almagest*, book 1, chapter 7.

276 *our knowledge . . . remembering*: Galileo returns to the Platonic theme of recollection; see p. 130 above.

284 *part of the tangent . . . point of contact*: a secant line of a circle is a line that intersects two points on the curve. If the secant is defined by two points, P and Q, with P fixed and Q variable, as Q approaches P along the curve, the secant becomes closer and closer to being the tangent at P, namely it 'just touches' the curve at that point.

the modern author just cited: Galileo is referring to the Jesuit Christoph Scheiner.

canna: a unit of length whose value varied in different regions of Italy, but was usually approximately 2 metres.

287 *56 times the semi-diameter of the Earth*: Galileo has modified the terminology that he used in the *Sidereal Message*, when he gave the distance of the Moon from the Earth as 'almost sixty terrestrial diameters'; see the note on p. 7 above.

296 *Quandoque bonus*: Simplicio quotes the first two words of the Roman poet Horace's famous phrase, 'quandoque bonus dormitat Homerus' ('sometimes good Homer nods') in his *Ars Poetica*, line 359.

298 *either assisting or informing*: the assisting spirits were angels who guided the planets in their course; the informing spirits were the internal moving principles of animate beings.

THIRD DAY

299 *is finite . . . or is infinite and boundless*: Copernicus states that the size of the Earth compared to the size of the heavens is 'like a point compared to a solid body or a finite magnitude compared to an infinite one', but he does not consider whether the universe itself is finite or infinite (*On the Revolution of the Heavenly Spheres*, book 1, chapter 6).

finite, bounded, and spherical: see Aristotle, *On the Heavens*, book 1, chapters 5–8, section 271b–277b, 27.

300 *would refuse even to look . . . accept them*: the field of view of Galileo's telescope was so narrow that only one-quarter of the Moon could be seen at a time, and people found this frustrating because they expected to see it whole. His friend, the philosopher Cesare Cremonini, was candid about his failure: 'Looking through those lenses', he told a friend, 'just makes me dizzy' (as reported by Paolo Gualdo in his letter to Galileo, 20 July 1611, *Opere*, 11, p. 165). Another professor, Giulio Libri, who had been Galileo's colleague in Padua before moving to Pisa, had the same trouble. When he died at the end of 1610 Galileo joked that although he had failed to see the new stars while on Earth, he might see them on his way to heaven (letter of Galileo to Paolo Gualdo in Padua, 17 December 1610, *Opere*, 10, p. 484). Galileo's bugbears were the pedants who swore by their books instead of looking through the telescope. 'This kind of person', he wrote to Kepler, 'thinks philosophy [used here in the sense of natural philosophy or natural science] is a book like the *Aeneid* or the *Odyssey*, and that truth is to be discovered, not in the world or in nature, but by *comparing texts* (I use their own words)' (letter to Kepler, 19 August 1610, *Opere*, 10, p. 422).

307 *on another occasion*: Salviati is referring to a passage in the Second Day (p. 295 above) where he criticized a student of the Jesuit Christoph Scheiner for failing to grasp that the year would last one natural day if the Earth did not revolve on its axis.

FOURTH DAY

310 *our own Tyrrhenian sea*: Salviati is referring to the coast of his (and Galileo's) native Tuscany.

311 *three intervals . . . tides*: at the beginning of the seventeenth century it was generally recognized that the tides went through four different cycles: the *daily cycle* with high and low tide recurring at intervals of 12 hours; the *monthly cycle* whereby the tides lag behind 50 minutes each day until they have gone round the clock and are back to their original position; the *half-monthly cycle* with high tides at new and full moon and low tides at quadrature and, finally, the *half-yearly cycle* with greater tides at the equinoxes than at the solstices. Galileo enumerates three periods: the diurnal, whose intervals 'in the Mediterranean . . . are of roughly six hours each', the monthly, which 'appears to derive from the motion of the Moon', and the annual, which 'appears to derive from the Sun'. But he fails to state the differences in the tides when the Moon is new, full, or at quadrature, and although he mentions that the tides at the solstices vary in size from those at the equinoxes, it is only some forty pages later (see pp. 351–2 below) that he states, erroneously, that they are greater at the solstices.

312 *Scylla and Charybdis*: in classical mythology, two monsters on either side of the Strait of Messina who embodied the danger to sailors of passing through the strait.

a prelate: Marcantonio de Dominis, a former Jesuit who was made bishop of Split in Croatia, left the Roman Church to become an Anglican, recanted, and returned to Rome, where he died in prison in 1624, the year of the publication of his book on the tides.

313 *the Moon with its temperate heat ... rarefied*: Girolamo Borro, who lectured at Pisa when Galileo was a student there, invokes the 'temperate heat' of the Moon that acts as an attractive force on the analogy of fire causing water to rise as it nears the boiling-point. Bernardino Telesio had suggested a rather more vague relationship between the Sun, the Moon, and the tides. He assumed that the sea rises and tends to boil over when it is heated by the Sun, and that it sets itself in motion to avoid evaporation, thus producing the flow and ebb of the tides (Girolamo Borro's *On the Tides in the Sea and the Inundation of the Nile* appeared in 1577 and was reprinted as least twice, the third time in 1583; Bernardino Telesio's views are to be found in his *On the Nature of Things*, book 1, chapter 12, a work that was published in 1565 and reprinted several times).

the water only rises ... where we are: to account for the tides, Marcantonio de Dominis postulated that an attractive force acted from the Moon on the ocean. A common objection to de Dominis' explanation was that high tide does not occur once a day when the Moon is directly above the sea but twice, the second time when the Moon is below the horizon. His theory, like Galileo's own hypothesis, entailed a 24-hour cycle and was rejected for failing to agree with experience. Galileo, of course, could not level this criticism at de Dominis, and he attacked him for failing to realize that water rises and falls only at the extremities and not at the centre of the Mediterranean. De Dominis can hardly be blamed for failing to detect this phenomenon: it only exists as a consequence of Galileo's own theory.

316 *like the breath ... whale*: this animistic interpretation of the tides on the analogy with respiration is set forth in Antonio Ferrari (known as Galateo), *On the Location of the Elements* (Basel, 1558).

Ancona, Ragusa, or Corfu: Ancona is on the western shore of the Adriatic; Ragusa (modern Dubrovnik) and Corfu are on its eastern shore.

323 *This will be easier to understand ... just now*: of the many ways that water can be made to flow, Galileo considered particularly suggestive the to-and-fro motion of water at the bottom of a boat that is alternately speeded up and slowed down. He likened the piling of the water now at one end and now at the other to the action of the tide. The analogy is not entirely satisfactory, however, since the acceleration or retardation is shared uniformly by the whole boat whereas, for the flux and reflux of the tides, it is not uniform throughout the sea basins in which they occur. Galileo parries this criticism by introducing a more sophisticated model

familiar to contemporary mathematicians and astronomers. He asks his readers to imagine that the ecliptic and the equator coincide. A point on the surface of the Earth can be considered to move on an epicycle attached to a deferent representing the Earth's orbit, as in the figure. The epicycle revolves once daily. For half the day the speed of the point is greater than that of the epicycle's centre (the centre of the Earth); for the other half the speed is less. Maximum and minimum velocities occur when a given point is collinear with the centre of both epicycle and deferent. This means that the greatest speed is at midnight and the lesser at noon, and thus entails a twelve-hour period for the tides, not a six-hour period as is actually the case since there are two high and two low tides every day.

324 *a mechanical model*: in the *Dialogue on the Tides* written in 1616 Galileo had written, 'I have a mechanical model that I will disclose at the appropriate time in which the effects of this marvellous composition of movements can be observed in detail' *(Opere*, 5, p. 386). Here the words 'that I will disclose at the appropriate time' have disappeared. This would seem to indicate that he had not been able, in the interval of sixteen years, to translate his idea into practice.

326 *I come, secondly . . . maximum slowness*: see the note to p. 323 above.

Aristotle . . . threw himself into the sea and drowned: the legend that Aristotle would have drowned himself in despair because he could not understand the cause of the tides goes back to the early Church Fathers and is found in Justin Martyr, a second-century Christian apologist, and Gregory of Nanzianus, a fourth-century archbishop of Constantinople. See I. During, *Aristotle and the Ancient Biographical Tradition* (Gothenburg, 1957).

327 *the strait between Scylla and Charybdis*: the strait of Messina; see the note to p. 312 above.

Atlantic: Galileo writes 'Ethiopian' *(Etiopico)* for the southern part of the Atlantic Ocean. The ambiguity surrounding the word 'Ethiopian' goes back to antiquity.

331 *'Even as the Aegean Sea', etc.*: Galileo is quoting Torquato Tasso, *Gerusalemme liberata*, 12.63. In Max Wickert's translation (*The Liberation of Jerusalem*, Oxford World's Classics (Oxford: Oxford University Press, 2009)): 'Even as the Aegean Sea, when Aquilo | and Notus cease to blow and churn and pound, | does not fall still, but in the heave and throe | of waves retains the motion and the sound . . .'. This flattering reference to Tasso as 'the divine poet' contrasts with Galileo's earlier dismissive attitude when he compared him unfavourably to Ariosto; see the note to p. 245 above.

as we have said on another occasion: see pp. 232–4 above.

334 *stronger than those coming from the west*: Galileo's claim to recorded evidence for the Mediterranean is puzzling for the prevailing wind east of Italy is a west wind in all seasons, and not an east wind as Sagredo vouches for.

341 *the wretched Orlando*: in Ariosto's *Orlando furioso*, Orlando tries to deny
     the unmistakeable evidence that Angelica, the object of his love, has
     married his rival Medoro. He is driven mad by jealousy when he can no
     longer deny the truth; hence Sagredo's reply. On Ariosto's poem see the
     note to p. 245 above.

     told of Aristotle: see p. 326 above.

349 *Mars . . . and even the Moon*: Galileo is aware of the fact that his theory
     of the tides does not sit easily on the known astronomical motion of
     the Sun, and he is anxious to remind his reader that the motions of the
     planets, for instance Mars and the Moon, are not perfectly understood.
     Kepler had found that the path of Mars is elliptical but Galileo would
     have nothing of this.

350 *a marked difference . . . divided by the equinoctial points*: the average
     apparent speed of the Sun around the Earth is about one degree per day
     (1/365.25 of the circle), but it speeds up and slows down. It takes 187
     days to move from the vernal to the autumn equinox (21 March–
     21 September) and 178 days from the autumn back to the vernal equinox
     (21 September–21 March).

351 *The additions are equal . . . are the smallest*: Galileo wishes to argue that the
     inclination of the Earth's axis with respect to its orbit around the Sun
     entails a modification of his original model. The annual and the diurnal
     motions are in the same line only at the solstices when their combination
     produces the greatest acceleration and the greatest retardation. At the
     equinoxes the two motions are inclined at their maximum angle, and the
     effect of their combination is consequently least. This is what Galileo's
     theory entails, but the reverse holds true: the equinoctial tides are the
     most extreme because they receive the maximum effect of the sun's
     gravitational pull, something Galileo did not consider.

352 *solstitial colure*: the meridian of the celestial sphere that passes through
     the two poles and the two solstices.

356 *An ancient mathematician*: this is Seleucus, a Hellenized Babylonian
     astronomer who flourished in the second century BC.

     similar childish ideas: in the Introduction to his *Astronomia Nova*,
     published in 1609, the German astronomer Johann Kepler had con-
     jectured that the Moon causes the tides.

357 *The fourth argument . . . says are imperceptible*: Galileo hoped that better
     telescopes would reveal that the motion of the Earth around the Sun had
     as a consequence a shift in the observed positions of the fixed stars at an
     interval of six months. This apparent displacement or difference in the
     apparent position of an object viewed along two different lines of sight is
     called parallax (see note to p. 35 above). The stars are actually too far for
     such a displacement to be noted.

     signor Cesare . . . Lincean Academy: in 1631, a year before the publication
     of the *Dialogue*, Galileo received an essay by Cesare Marsili (1552–1633)
     in which he declared that he had detected a shift in the meridian that had

been traced on the floor of the church of St Petronio in Bologna, where it can still be seen. Marsili's observations were not conclusive of the motion of the Earth, but Galileo had great hopes that they would be.

358 *a most learned and eminent person . . . submit*: this person is none less than the Pope, Urban VIII, and it was unfortunate that the argument should have been made by Simplicio, who cut such a pitiful figure in the *Dialogue*. Worse still, it is immediately ridiculed by Salviati, who had acted as Galileo's spokesman throughout the entire discussion.

THE TRIAL

360 *eventually printed in Florence*: the *Dialogue* was published in Florence on 21 February 1632 but an outbreak of the plague rendered communications difficult and copies of the book did not reach Rome before April. When questions were raised about the way the book had been authorized as well as about its content, the Pope suspended further distribution and appointed a special Commission of Enquiry. The members were Niccolò Riccardi, the Master of the Sacred Palace, an office which included the responsibility of licensing books to be printed; Agostino Oreggi, who was the papal theologian; and, in all likelihood, a Jesuit by the name of Melchior Inchofer, who had had one of his own books placed on the Index.

364 *Galileo's First Deposition*: Galileo was summoned to Rome in September 1632 but he pleaded ill-health and managed to postpone his trip for another five months. He arrived in Rome on 13 February 1633, and while his case was being studied by the Vatican officials he stayed with Francesco Niccolini, the Tuscan Ambassador and an old friend. On 12 April he was driven to the headquarters of the Holy Office, where he was not placed in a cell but provided with a three-room suite. He was kept in the Vatican until 30 April, when he was allowed to return to the Florentine Embassy.

the Revd Brother . . . Procurator Fiscal of the Holy Office: Galileo's trial was not conducted before the Cardinals of the Holy Office but before two officials: Vincenzo Maculano, a Dominican scholar and engineer who had recently been appointed Commissioner-General, and Carlo Sinceri, who had joined the Holy Office as early as 1606. The 'court' met only four times: on 12 April, 30 April, 10 May, and 21 June 1633. No one else was present.

he was questioned as follows: the introductory and concluding text of this document, and the questions put to Galileo, are in Latin; Galileo's replies are given in Italian.

365 *several Cardinals . . . d'Ascoli*: Cardinals [Robert] Bellarmine and [Giovanni Battista] Bonsi are mentioned by their surname, the others are referred to by the names of the church of which they were the titular: 'Aracoeli' stands for Agostino Galamani, 'Sant'Eusebio' for Ferdinando Taverna, and 'd'Ascoli' for Felice Centini.

366 *the Cardinal's reply to a letter*: see Bellarmine's letter to Foscarini in this volume, pp. 94–6.

367 *Document B*: see p. 370 below.

the injunction he was given: acting on the orders of Pope Paul V, Cardinal Bellarmine met Galileo on 26 February 1616 and enjoined him to cease teaching that the Earth moved. See next note.

368 *that he could not in any way . . . the said theory*: an unsigned memorandum reads: 'In the residence of his Eminence Cardinal Bellarmine, the above-mentioned Galileo was summoned and was admonished by his Eminence, in the presence of the Revd Father Michelangelo Seghizzi, of the Dominican Order, the Commissioner-General of the Holy Office, that the above opinion was an error and that he should abandon it. And immediately thereafter, in my presence and that of witnesses, the Cardinal still being present, the said Commissioner enjoined and commanded the said Galileo, in the name of his Holiness the Pope and of the whole Congregation of the Holy Office, that he should relinquish altogether the above opinion, that the Sun is the centre of the world and at rest and that the Earth moves; and that he should not henceforth hold, teach, or defend it in any way, verbally or in writing. Otherwise, proceedings would be taken against him by the Holy Office. The said Galileo acquiesced in this ruling and promised to obey it.'

370 *Galileo . . . having requested a hearing*: Galileo had a private meeting outside the courtroom with Maculano on 27 April, when he agreed to a plea-bargain of sorts. Galileo agreed to admit that he had erred in arguing too strongly for the Copernican position, and Maculano promised to recommend leniency. Hence Galileo's request for a second hearing.

371 *I am more desirous of glory than is seemly*: Galileo quotes (in Latin) Cicero, *Letters to his Friends*, 9.14.

377 *I, Galileo . . . whole Christian Commonwealth*: Galileo appeared before the Holy Office in their office in the convent adjoining the church of Santa Maria sopra Minerva on 22 June 1633. Of the ten Cardinals who were members, seven were in attendance. This was the average number at most meetings. The text that Galileo read was not his own but had been prepared by officials of the Holy Office.

TWO NEW SCIENCES

379 *Salviati, Sagredo, and Simplicio*: the three interlocutors who appear in the *Two New Sciences* that was published in 1638 bear the same names as those in the *Dialogue* of 1632. See the notes to pp. 125–6 above. Salviati remains a spokesman for Galileo but Sagredo occasionally takes positions that Galileo had once considered and then rejected. Simplicio is no longer the stubborn Aristotelian philosopher but an earnest enquirer.

Arsenal: the name of the Venetian shipyard.

383 *exactly that amount of force . . . state of rest*: what Sagredo presents here is the view that Galileo held when he first examined the question of natural acceleration in his early work by seeking its cause in terms of an 'impressed force'. He came to reject this approach as fruitless.

our Author's purpose: during the Third Day, Salviati has been reading out loud a Latin treatise on motion composed by the author (Galileo) when he was a professor at the University of Padua.

384 *impact . . . just one braccio*: it is true that impact is proportional to the height of fall, but this does not apply to the speed acquired as Sagredo suggests, making an assumption that Galileo had accepted in 1604 and abandoned later.

vires acquirat eundo: Virgil, *Aeneid*, 4.175, where the reference is to rumour.

385 *the Academician*: Galileo; see the note to p. 139 above.

389 *the reading*: Salviati had resumed reading Galileo's Latin treatise on motion.

390 *from unity, 1, 3, 5, etc.*: the 'odd number' rule is equivalent to the 'squared times' law: when the distance increases over three successive equal periods of time, the total distance covered in the first interval of time is 1 or 1^2; in the second interval of time, $1 + 3 = 4$ or 2^2, and in the third interval of time, $1 + 3 + 5 = 9$ or 3^2.

INDEX

The Oxford World's Classics Website

www.worldsclassics.co.uk

- Browse the full range of Oxford World's Classics online

- Sign up for our monthly e-alert to receive information on new titles

- Read extracts from the Introductions

- Listen to our editors and translators talk about the world's greatest literature with our Oxford World's Classics audio guides

- Join the conversation, follow us on Twitter at OWC_Oxford

- Teachers and lecturers can order inspection copies quickly and simply via our website

www.worldsclassics.co.uk

American Literature

British and Irish Literature

Children's Literature

Classics and Ancient Literature

Colonial Literature

Eastern Literature

European Literature

Gothic Literature

History

Medieval Literature

Oxford English Drama

Poetry

Philosophy

Politics

Religion

The Oxford Shakespeare

A complete list of Oxford World's Classics, including Authors in Context, Oxford English Drama, and the Oxford Shakespeare, is available in the UK from the Marketing Services Department, Oxford University Press, Great Clarendon Street, Oxford OX2 6DP, or visit the website at www.oup.com/uk/worldsclassics.

In the USA, visit www.oup.com/us/owc for a complete title list.

Oxford World's Classics are available from all good bookshops. In case of difficulty, customers in the UK should contact Oxford University Press Bookshop, 116 High Street, Oxford OX1 4BR.

Bhagavad Gita

The Bible Authorized King James Version
With Apocrypha

Dhammapada

Dharmasūtras

The Koran

The Pañcatantra

The Sauptikaparvan (from the
Mahabharata)

The Tale of Sinuhe and Other Ancient
Egyptian Poems

The Qur'an

Upaniṣads

ANSELM OF CANTERBURY	The Major Works
THOMAS AQUINAS	Selected Philosophical Writings
AUGUSTINE	The Confessions On Christian Teaching
BEDE	The Ecclesiastical History
HEMACANDRA	The Lives of the Jain Elders
KĀLIDĀSA	The Recognition of Śakuntalā
MANJHAN	Madhumalati
ŚĀNTIDEVA	The Bodhicaryàvatàra